Communications and Control Engineering

Springer
London
Berlin
Heidelberg
New York
Barcelona
Budapest
Hong Kong
Milan
Paris
Santa Clara
Singapore
Tokyo

Published titles include:

Sampled-Data Control Systems
J. Ackermann

Interactive System Identification
T. Bohlin

The Riccatti Equation
S. Bittanti, A.J. Laub and J.C. Willems (Eds)

Analysis and Design of Stream Ciphers
R.A. Rueppel

Sliding Modes in Control Optimization
V.I. Utkin

Fundamentals of Robotics
M. Vukobratović

Parametrizations in Control, Estimation and Filtering Problems: Accuracy Aspects
M. Gevers and G. Li

Parallel Algorithms for Optimal Control of Large Scale Linear Systems
Zoran Gajić and Xuemin Shen

Loop Transfer Recovery: Analysis and Design
A. Saberi, B.M. Chen and P. Sannuti

Markov Chains and Stochastic Stability
S.P. Meyn and R.L. Tweedie

Robust Control: Systems with Uncertain Physical Parameters
J. Ackermann in co-operation with A. Bartlett, D. Kaesbauer, W. Sienel and R. Steinhauser

Optimization and Dynamical Systems
U. Helmke and J.B. Moore

Optimal Sampled-Data Control Systems
Tongwen Chen and Bruce Francis

Nonlinear Control Systems (3rd edition)
Alberto Isidori

Theory of Robot Control
C. Canudas de Wit, B. Siciliano and G. Bastin (Eds)

María M. Seron, Julio H. Braslavsky and
Graham C. Goodwin

Fundamental Limitations in Filtering and Control

With 114 Figures

Springer

María M. Seron, PhD
Julio H. Braslavsky, PhD
Graham C. Goodwin, Professor

Department of Electrical and Computer Engineering, University of Newcastle, Newcastle, New South Wales 2308, Australia

ISBN 3-540-76126-8 Springer-Verlag Berlin Heidelberg New York

British Library Cataloguing in Publication Data
A catalogue record for this book is available from the British Library

Printed in Great Britain

Typesetting: Camera ready by authors
Printed and bound at the Athenæum Press Ltd, Gateshead
69/3830-543210 Printed on acid-free paper

Preface

This book deals with the issue of fundamental limitations in filtering and control system design. This issue lies at the very heart of feedback theory since it reveals what is achievable, and conversely what is not achievable, in feedback systems.

The subject has a rich history beginning with the seminal work of Bode during the 1940's and as subsequently published in his well-known book *Feedback Amplifier Design* (Van Nostrand, 1945). An interesting fact is that, although Bode's book is now fifty years old, it is still extensively quoted. This is supported by a science citation count which remains comparable with the best contemporary texts on control theory.

Interpretations of Bode's results in the context of control system design were provided by Horowitz in the 1960's. For example, it has been shown that, for single-input single-output stable open-loop systems having relative degree greater than one, the integral of the logarithmic sensitivity with respect to frequency is zero. This result implies, among other things, that a reduction in sensitivity in one frequency band is necessarily accompanied by an increase of sensitivity in other frequency bands. Although the original results were restricted to open-loop stable systems, they have been subsequently extended to open-loop unstable systems and systems having nonminimum phase zeros.

The original motivation for the study of fundamental limitations in feedback was control system design. However, it has been recently realized that similar constraints hold for many related problems including filtering and fault detection. To give the flavor of the filtering results, consider the frequently quoted problem of an inverted pendulum. It is well known that this system is completely observable from measurements of the carriage position. What is less well known is that it is fundamentally difficult to estimate the pendulum angle from measurements of the carriage position due to the location of open-loop nonminimum phase zeros and unstable poles. Minimum sensitivity peaks of 40 dB are readily predictable using Poisson integral type formulae *without* needing to carry out a specific design. This clearly suggests that a change in the instrumentation is called for, i.e., *one should measure the angle directly*. We see, in this example, that the fundamental limitations point directly to the inescapable nature of the difficulty and thereby eliminate the possibility of expending effort on various filter design strategies that we know, ab initio, are doomed to failure.

Recent developments in the field of fundamental design limitations include extensions to multivariable linear systems, sampled-data systems, and nonlinear systems.

At this point in time, a considerable body of knowledge has been assembled on the topic of fundamental design limitations in feedback systems. It is thus timely to summarize the key developments in a modern and comprehensive text. This has been our principal objective in writing this book. We aim to cover all necessary background and to give new succinct treatments of Bode's original work together with all contemporary results.

The book is organized in four parts. The first part is introductory and it contains a chapter where we cover the significance and history of design limitations, and motivate future chapters by analyzing design limitations arising in the time domain.

The second part of the book is devoted to design limitations in feedback control systems and is divided in five chapters. In Chapter 2, we summarize the key concepts from the theory of control systems that will be needed in the sequel. Chapter 3 examines fundamental design limitations in linear single-input single-output control, while Chapter 4 presents results on multi-input multi-output control. Chapters 5 and 6 develop corresponding results for periodic and sampled-data systems respectively.

Part III deals with design limitations in linear filtering problems. After setting up some notation and definitions in Chapter 7, Chapter 8 covers the single-input single-output filtering case, while Chapter 9 studies the multivariable case. Chapters 10 and 11 develop the extensions to the related problems of prediction and fixed-lag smoothing.

Finally, Part IV presents three chapters with very recent results on sensitivity limitations for nonlinear filtering and control systems. Chapter 12 introduces notation and some preliminary results, Chapter 13 covers feedback control systems, and Chapter 14 the filtering case.

In addition, we provide an appendix with an almost self-contained review of complex variable theory, which furnishes the necessary mathematical background required in the book.

Because of the pivotal role played by design limitations in the study of feedback systems, we believe that this book should be of interest to research and practitioners from a variety of fields including Control, Communications, Signal Processing, and Fault Detection. The book is self-contained and includes all necessary background and mathematical preliminaries. It would therefore also be suitable for junior graduate students in Control, Filtering, Signal Processing or Applied Mathematics.

The authors wish to deeply thank several people who, directly or indirectly, assisted in the preparation of the text. Our appreciation goes to Greta Davies for facilitating the authors the opportunity to complete this project in Australia. In the technical ground, input and insight were obtained from Gjerrit Meinsma, Guillermo Gómez, Rick Middleton and Thomas Brinsmead. The influence of Jim Freudenberg in this work is immense.

Contents

Part I

Introduction

1

A Chronicle of System Design Limitations

1.1 Introduction

This book is concerned with fundamental limits in the design of feedback control systems and filters. These limits tell us what is feasible and, conversely, what is infeasible, in a given set of circumstances. Their significance arises from the fact that they subsume *any particular* solution to a problem by defining the characteristics of *all possible* solutions.

Our emphasis throughout is on system analysis, although the results that we provide convey strong implications in system synthesis. For a variety of dynamical systems, we will derive relations that represent fundamental limits on the achievable performance of all possible designs. These relations depend on both constitutive and structural properties of the system under study, and are stated in terms of functions that quantify system performance in various senses.

Fundamental limits are actually at the core of many fields of engineering, science and mathematics. The following examples are probably well known to the reader.

Example 1.1.1 (The Cramér-Rao Inequality). In *Point Estimation Theory*, a function $\hat{\theta}(Y)$ of a random variable Y — whose distribution depends on an unknown parameter θ — is an *unbiased estimator for* θ if its expected value satisfies

$$E_\theta\{\hat{\theta}(Y)\} = \theta \,, \qquad (1.1)$$

where E_θ denotes expectation over the parametrized density function $p(\cdot;\theta)$ for the data.

A natural measure of performance for a parameter estimator is the covariance of the estimation error, defined by $E_\theta\{(\hat{\theta}-\theta)^2\}$. Achieving a small covariance of the error is usually considered to be a good property of an unbiased estimator. There is, however, a limit on the minimum value of covariance that can be attained. Indeed, a relatively straightforward mathematical derivation from (1.1) leads to the following inequality, which holds for *any* unbiased estimator,

$$E_\theta\{(\hat{\theta}-\theta)^2\} \geq \left(E_\theta\left\{\left(\frac{\partial \log p(y;\theta)}{\partial\theta}\right)^2\right\}\right)^{-1},$$

where $p(\cdot;\theta)$ defines the density function of the data $y \in Y$.

The above relation is known as the *Cramér-Rao Inequality*, and the right hand side (RHS) the *Cramér-Rao Lower Bound* (Cramér 1946). This plays a fundamental role in Estimation Theory (Caines 1988). Indeed, an estimator is considered to be efficient if its covariance is equal to the Cramér-Rao Lower Bound. Thus, this bound provides a benchmark against which all practical estimators can be compared. ◦

Another illustration of a relation expressing fundamental limits is given by Shannon's Theorem of Communications.

Example 1.1.2 (The Shannon Theorem). A celebrated result in *Communication Theory* is the *Shannon Theorem* (Shannon 1948). This crucial theorem establishes that given an information source and a communication channel, there exists a coding technique such that the information can be transmitted over the channel at any rate R less than the channel capacity C and with arbitrarily small frequency of errors despite the presence of noise (Carlson 1975). In short, the probability of error in the received information can be made arbitrarily small provided that

$$R \leq C\,. \tag{1.2}$$

Conversely, if $R > C$, then reliable communication is impossible. When specialized to continuous channels,[1] a complementary result (known as the Shannon-Hartley Theorem) gives the channel capacity of a band-limited channel corrupted by white gaussian noise as

$$C = B\log_2(1+S/N) \quad \text{bits/sec},$$

[1] A continuous channel is one in which messages are represented as waveforms, i.e., continuous functions of time, and the relevant parameters are the bandwidth and the signal-to-noise ratio (Carlson 1975).

where the bandwidth, B, and the signal-to-noise ratio, S/N, are the relevant channel parameters.

The Shannon-Hartley law, together with inequality (1.2), are fundamental to communication engineers since they (i) represent the absolute best that can be achieved in the way of reliable information transmission, and (ii) they show that, for a specified information rate, one can reduce the signal power provided one increases the bandwidth, and vice versa (Carlson 1975). Hence these results both provide a benchmark against which practical communication systems can be evaluated, and capture the inherent trade-offs associated with physical communication systems. ◦

Comparing the fundamental relations in the above examples, we see that they possess common qualities. Firstly, they evolve from basic axioms about the nature of the universe. Secondly, they describe inescapable performance bounds that act as benchmarks for practical systems. And thirdly, they are recognized as being central to the design of real systems.

The reader may wonder why it is important to know the existence of fundamental limitations before carrying out a particular design to meet some desired specifications. Åström (1995) quotes an interesting example of the latter issue. This example concerns the design of the flight controller for the X-29 aircraft. Considerable design effort was recently devoted to this problem and many different optimization methods were compared and contrasted. One of the design criteria was that the phase margin should be greater than 45° for all flight conditions. At one flight condition the model contained an unstable pole at 6 and a nonminimum phase zero at 26. A relatively simple argument based on the fundamental laws applicable to feedback loops (see Example 2.3.2 in Chapter 2) shows that a phase margin of 45° is infeasible! It is interesting to note that many design methods were used in a futile attempt to reach the desired goal.

As another illustration of inherently difficult problems, we learn from virtually every undergraduate text book on control that the states of an inverted pendulum are completely observable from measurements of the carriage position. However, the system has an open right half plane (ORHP) zero to the left of a real ORHP pole. A simple calculation based on integral sensitivity constraints (see §8.5 in Chapter 8) shows that sensitivity peaks of the order of 50:1 are unavoidable in the estimation of the pendulum angle when only the carriage position is measured. This, in turn, implies that relative input errors of the order of 1% will appear as angle relative estimation errors of the order of 50%. Note that this claim can be made *before any particular estimator is considered.* Thus much wasted effort can again be avoided. The inescapable conclusion is that we should redirect our efforts to building angle measuring transducers rather than attempting to estimate the angle by an inherently sensitive procedure.

In the remainder of the book we will expand on the themes outlined above. We will find that the fundamental laws divide problems into those that are essentially easy (in which case virtually any sensible design method will give a satisfactory solution) and those that are essentially hard (in which case no design method will give a satisfactory solution). We believe that understanding these inherent design difficulties readily justifies the effort needed to appreciate the results.

1.2 Performance Limitations in Dynamical Systems

In this book we will deal with very general classes of *dynamic systems*. The dynamic systems that we consider are characterized by three key attributes, namely:

(i) they consist of particular interconnections of a "known part" — the plant — and a "design part" — the controller or filter — whose structure is such that certain signals interconnecting the parts are indicators of the performance of the overall system;

(ii) the parts of the interconnection are modeled as input-output operators[2] with causal dynamics, i.e., an input applied at time t_0 produces an output response for $t > t_0$; and

(iii) the interconnection regarded as a whole system is stable, i.e., a bounded input produces a bounded response (the precise definition will be given later).

We will show that, when these attributes are combined within an appropriate mathematical formalism, we can derive fundamental relations that may be considered as being systemic versions of the Cramér-Rao Lower Bound of Probability and the Channel Capacity Limit of Communications. These relations are fundamental in the sense that they describe achievable — or non achievable — properties of the overall system only in terms of the known part of the system, i.e., they hold for *any* particular choice of the design part.

As a simple illustrative example, consider the unity feedback control system shown in Figure 1.1.

To add a mathematical formalism to the problem, let us assume that the plant and controller are described by finite dimensional, linear time-invariant (LTI), scalar, continuous-time dynamical systems. We can thus use Laplace transforms to represent signals. The plant and controller can

[2] It is sufficient here to consider an input-output operator as a mapping between input and output signals.

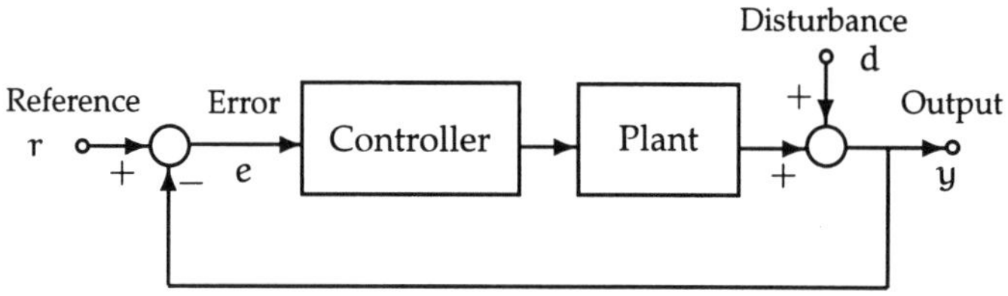

FIGURE 1.1. Feedback control system.

be described in transfer function form by $G(s)$ and $K(s)$, where

$$G(s) = \frac{N_G(s)}{D_G(s)}, \qquad \text{and} \qquad K(s) = \frac{N_K(s)}{D_K(s)}. \tag{1.3}$$

The reader will undoubtedly know[3] that the transfer functions from reference input to output and from disturbance input to output are given respectively by T and S, where

$$T = \frac{N_G N_K}{N_G N_K + D_G D_K}, \tag{1.4}$$

$$S = \frac{D_G D_K}{N_G N_K + D_G D_K}. \tag{1.5}$$

Note that these are dimensionless quantities since they represent the transfer function (ratio) between like quantities that are measured in the same units. Also $T(j\omega)$ and $S(j\omega)$ describe the response to inputs of a particular type, namely pure sinusoids. Since $T(j\omega)$ and $S(j\omega)$ are dimensionless, it is appropriate to compare their respective amplitudes to benchmark values. At each frequency, the usual value chosen as a benchmark is unity, since $T(j\omega_0) = 1$ implies that the magnitude of the output is equal to the magnitude of the reference input at frequency ω_0, and since $S(j\omega_0) = 1$ implies that the magnitude of the output is equal to the magnitude of the disturbance input at frequency ω_0. More generally, the frequency response of T and S can be used as measures of stability robustness with respect to modeling uncertainties, and hence it is sensible to compare them to "desired shapes" that act as benchmarks.

Other domains also use dimensionless quantities. For example, in Electrical Power Engineering it is common to measure currents, voltages, etc., as a fraction of the "rated" currents, voltages, etc., of the machine. This system of units is commonly called a "per-unit" system. Similarly, in Fluid Dynamics, it is often desirable to determine when two different flow situations are similar. It was shown by Osborne Reynolds (Reynolds 1883)

[3]See Chapter 2 for more details.

that two flow scenarios are dynamically similar when the quantity

$$R = \frac{ul\rho}{\mu},$$

(now called the Reynolds number) is the same for both problems.[4] The Reynolds number is the ratio of inertial to viscous forces, and high values of R invariable imply that the flow will be turbulent rather than laminar. As can be seen from these examples, dimensionless quantities facilitate the comparison of problems with critical (or benchmark) values.

The key question in scalar feedback control synthesis is how to find a *particular* value for the design polynomials N_K and D_K in (1.3) so that the feedback loop satisfies certain desired properties. For example, it is usually desirable (see Chapter 2) to have $T(j\omega) = 1$ at low frequencies and $S(j\omega) = 1$ at high frequencies. These kinds of design goals are, of course, important questions; but we seek deeper insights. Our aim is to examine the fundamental and unavoidable constraints on T and S that hold *irrespective* of which controller K is used — provided only that the loop is stable, linear, and time-invariant (actually, in the text we will relax these latter restrictions and also consider nonlinear and time-varying loops).

In the linear scalar case, equations (1.5) and (1.4) encapsulate the key relationships that lead to the constraints. The central observation is that we require the loop to be stable and hence we require that, whatever value for the controller transfer function we choose, the resultant closed loop *characteristic polynomial* $N_G N_K + D_G D_K$ must have its zeros in the open left half plane.

A further observation is that the two terms $N_G N_K$ and $D_G D_K$ of the characteristic polynomial appear in the numerator of T and S respectively. These observations, in combination, have many consequences, for example we see that

(i) $S(s) + T(s) = 1$ for all s (called the complementarity constraint);

(ii) if the characteristic polynomial has all its zeros to the left of $-\alpha$, where α is some nonnegative real number, then the functions S and T are analytic in the half plane to the right of $-\alpha$ (called analyticity constraint);

(iii) if q is a zero of the plant numerator N_G (i.e., a plant zero), such that $\operatorname{Re} q > -\alpha$ (here $\operatorname{Re} s$ denotes real part of the complex number s), then $T(q) = 0$ and $S(q) = 1$; similarly, if p is a zero of the plant denominator D_G (i.e., a plant pole), such that $\operatorname{Re} q > -\alpha$, then $T(p) = 1$ and $S(p) = 0$ (called interpolation constraints).

[4]Here u is a characteristic velocity, l a characteristic length, ρ the fluid density, and μ the viscosity.

The above seemingly innocuous constraints actually have profound implications on the achievable performance as we will see below.

1.3 Time Domain Constraints

In the main body of the book we will carry out an in-depth treatment of constraints for interconnected dynamic systems. However, to motivate our future developments we will first examine some preliminary results that follow very easily from the use of the Laplace transform formalism. In particular we have the following result.

Lemma 1.3.1. Let $H(s)$ be a strictly proper transfer function that has all its poles in the half plane $\operatorname{Re} s \leq -\alpha$, where α is some finite real positive number (i.e., $H(s)$ is analytic in $\operatorname{Re} s > -\alpha$). Also, let $h(t)$ be the corresponding time domain function, i.e.,

$$H(s) = \mathcal{L}h(t) ,$$

where $\mathcal{L}\cdot$ denotes the Laplace transformation. Then, for any s_0 such that $\operatorname{Re} s_0 > -\alpha$, we have

$$\int_0^{\infty} e^{-s_0 t} h(t)\, dt = \lim_{s \to s_0} H(s) .$$

Proof. From the definition of the Laplace transform we have that, for all s in the region of convergence of the transform, i.e., for $\operatorname{Re} s > -\alpha$,

$$H(s) = \int_0^{\infty} e^{-st} h(t)\, dt .$$

The result then follows since s_0 is in the region of convergence of the transform. □

In the following subsection, we will apply the above result to examine the properties of the step responses of the output and error in Figure 1.1.

1.3.1 *Integrals on the Step Response*

We will analyze here the impact on the step response of the closed-loop system of open-loop poles at the origin, unstable poles, and nonminimum phase zeros. We will then see that the results below quantify limits in performance as constraints on transient properties of the system such as *rise time, settling time, overshoot* and *undershoot*.

Throughout this subsection, we refer to Figure 1.1, where the plant and controller are as in (1.3), and where e and y are the time responses to a unit step input (i.e., $r(t) - d(t) = 1, \forall t$).

We then have the following results relating open-loop poles and zeros with the step response.

Theorem 1.3.2 (Open-loop integrators). Suppose that the closed loop in Figure 1.1 is stable. Then,

(i) for $\lim_{s\to 0} sG(s)K(s) = c_1, 0 < |c_1| < \infty$, we have that

$$\lim_{t\to\infty} e(t) = 0\,,$$
$$\int_0^\infty e(t)\,dt = \frac{1}{c_1}\,;$$

(ii) for $\lim_{s\to 0} s^2G(s)K(s) = c_2, 0 < |c_2| < \infty$, we have that

$$\lim_{t\to\infty} e(t) = 0\,,$$
$$\int_0^\infty e(t)\,dt = 0\,.$$

Proof. Let E, Y, R and D denote the Laplace transforms of e, y, r and d, respectively. Then,

$$E(s) = S(s)[R(s) - D(s)]\,, \tag{1.6}$$

where S is the sensitivity function defined in (1.5), and $R(s) - D(s) = 1/s$ for a unit step input. Next, note that in case (i) the open-loop system GK has a simple pole at $s = 0$, i.e., $G(s)K(s) = \tilde{L}(s)/s$, where $\lim_{s\to 0}\tilde{L}(s) = c_1$. Accordingly, the sensitivity function has the form

$$S(s) = \frac{s}{s + \tilde{L}(s)}\,,$$

and thus, from (1.6),

$$\lim_{s\to 0} E(s) = \frac{1}{c_1}\,. \tag{1.7}$$

From (1.7) and the Final Value Theorem (e.g., Middleton & Goodwin 1990), we have that

$$\begin{aligned}\lim_{t\to\infty} e(t) &= \lim_{s\to 0} sE(s)\\ &= 0\,.\end{aligned}$$

Similarly, from (1.7) and Lemma 1.3.1,

$$\begin{aligned}\int_0^\infty e(t)\,dt &= \lim_{s\to 0} E(s)\\ &= \frac{1}{c_1}\,.\end{aligned}$$

This completes the proof of case (i).

Case (ii) follows in the same fashion, on noting that here the open-loop system GK has a double pole at $s = 0$. □

Theorem 1.3.2 states conditions that the error step response has to satisfy provided the open-loop system has poles at the origin, i.e., it has pure integrators. The following result gives similar constraints for ORHP open-loop poles.

Theorem 1.3.3 (ORHP open-loop poles). Consider Figure 1.1, and suppose that the open-loop plant has a pole at $s = p$, such that $\operatorname{Re} p > 0$. Then, if the closed loop is stable,

$$\int_0^\infty e^{-pt} e(t)\, dt = 0 \,, \tag{1.8}$$

and

$$\int_0^\infty e^{-pt} y(t)\, dt = \frac{1}{p} \,. \tag{1.9}$$

Proof. Note that, by assumption, $s = p$ is in the region of convergence of $E(s)$, the Laplace transform of the error. Then, using (1.6) and Lemma 1.3.1, we have that

$$\begin{aligned}\int_0^\infty e^{-pt} e(t)\, dt &= E(p) \\ &= \frac{S(p)}{p} \\ &= 0 \,,\end{aligned}$$

where the last step follows since $s = p$ is a zero of S, by the interpolation constraints. This proves (1.8). Relation (1.9) follows easily from (1.8) and the fact that $r - d = 1$, i.e.,

$$\begin{aligned}\int_0^\infty e^{-pt} y(t)\, dt &= \int_0^\infty e^{-pt} \left(r(t) - d(t) - e(t)\right)\, dt \\ &= \int_0^\infty e^{-pt}\, dt \\ &= \frac{1}{p} \,.\end{aligned}$$

□

A result symmetric to that of Theorem 1.3.3 holds for plants with nonminimum phase zeros, as we see in the following theorem.

Theorem 1.3.4 (ORHP open-loop zeros). Consider Figure 1.1, and suppose that the open-loop plant has a zero at $s = q$, such that $\operatorname{Re} q > 0$. Then, if the closed loop is stable,

$$\int_0^\infty e^{-qt}\, e(t)\, dt = \frac{1}{q} \,, \tag{1.10}$$

and

$$\int_0^\infty e^{-qt}\, y(t)\, dt = 0\,. \tag{1.11}$$

Proof. Similar to that of Theorem 1.3.3, except that here $T(q) = 0$. □

The above theorems assert that if the plant has an ORHP open-loop pole or zero, then the error and output time responses to a step must satisfy integral constraints that hold for *all* possible controller giving a stable closed loop. Moreover, if the plant has *real* zeros or poles in the ORHP, then these constraints display a balance of exponentially weighted areas of positive and negative error (or output). It is evident that the same conclusions hold for ORHP zeros and/or poles of the controller. Actually, equations (1.8) and (1.10) hold for open-loop poles and zeros that lie to the right of all closed-loop poles, provided the open-loop system has an integrator. Hence, *stable* poles and *minimum phase* zeros also lead to limitations in certain circumstances.

The time domain integral constraints of the previous theorems tell us fundamental properties of the resulting performance. For example, Theorem 1.3.2 shows that a plant-controller combination containing a double integrator will have an error step response that necessarily overshoots (changes sign) since the integral of the error is zero. Similarly, Theorem 1.3.4 implies that if the open-loop plant (or controller) has real ORHP zeros then the closed-loop transient response can be arbitrarily poor (depending only on the location of the closed-loop poles relative to q), as we show next. Assume that the closed-loop poles are located to the left of $-\alpha$, $\alpha > 0$. Observe that the time evolution of e is governed by the closed-loop poles. Then as q becomes much smaller than α, the weight inside the integral, e^{-qt}, can be approximated to 1 over the transient response of the error. Hence, since the RHS of (1.10) grows as q decreases, we can immediately conclude that real ORHP zeros much smaller than the magnitude of the closed-loop poles will produce large transients in the step response of a feedback loop. Moreover this effect gets worse as the zeros approach the imaginary axis.

The following example illustrates the interpretation of the above constraints.

Example 1.3.1. Consider the plant

$$G(s) = \frac{q - s}{s(s+1)}\,,$$

where q is a positive real number. For this plant we use the internal model control paradigm (Morari & Zafiriou 1989) to design a controller in Figure 1.1 that achieves the following complementarity sensitivity function

$$T(s) = \frac{q - s}{q(0.2s+1)^2}\,.$$

This design has the properties that, for every value of the ORHP plant zero, q, (i) the two closed-loop poles are fixed at $s = -5$, and (ii) the error goes to zero in steady state. This allows us to study the effect in the transient response of q approaching the imaginary axis. Figure 1.2 shows the time responses of the error and the output for decreasing values of q. We can see from this figure that the amplitude of the transients indeed becomes larger as q becomes much smaller than the magnitude of the closed-loop poles, as already predicted from our previous discussion.

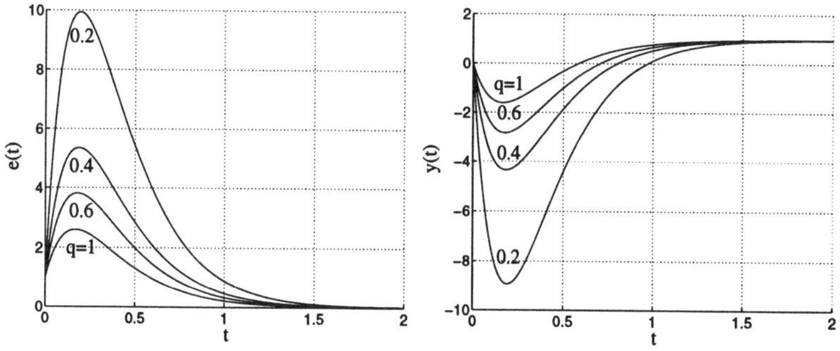

FIGURE 1.2. Error and output time responses of a nonminimum phase plant.

○

1.3.2 *Design Interpretations*

The results of the previous section have straightforward implications concerning standard quantities used as figures of merit of the system's ability to reproduce step functions. We consider here the *rise time*, the *settling time*, the *overshoot* and the *undershoot*.

The *rise time* approximately quantifies the minimum time it takes the system to reach the vicinity of its new set point. Although this term has intuitive significance, there are numerous possibilities to define it rigorously (cf. Bower & Schultheiss 1958). We define it by

$$t_r \triangleq \sup_{\delta} \left\{ \delta : y(t) \leq \frac{t}{\delta} \quad \text{for all } t \text{ in } [0, \delta] \right\} . \tag{1.12}$$

The *settling time* quantifies the time it takes the transients to decay below a given settling level, say ϵ, commonly between 1 and 10%. It is defined by

$$t_s \triangleq \inf_{\delta} \left\{ \delta : |y(t) - 1| \leq \epsilon \quad \text{for all } t \text{ in } [\delta, \infty) \right\} . \tag{1.13}$$

Here, the step response of the system has been normalized to have unitary final value, which is also assumed throughout this section.

Finally, the *overshoot* is the maximum value by which the output exceeds its final set point value, i.e.,

$$y_{os} \triangleq \sup_t \{-e(t)\} \; ;$$

and the *undershoot* is the maximum negative peak of the system's output, i.e.,

$$y_{us} \triangleq \sup_t \{-y(t)\} \; .$$

Figure 1.3 shows a typical step response and illustrates these quantities.

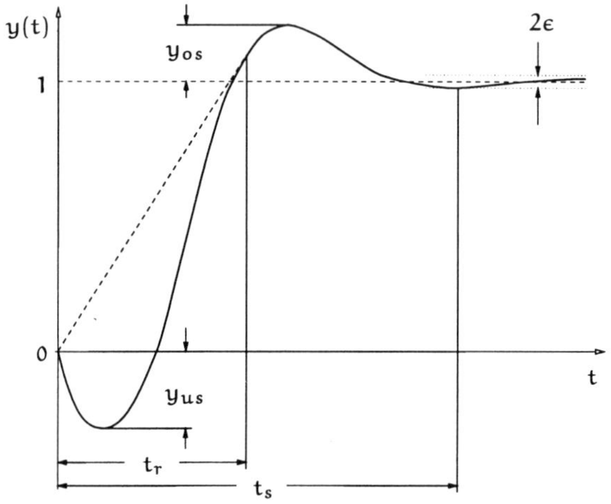

FIGURE 1.3. Time domain specifications.

Corollary 1.3.5 (Overshoot and real ORHP poles). A stable unity feedback system with a real ORHP open-loop pole, say at $s = p$, must have overshoot in its step response. Moreover, if t_r is the rise time defined by (1.12), then

$$\begin{aligned} y_{os} &\geq \frac{(pt_r - 1)e^{pt_r} + 1}{pt_r} \\ &\geq \frac{pt_r}{2} \, . \end{aligned} \tag{1.14}$$

Proof. The existence of overshoot follows immediately from Theorem 1.3.3, since $e(t)$ cannot have a single sign unless it is zero for all t. From the definition of rise time in (1.12) we have that $y(t) \leq t/t_r$ for $t \leq t_r$, i.e., $e(t) \geq 1 - t/t_r$. Using this, we can write from the integral equality (1.8)

$$-\int_{t_r}^{\infty} e^{-pt} e(t)\, dt \geq \int_0^{t_r} e^{-pt} \left(1 - \frac{t}{t_r}\right) dt \, . \tag{1.15}$$

From (1.15) and the definition of overshoot, it follows that

$$y_{os}\frac{e^{-pt_r}}{p} = y_{os}\int_{t_r}^{\infty} e^{-pt}\,dt \tag{1.16}$$

$$\geq \int_0^{t_r} e^{-pt}\left(1-\frac{t}{t_r}\right)dt$$

$$= \frac{(pt_r - 1) + e^{-pt_r}}{p^2 t_r}\,. \tag{1.17}$$

Equation (1.14) is then obtained from (1.16) - (1.17). □

Corollary 1.3.5 shows that if the closed-loop system is "slow", i.e., it has a large rise time, the step response will present a large overshoot if there are open-loop unstable real poles[5]. Intuitively, we can deduce from this result that unstable poles will demand a "fast" closed-loop system — or equivalently, a larger closed-loop bandwidth — to keep an acceptable performance. The farther from the $j\omega$-axis the poles are, the more stringent this bandwidth demand will be.

An analogous situation is found in relation with real nonminimum phase zeros and undershoot in the system's response, as we see in the next corollary.

Corollary 1.3.6 (Undershoot and real ORHP zeros). A stable unity feedback system with a real ORHP open-loop zero, say at $s = q$, must have undershoot in its step response. Moreover, if t_s and ϵ are the settling time and level defined by (1.13),

$$y_{us} \geq \frac{1-\epsilon}{e^{qt_s}-1}\,. \tag{1.18}$$

Proof. Similar to Corollary 1.3.5, this time using (1.11) and the definition of settling time and undershoot. □

The interpretation for Corollary 1.3.6 is that if the system has real nonminimum phase zeros, then its step response will display large undershoots as the settling time is reduced, i.e., the closed-loop system is made "faster". Notice that this situation is quite the opposite to that for real unstable poles, for now real nonminimum phase zeros will demand a *short* closed-loop bandwidth for good performance. Moreover, here the *closer* to the imaginary axis the zeros are, the *stronger* the demand for a short bandwidth will be.

Evidently from the previous remarks, a clear trade-off in design arises when the open-loop system is both unstable and nonminimum phase,

[5] This is in contrast with the case of open-loop *stable* systems, where large overshoots normally arise from *short* rise times.

since, depending on the relative position of these poles and zeros, a completely satisfactory performance may not be possible. The following result considers such a case.

Corollary 1.3.7. Suppose a stable unity feedback system has a real ORHP open-loop zero at $s = q$ and a real ORHP open-loop pole at $s = p$, $p \neq q$. Then,

(i) if $p < q$, the overshoot satisfies

$$y_{os} \geq \frac{p}{q-p},$$

(ii) if $p > q$, the undershoot satisfies

$$y_{us} \geq \frac{q}{p-q}.$$

Proof. For case (i) combine (1.8) and (1.10) to obtain

$$\int_0^\infty \left(e^{-pt} - e^{-qt}\right)[-e(t)]\,dt = \frac{1}{q}.$$

Using the definition of overshoot yields

$$\begin{aligned} \frac{1}{q} &\leq y_{os} \int_0^\infty \left(e^{-pt} - e^{-qt}\right) dt \\ &= y_{os} \frac{q-p}{pq}. \end{aligned} \tag{1.19}$$

The result then follows from (1.19) by using the fact that $q > p$.

Case (ii) can be shown similarly by combining (1.9) and (1.11) and using the fact that $q < p$. □

In the following subsection, we illustrate the previous results by analyzing time domain limitations arising in the control of an inverted pendulum. This example will be revisited in Chapter 3, where we study frequency domain limitations in the context of feedback control, and in Chapter 8, where we analyze frequency domain limitations from a filtering point of view.

1.3.3 *Example: Inverted Pendulum*

Consider the inverted pendulum shown in Figure 1.4. The linearized model for this system about the origin (i.e., $\theta = \dot{\theta} = y = \dot{y} = 0$) has the following transfer function from force, u, to carriage position, y

$$\frac{Y(s)}{U(s)} = \frac{(s-q)(s+q)}{M\,s^2(s-p)(s+p)}, \tag{1.20}$$

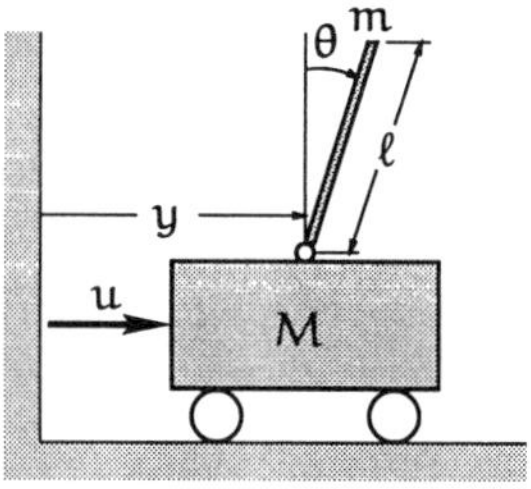

FIGURE 1.4. Inverted pendulum.

where

$$q = \sqrt{g/\ell}\,,$$
$$p = \sqrt{\frac{(M+m)g}{M\ell}}\,.$$

In the above definitions, g is the gravitational constant, m is the mass at the end of the pendulum, M is the carriage's mass, and ℓ is the pendulum's length.

We readily see that this system satisfies the conditions discussed in Corollary 1.3.7, part (ii). Say that we normalize so that $q = 1$ and take $m/M = 0.1$, so that $p = 1.05$. Corollary 1.3.7 then predicts an undershoot greater than 20!

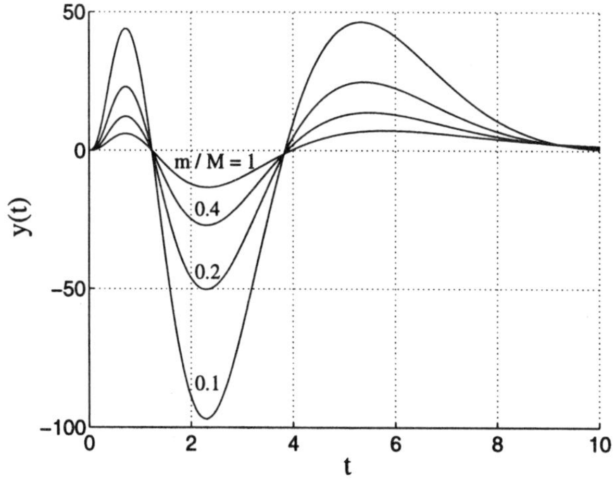

FIGURE 1.5. Position time response of the inverted pendulum.

To test the results, we designed an LQG-LQR controller[6] that fixes the closed-loop poles at $s = -1, -2, -3, -4$. Figure 1.5 shows the step response of the carriage position for fixed $q = 1$, and for different values of the mass ratio m/M (which imply, in turn, different locations of the open-loop poles of the plant). We can see from this figure that (i) the lower bound on the undershoot predicted by Corollary 1.3.7 is conservative (this is due to the approximation $-y \approx y_{us}$ implicitly used to derived this bound), and (ii) the bound correctly predicts an increase of the undershoot as the difference $p - q$ decreases.

1.4 Frequency Domain Constraints

The results presented in §1.3 were expressed in the time domain using Laplace transforms. However, one might expect that corresponding results hold in the frequency domain. This will be a major theme in the remainder of the book. To give the flavor of the results, we will briefly discuss constraints induced by zeros on the imaginary axis, or ORHP zeros arbitrarily close to the imaginary axis. Analogous conclusions hold for poles on the imaginary axis.

Note that, assuming closed-loop stability, then an open-loop zero on the imaginary axis at $j\omega_q$ implies that

$$T(j\omega_q) = 0, \qquad \text{and} \qquad S(j\omega_q) = 1. \tag{1.21}$$

We have remarked earlier that a common design objective is to have $S(j\omega) \ll 1$ at low frequencies, i.e., for $\omega \in [0, \omega_1]$ for some ω_1. Clearly, if $\omega_q < \omega_1$, then this goal is *inconsistent* with (1.21). Now say that the open-loop plant has a zero at $q = \epsilon + j\omega_q$, where ϵ is small and positive. Then we might expect (by continuity) that $|S(j\omega)|$ would have a tendency to be near 1 in the vicinity of $\omega = \omega_q$. Actually, it turns out to be possible to force $|S(j\omega)|$ to be small for frequencies $\omega \in [0, \omega_1]$ where $\omega_1 > \omega_q$. However, one has to pay a heavy price for trying to defeat "the laws of nature" by not allowing $|S(j\omega)|$ to approach 1 near ω_q. Indeed, it turns out that the "price" is an even larger peak in $|S(j\omega)|$ for some other value of ω. We will show this using the continuity (analyticity) properties of functions of a complex variable. Actually, we will see that many interesting properties of linear feedback systems are a direct consequence of the properties of analytic functions.

[6] See e.g., Kwakernaak & Sivan (1972).

1.5 A Brief History

Bode (1945) used analytic function theory to examine the properties of feedback loops in the frequency domain. At the time, Bode was working at Bell Laboratories. He used complex variable theory to show that there were restrictions on the type of frequency domain response that could be obtained from a stable feedback amplifier circuit. In particular, he showed — mutatis mutandis — that the sensitivity function, S, (defined in (1.5) for a particular case) must satisfy the following integral relation for a stable open-loop plant

$$\int_0^\infty \log|S(j\omega)|\,d\omega = 0\,. \tag{1.22}$$

This result shows that it is not possible to achieve arbitrary sensitivity reduction (i.e., $|S| < 1$) at all points of the imaginary axis. Thus, if $|S(j\omega)|$ is smaller than one over a particular frequency range then it must necessarily be greater than one over some other frequency range.

Bode also showed that, for stable minimum phase systems, it was not necessary to specify both the magnitude and phase response in the frequency domain since each was determined uniquely by the other.

Horowitz (1963) applied Bode's theorems to the feedback control problem, and also obtained some preliminary results for open-loop unstable systems. These latter extensions turned out to be in error due to a missing term, but the principle is sound.

Francis & Zames (1984) studied the feedback constraints imposed by ORHP zeros of the plant in the context of $\mathcal{H}_\infty$ optimization. They showed that if the plant has zeros in the ORHP, then the peak magnitude of the frequency response of $S(j\omega)$ necessarily becomes very large if $|S(j\omega)|$ is made small over frequencies which exceed the magnitude of the zeros. This phenomenon has become known as the "water-bed" or "push-pop" effect.

Freudenberg & Looze (1985) brought many of the results together. They also produced definitive results for the open-loop unstable case. For example, in the case of an unstable open-loop plant, (1.22) generalizes to (see Theorem 3.1.4 in Chapter 3)

$$\frac{1}{\pi}\int_0^\infty \log|S(j\omega)|\,d\omega \geq \sum_{i=1}^{n_p} p_i\,, \tag{1.23}$$

where $\{p_i : i = 1, \ldots, n_p\}$ is the set of ORHP poles of the open-loop plant. Equality is achieved in (1.23) if the set $\{p_i : i = 1, \ldots, n_p\}$ also includes all the ORHP poles of the controller.

In addition, Freudenberg & Looze expressed the integral constraints in various formats, both along the lines of Bode and in a different form using the related idea of Poisson integrals. In particular, the Poisson integrals permit the derivation of an insightful closed expression that displays the water-bed effect experienced by nonminimum phase systems. This expression, however, is not tight if the plant has more than one ORHP zero.

About the same time, O'Young & Francis (1985) used Nevanlinna-Pick theory to characterize the smallest upper bound on the norm of the multivariable sensitivity function over a frequency range, with the constraint that the norm remain bounded at all frequencies. This characterization can be used to show the water-bed effect in multivariable nonminimum phase systems, and is tight for any number of ORHP zeros of the plant. Yet, no closed expression is available for this characterization but rather it has to be computed iteratively for each given plant.

In 1987, Freudenberg & Looze extended the Bode integrals to scalar plants with time delays. In 1988 the same authors published a book that summarized the results for scalar systems, and also addressed the multivariable case using singular values.

In 1990, Middleton obtained Bode-type integrals for the complementary sensitivity function T. For example, the result equivalent to (1.23), for an open-loop system having at least two pure integrators, is (see Theorem 3.1.5 in Chapter 3)

$$\frac{1}{\pi}\int_0^\infty \log|T(j\omega)|\frac{d\omega}{\omega^2} \geq \frac{\tau}{2} + \sum_{i=1}^{n_q}\frac{1}{q_i}, \tag{1.24}$$

where $\{q_i : i = 1, \ldots, n_q\}$ is the set of ORHP zeros of the open-loop plant, and τ is the plant pure time delay.

Comparing (1.24) with (1.10) we see that (perhaps not unexpectedly) there is a strong connection between the time and frequency domain results. Indeed, this is reasonable since the only elements that are being used are the complementarity, interpolation and analyticity constraints introduced in §1.2.

Recent extensions of the results include multivariable systems, filtering problems, periodic systems, sampled-data systems and, very recently, nonlinear systems. We will cover all of these results in the remainder of the book.

1.6 Summary

This chapter has introduced the central topic of this book. We are concerned with fundamental limitations in the design of dynamical systems, limitations that are imposed by structural and constitutive characteristics of the system under study. As we have seen, fundamental limitations are central to other disciplines; indeed, we have provided as examples the Cramér-Rao Inequality of Estimation Theory, and the Shannon Theorem of Communications.

We have presented, through an example of scalar feedback control, two of the mappings that are central to this book, namely, the sensitivity and complementarity sensitivity functions. These mappings are indicators of closed-loop performance as well as stability robustness and, as such, it is natural to require that they meet certain desired design specifications. We argue that it is important to establish the limits that one faces when attempting to achieve these specifications before any design is carried out. For example, it is impossible for a stable closed-loop system to achieve sensitivity reduction over a frequency range where the open-loop system has a pure imaginary zero. More generally, ORHP zeros and poles of the open-loop system impose constraints on the achievable frequency response of the sensitivity and complementarity sensitivity functions.

As a further illustration of these constraints, we have studied the effect on the closed-loop step response of pure integrators and ORHP zeros and poles of the open-loop system. We have seen, inter-alia, that a stable unity feedback system with a real ORHP open-loop zero must have undershoot in its step response. A similar conclusion holds with ORHP open-loop poles and overshoot in the step response. These limitations are obviously worse for plants having both nonminimum phase zeros and unstable poles; the inverted pendulum example illustrates these complications.

Finally, we have provided an overview of the published work that focuses on systems design limitations, the majority of which build on the original work of Bode (1945).

Notes and References

Some of the studies of Bode seem to have been paralleled in Europe. For example, some old books refer to the Bode gain-phase relationship as the *Bayard-Bode* gain-phase relationship (e.g., Gille, Pelegrin & Decaulne (1959, pp. 154-155), Naslin (1965)), although precise references are not given.

§1.3 is mainly extracted from Middleton (1991).

Part II

Limitations in Linear Control

2

Review of General Concepts

This chapter collects some concepts related to linear, time-invariant systems, as well as properties of feedback control systems. It is mainly intended to introduce notation and terminology, and also to provide motivation and a brief review of the background material for Part II. The interested reader may find a more extensive treatment of the topics covered here in the books and papers cited in the Notes and References section at the end of the chapter.

Notation. As usual, $\mathbb{N}$, $\mathbb{R}$ and $\mathbb{C}$ denote the natural, real and complex numbers, respectively. $\mathbb{N}_0$ denotes the set $\mathbb{N} \cup \{0\}$. The *extended complex plane* is the set of all finite complex numbers (the complex plane $\mathbb{C}$) and the *point at infinity*, ∞. We denote the extended complex plane by $\mathbb{C}_e = \mathbb{C} \cup \{\infty\}$. The real and imaginary parts of a complex number, s, are denoted by $\operatorname{Re} s$ and $\operatorname{Im} s$ respectively.

We will denote by $\mathbb{C}^+$ and $\mathbb{C}^-$ the open right and left halves of the complex plane, and by $\overline{\mathbb{C}^+}$ and $\overline{\mathbb{C}^-}$ their corresponding closed versions. Sometimes, we will use the obvious acronyms ORHP, OLHP, CRHP, and CLHP. Similarly, the symbols $\mathbb{D}$ and $\overline{\mathbb{D}}^c$ denote the regions inside and outside the unit circle $|z| = 1$ in the complex plane, and $\overline{\mathbb{D}}$ and $\mathbb{D}^c$ their corresponding closed versions.

The Laplace and $\mathcal{Z}$ transforms of a function f are denoted by $\mathcal{L}f$ and $\mathcal{Z}f$, respectively. In general, the symbol s is used to denote variables when working with Laplace transforms, and z when working with $\mathcal{Z}$ transforms. Finally, we use lower case letters for time domain functions, and upper case letters for both constant matrices and transfer functions.

2.1 Linear Time-Invariant Systems

A common practice is to assume that the system under study is linear time-invariant (LTI), causal, and of finite-dimension.[1] If the signals are assumed to evolve in continuous time,[2] then an input-output model for such a system in the time domain has the form of a convolution equation,

$$y(t) = \int_{-\infty}^{\infty} h(t-\tau)\, u(\tau)\, d\tau \ , \tag{2.1}$$

where u and y are the system's input and output respectively. The function h in (2.1) is called the *impulse response* of the system, and causality means that $h(t) = 0$ for $t < 0$.

The above system has an equivalent *state-space* description

$$\begin{aligned} \dot{x}(t) &= Ax(t) + Bu(t) \ , \\ y(t) &= Cx(t) + Du(t) \ , \end{aligned} \tag{2.2}$$

where A, B, C, D are real matrices of appropriate dimensions.

An alternative input-output description, which is of special interest here, makes use of the *transfer function*,[3] corresponding to system (2.1). The transfer function, H say, is given by the Laplace transform of h in (2.1), i.e.,

$$H(s) = \int_{0}^{\infty} e^{-st} h(t)\, dt \ .$$

After taking Laplace transform, (2.1) takes the form

$$Y(s) = H(s)U(s) \ , \tag{2.3}$$

where U and Y are the Laplace transforms of the input and output signals respectively.

The transfer function is related with the state-space description as follows

$$H(s) = C(sI - A)^{-1}B + D \ ,$$

which is sometimes denoted as

$$H \overset{s}{=} \left[\begin{array}{c|c} A & B \\ \hline C & D \end{array}\right] \ . \tag{2.4}$$

---

[1] For an introduction to these concepts see e.g., Kailath (1980), or Sontag (1990) for a more mathematically oriented perspective.

[2] Chapters 3 and 4 assume LTI systems in continuous time, Chapter 5 deals with periodically time-varying systems in discrete time, and Chapter 6 with sampled-data systems, i.e., a combination of digital control and LTI plants in continuous time.

[3] Sometimes the name *transfer matrix* is also used in the multivariable case.

We next discuss some properties of transfer functions. The transfer function H in (2.4) is a matrix whose entries are scalar rational functions (due to the hypothesis of finite-dimensionality) with real coefficients. A scalar rational function will be said to be *proper* if its *relative degree*, defined as the difference between the degree of the denominator polynomial minus the degree of the numerator polynomial, is nonnegative. We then say that a transfer matrix H is proper if all its entries are proper scalar transfer functions. We say that H is *biproper* if both H and H^{-1} are proper. A square transfer matrix H is *nonsingular* if its determinant, det H, is not identically zero.

For a discrete-time system mapping a discrete input sequence, u_k, into an output sequence, y_k, an appropriate input-output model is given by

$$Y(z) = H(z)U(z) ,$$

where U and Y are the $\mathcal{Z}$ transforms of the sequences u_k and y_k, and are given by

$$U(z) = \sum_{k=0}^{\infty} u_k z^{-k} , \qquad \text{and} \qquad Y(z) = \sum_{k=0}^{\infty} y_k z^{-k} ,$$

and where H is the corresponding transfer matrix in the $\mathcal{Z}$-transform domain. All of the above properties of transfer matrices apply also to transfer functions of discrete-time systems.

2.1.1 Zeros and Poles

The *zeros* and *poles* of a scalar, or single-input single-output (SISO), transfer function H are the roots of its numerator and denominator polynomials respectively. Then H is said to be *minimum phase* if all its zeros are in the OLHP, and *stable* if all its poles are in the OLHP. If H has a zero in the CRHP, then H is said to be *nonminimum phase*; similarly, if H has a pole in the CRHP, then H is said to be *unstable*.

Zeros and poles of multivariable, or multiple-input multiple-output (MIMO), systems are similarly defined but also involve directionality properties. Given a proper transfer matrix H with the minimal realization[4] (A, B, C, D) as in (2.2), a point $q \in \mathbb{C}$ is called a *transmission zero*[5] of H if there exist complex vectors x and Ψ_o such that the relation

$$\begin{bmatrix} x^* & \Psi_o^* \end{bmatrix} \begin{bmatrix} qI - A & -B \\ -C & -D \end{bmatrix} = 0 \tag{2.5}$$

[4] A minimal realization is a state-space description that is both controllable and observable.

[5] Since transmission zeros are the only type of multivariable zeros that we will deal with, we will often refer to them simply as "zeros". See MacFarlane & Karcanias (1976) for a complete characterization.

holds, where $\Psi_o^*\Psi_o=1$ (the superscript '$*$' indicates conjugate transpose). The vector Ψ_o is called the *output zero direction* associated with q and, from (2.5), it satisfies $\Psi_o^* H(q) = 0$. Transmission zeros verify a similar property with *input zero directions*, i.e., there exists a complex vector Ψ_i, $\Psi_i^*\Psi_i = 1$, such that $H(q)\Psi_i = 0$. A zero direction is said to be *canonical* if it has only one nonzero component.

For a given zero at $s = q$ of a transfer matrix H, there may exist more than one input (or output) direction. In fact, there exist as many input (or output) directions as the drop in rank of the matrix $H(q)$. This deficiency in rank of the matrix $H(s)$ at $s = q$ is called the *geometric multiplicity* of the zero at frequency q.

The *poles* of a transfer matrix H are the eigenvalues of the evolution matrix of any minimal realization of H. We will assume that the sets of ORHP zeros and poles of H are disjoint. Then, as in the scalar case, H is said to be *nonminimum phase* if it has a transmission zero at $s = q$ with q in the CRHP. Similarly, H is said to be *unstable* if it has a pole at $s = p$ with p in the CRHP. By extension, a pole in the CRHP is said to be unstable, and a zero in the CRHP is called nonminimum phase.

It is known (e.g., Kailath 1980, p. 467) that if H admits a left or right inverse, then a pole of H will be a zero of H^{-1}. In this case we will refer to the input and output directions of the pole as those of the corresponding zero of H^{-1}.

With a slight abuse of terminology, the above notions of zeros and poles will be used also for nonproper transfer functions, without of course the state-space interpretation.

Finally, poles and zeros of discrete-time systems are defined in a similar way, the stability region being then the open unit disk instead of the OLHP. In particular, a transfer function is nonminimum phase if it has zeros outside the open unit disk, $\mathbb{D}$, and it is unstable if it has poles outside $\mathbb{D}$.

For certain applications, it will be convenient to factorize transfer functions of discrete systems in a way that their zeros at infinity are explicitly displayed.

Example 2.1.1. A proper transfer function corresponding to a scalar discrete-time system has the form

$$H(z) = \frac{b_0 z^m + \cdots + b_m}{z^n + a_1 z^{n-1} + \cdots + a_n}, \tag{2.6}$$

where $n \geq m$. Let $\delta = n - m$ be the relative degree of H given above. Then H can be equivalently written as

$$H(z) = \tilde{H}(z) z^{-\delta}, \tag{2.7}$$

where $\tilde{H}$ is a biproper transfer function, i.e., it has relative degree zero. Note that (2.7) explicitly shows the zeros at infinity of H.[6] ○

2.1.2 *Singular Values*

At a fixed point $s \in \mathbb{C}$, let the singular value decomposition (Golub & Van Loan 1983) of a transfer matrix $H \in \mathbb{C}^{n\times n}$ be given by

$$H(s) = \sum_{i=1}^{n} \sigma_i(H(s)) v_i^*(H(s)) u_i(H(s)) \ ,$$

where $\sigma_i(H(s))$ are the *singular values* of $H(s)$, and are ordered so that $\sigma_1 \geq \sigma_2 \geq \cdots \geq \sigma_n$. Each set of vectors v_i and u_i form an orthonormal basis of the space $\mathbb{C}^n$ and are termed the *left* and *right singular vectors*, respectively. When the singular values are evaluated on the imaginary axis, i.e., for $s = j\omega$, then they are called *principal gains* of the transfer matrix, and the corresponding singular vectors are the *principal directions*. Principal gains and directions are useful in characterizing directionality properties of matrix transfer functions (Freudenberg & Looze 1987).

It is well-known that the singular values of H can be alternatively determined from the relation

$$\sigma_i^2(H(s)) = \lambda_i(H^*(s)H(s)) \ , \tag{2.8}$$

where $\lambda_i(H^*H)$ denotes the i-th eigenvalue of the matrix H^*H. We will denote the largest singular value of H by $\overline{\sigma}(H)$, and its smallest singular value by $\underline{\sigma}(H)$.

2.1.3 *Frequency Response*

The *frequency response* of a stable system, is defined as the response in steady-state (i.e., after the natural response has died out) to complex sinusoidal inputs of the form $u = u_0\, e^{j\omega t}$, where u_0 is a constant vector. It is well known that this response, denoted by y_{ss}, is given by

$$y_{ss}(t) = H(j\omega)\, u_0\, e^{j\omega t} \ .$$

Hence, the steady-state response of a stable transfer function H to a complex sinusoid of frequency ω is given by the input scaled by a "complex gain" equal to $H(j\omega)$.

For scalar systems, note that $H(j\omega) = |H(j\omega)| e^{j \arg H(j\omega)}$. It is usual to call $H(j\omega)$ the frequency response of the system, and $|H(j\omega)|$ and $\arg H(j\omega)$ the

[6]Note that the transfer function $H(z)$ has, including those at ∞, the same number of zeros and poles, i.e., n. In fact, a rational function assumes *every value* the same number of times (e.g., Markushevich 1965, p. 163), 0 and ∞ being just two particular values of interest.

magnitude and *phase* frequency responses, respectively. Note that the magnitude frequency response gives the "gain" of H at each frequency, i.e., the ratio of output amplitude to input amplitude. It is a common practice to plot the logarithm of the magnitude response,[7] and the phase response versus ω on a logarithmic scale. These are called the *Bode plots*.

For multivariable systems, the extension of these concepts is not unique. One possible characterization of the gain of a MIMO system is by means of its principal gains, defined in §2.1.2. In particular, the smallest and largest principal gains are of special interest, since

$$\underline{\sigma}(H(j\omega)) \leq \frac{|H(j\omega)u(j\omega)|}{|u(j\omega)|} \leq \overline{\sigma}(H(j\omega)) , \tag{2.9}$$

where $|\cdot|$ denotes the Euclidean norm. Hence, the gain of H (understood here as the ratio of output norm to input norm) is always between its smallest and largest principal gains.

A useful measure of the gain of a system is obtained by taking the supremum over all frequencies of its largest principal gain. Let H be proper transfer function with no poles on the imaginary axis; then the *infinity norm* of H, denoted by $\|H\|_\infty$, is defined as

$$\|H\|_\infty = \sup_\omega \overline{\sigma}(H(j\omega)) . \tag{2.10}$$

For scalar systems $\overline{\sigma}(H(j\omega)) = |H(j\omega)|$, and hence the infinity norm is simply the peak value of the magnitude frequency response.

2.1.4 *Coprime Factorization*

Coprime factorization of transfer matrices is a useful way of describing multivariable systems. It consists of expressing the transfer matrix in question as a "ratio" between stable transfer matrices. Due to the noncommutativity of matrices, there exist left and right coprime factorizations. We will use the notation "lcf" and "rcf" to stand for left and right coprime factorization, respectively.

The following definitions are reviewed from Vidyasagar (1985).

Definition 2.1.1 (Coprimeness). Two stable and proper transfer matrices $\tilde{D}$, $\tilde{N}$ (N, D) having the same number of rows (columns) are left (right) *coprime* if and only if there exist stable and proper transfer matrices $\tilde{Y}$, $\tilde{X}$ (X, Y) such that

$$\tilde{N}\tilde{X} + \tilde{D}\tilde{Y} = I . \qquad (XN + YD = I .)$$

◦

[7] Frequently in *decibels (dB)*, where $|H|_{dB} = 20 \log_{10} |H|$.

Definition 2.1.2 (Lcf, Rcf). Suppose H is a proper transfer matrix. An ordered pair, $(\tilde{D}, \tilde{N})$, of proper and stable transfer matrices is a *lcf* of H if $\tilde{D}$ is square and $\det(\tilde{D}) \neq 0$, $H = \tilde{D}^{-1}\tilde{N}$, and $\tilde{D}$, $\tilde{N}$ are right coprime.

Similarly, an ordered pair (N, D), of proper and stable transfer matrices is a *rcf* of H if D is square and $\det(D) \neq 0$, $H = ND^{-1}$, and N, D are right coprime. ◦

$\tilde{N}$ and N will be called the *numerators* of the lcf or rcf, respectively. Similarly, $\tilde{D}$ and D will be called the *denominators* of the lcf or rcf, respectively.

Every proper transfer matrix admits left and right coprime factorizations. Also, if $H = \tilde{D}^{-1}\tilde{N} = ND^{-1}$, then (Kailath 1980, Chapter 6)

- q is a zero of H if and only if $N(s)$ ($\tilde{N}(s)$) loses rank at $s = q$.
- p is a pole of H if and only if $D(s)$ ($\tilde{D}(s)$) loses rank at $s = p$.

Note that all of the concepts defined above apply to both continuous and discrete-time systems. In particular, for continuous-time systems, the factorizations are performed over the ring of proper transfer matrices with poles in the OLHP; on the other hand, for discrete-time systems, the factorizations are performed over the ring of proper transfer matrices with poles in the open unit disk.

2.2 Feedback Control Systems

Most of Part II is concerned with the unity feedback configuration of Figure 2.1, where the *open-loop system*, L, is formed of the series connection of the *plant*, G, and *controller*, K, i.e.,

$$L = GK .$$

In broad terms, the general feedback control problem is to design the controller for a given plant such that the *output*, y, follows the *reference input*, r, in some specified way. This task has to be accomplished, in general, in the presence of disturbances affecting the loop. Two common disturbances are *output disturbances*, d, and *sensor* or *measurement noise*, w. The design generally assumes a *nominal* model for the plant, and then takes additional precautions to ensure that the system continues to perform in a reasonable fashion under perturbations of this nominal model.

We consider throughout the rest of this chapter that the (nominal) plant and controller are continuous-time, linear, time-invariant systems, described by transfer functions G and K respectively. Some of these assumptions will be relaxed in other chapters.

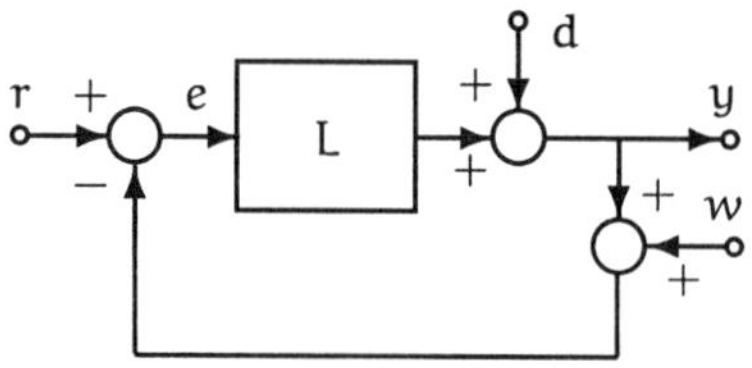

FIGURE 2.1. Feedback control system.

2.2.1 *Closed-Loop Stability*

A basic requirement for a feedback control loop is that of *internal stability*, or simply closed-loop stability, according to the following definition.

Definition 2.2.1 (Internal Stability). Let the open-loop system in Figure 2.1 be given by $L = GK$. Then the closed loop is *internally stable* if and only if $I + GK$ is nonsingular and the four transfer functions

$$\begin{bmatrix} (I+GK)^{-1} & -(I+GK)^{-1}G \\ K(I+GK)^{-1} & I-K(I+GK)^{-1}G \end{bmatrix},$$

have all of their poles in the OLHP. ∘

We say that L is *free of unstable hidden modes* if there are no cancelations of CRHP zeros and poles between the plant and controller whose cascade connection forms L. In the multivariable case, cancelations involve both location and directions of zeros and poles. For example, if the plant and controller are expressed using coprime factorizations as

$$G = \tilde{D}_G^{-1}\tilde{N}_G = N_G D_G^{-1},$$
$$K = \tilde{D}_K^{-1}\tilde{N}_K = N_K D_K^{-1},$$

then $L = GK$ has an unstable hidden mode if $\tilde{D}_G$ and $\tilde{N}_K$ share an ORHP zero with the same input direction (Gómez & Goodwin 1995). It is easy to see that the closed loop is not internally stable if L has unstable hidden modes.

For discrete-time systems, the above concepts apply with the corresponding definition of the stability region.

2.2.2 *Sensitivity Functions*

From Figure 2.1 we see that

$$Y = D + L[R - W - Y].$$

Solving for Y we have

$$Y = (I+L)^{-1}D + L(I+L)^{-1}[R - W]. \tag{2.11}$$

The above expression suggests that two functions of central interest in the design of feedback systems are the *sensitivity function*, S, and *complementary sensitivity function*, T, given respectively by

$$S(s) = [I + L(s)]^{-1}\,, \qquad \text{and} \qquad T(s) = L(s)\,[I + L(s)]^{-1}\,. \tag{2.12}$$

We see that S maps output disturbances to the output, and T maps both reference and sensor noise to the output. It is also straightforward to check that S also maps the reference input to the error in Figure 2.1.[8] In fact, S and T are intimately connected with closed-loop performance and robustness properties, as we show in the following subsections.

2.2.3 Performance Considerations

If the closed loop is internally stable, S and T are stable transfer functions. Then, the steady-state response of the system output to a disturbance $d = d_0 e^{j\omega t}$, for $d_0 \in \mathbb{C}^n$, is given by (cf. §2.1.3)

$$y_d(t) = S(j\omega) d_0 e^{j\omega t}\,.$$

Thus, the response to d can be made small by requiring $|S(j\omega)d_0| \ll 1$. Clearly, from (2.9), the response of the system to output disturbances of any direction and frequency ω can be made small if

$$\overline{\sigma}(S(j\omega)) \ll 1\,. \tag{2.13}$$

A similar analysis for T shows that the response of the system to sensor noise of any direction and frequency ω can be made small if

$$\overline{\sigma}(T(j\omega)) \ll 1\,. \tag{2.14}$$

Recall that T also maps the reference input to the system output. It is thus clear that the feedback loop will have poor performance unless the frequency content of the reference input and the measurement noise are disjoint. This shows that there is an inherent trade-off between reference tracking and noise attenuation.

Another trade-off arises between attenuation of output disturbances and sensor noise. Namely, the relationship $S + T = I$ implies that the specifications (2.13) and (2.14) cannot be both satisfied over the same frequency range.

The above discussion suggests that a sensible design should focus on achievable specifications. Consider, for example, the scalar case. If we assume that the reference input has low frequency content (which is typically the case), then it is reasonable to require

$$|T(j\omega)| \approx 1, \qquad \forall \omega \in [0, \omega_1]\,,$$

---

[8] In the following chapters, d and w in this figure are frequently set to zero; the reader is asked to keep in mind that S and T also lead the output response to these signals.

for some ω_1. This is equivalent to

$$|S(j\omega)| \ll 1, \qquad \forall \omega \in [0, \omega_1] . \tag{2.15}$$

Note that the above specification implies that output disturbances having frequency content in the range $[0, \omega_1]$ will be attenuated at the system output. However, making ω_1 too large may result in large magnitudes at the *plant input*. To see this, note that $|S| = 1/|1 + L|$ can only be made small by making the magnitude of the open-loop system $L = GK$ large. However, making the open-loop gain large over a frequency range where the gain of the plant $|G|$ is small requires a high controller gain; hence, the response of the plant input to disturbances in this range will be very large, usually leading to saturation. We conclude that the range of sensitivity reduction (and thus of reference tracking) is limited by the plant's input response to disturbances.

It is then common to aim at a design that achieves specifications of the form

$$\begin{aligned} |S(j\omega)| &\ll 1, \qquad \forall \omega \in [0, \omega_1] , \\ |T(j\omega)| &\ll 1, \qquad \forall \omega \in [\omega_2, \infty) , \end{aligned}$$

for some $\omega_2 > \omega_1$. Typical shapes for S and T satisfying specifications of this kind are shown in Figure 2.2.

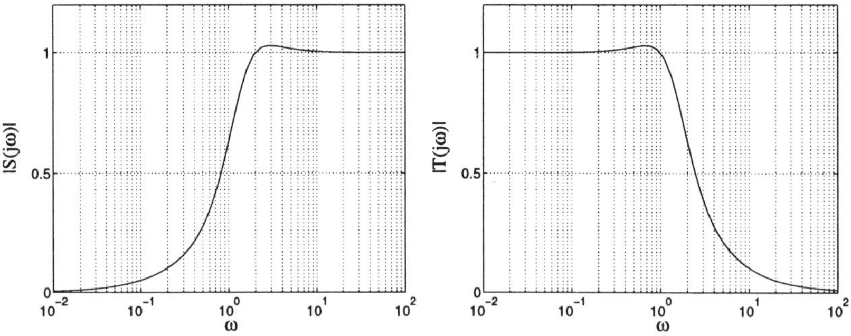

FIGURE 2.2. Typical shapes for $|S(j\omega)|$ and $|T(j\omega)|$.

If the above requirements are satisfied, then the peak values of S and T will occur in the intermediate range (ω_1, ω_2). It is desirable to keep these peaks as small as possible in order to avoid overly large sensitivity to disturbances and excessive influence of sensor noise. A key point to emerge later in the book, however, is that nonminimum phase zeros and unstable poles of the open-loop system impose lower limits on the achievable peaks of the sensitivity and complementary sensitivity functions.

2.2.4 *Robustness Considerations*

If the actual plant, $\tilde{G}$ say, differs from the nominal model G, then additional requirements are needed to preserve stability and good performance. The usual procedure is to consider a particular model for the allowable perturbations (usually based on practical considerations), and then to impose a condition on the design that will guarantee robust stability (and/or performance) for all perturbed plants within the set of models having the allowable perturbations.

Two common models used to describe plant uncertainty are *divisive* and *multiplicative* perturbation models. The (input) divisive perturbation model, or perturbation of the plant inverse, assumes that

$$\tilde{G} = G(I + \Delta)^{-1} , \tag{2.16}$$

where Δ is a stable transfer function satisfying a frequency dependent magnitude bound

$$\overline{\sigma}(\Delta(j\omega)) \leq W(\omega), \qquad \forall\omega . \tag{2.17}$$

The (output) multiplicative perturbation model assumes that

$$\tilde{G} = (I + \Delta)G , \tag{2.18}$$

where Δ is a stable transfer function satisfying a frequency dependent magnitude bound similar to (2.17). If this is the only information available about the uncertainty, then Δ is termed *unstructured* uncertainty.

It turns out that the sensitivity and complementary sensitivity functions each characterize stability robustness of the system against divisive and multiplicative plant uncertainty respectively. Indeed, the feedback system will be stable for all plants described by (2.16), (2.17), with Δ stable, if and only if the system is stable when $\Delta = 0$ and

$$\overline{\sigma}(S(j\omega)) < 1/W(\omega) , \qquad \forall\omega , \tag{2.19}$$

where W is the bound in (2.17).

Similarly, if the nominal closed loop is stable, then the perturbed closed loop will remain stable for all plants described by (2.18), (2.17), with Δ stable, if and only if

$$\overline{\sigma}(T(j\omega)) < 1/W(\omega) , \qquad \forall\omega . \tag{2.20}$$

We remark that the same condition (2.19) is also necessary and sufficient for robust stability against *additive* perturbations of the open-loop of the form $\tilde{L} = L + \Delta$, where Δ is stable and satisfies (2.17).

Note that specifications such as (2.19) and (2.20) give more insights into the desirable shapes for S and T. For example, in the scalar case, a typical bound $W(\omega)$ for multiplicative perturbation grows at high frequencies.[9]

---

[9]The multiplicative perturbation model is useful to describe high frequency modelling inaccuracy, common in practice (Doyle & Stein 1981).

Thus, robust stability against multiplicative uncertainty requires that $|T(j\omega)|$ be small at high frequencies.

Finally, we briefly turn to performance robustness considerations. We have seen in §2.2.3 that S and T are indicators of feedback system performance. We will next consider how these functions are affected by plant variations. We focus on the scalar case but similar conclusions hold, mutatis mutandis, for multivariable systems.

Assume that the loop gain changes from its nominal value L to its actual value $\tilde{L}$. It is not difficult to show that the relative changes in the sensitivity and complementary sensitivity functions are given by

$$\frac{\tilde{S} - S}{\tilde{S}} = -T\,\frac{\tilde{L} - L}{L} ,$$

$$\frac{\tilde{T} - T}{\tilde{T}} = S\,\frac{\tilde{L} - L}{\tilde{L}} .$$

These relations show that the sensitivity function will be robust with respect to changes in the loop gain in those frequency ranges where the nominal complementary sensitivity is small (typically at high frequencies); conversely the complementary sensitivity will be robust with respect to changes in the loop gain in those frequency ranges where the nominal sensitivity is small (typically at low frequencies).

2.3 Two Applications of Complex Integration

In the previous section we discussed the importance of attaining a desired shape for the frequency response of relevant transfer functions of feedback control loops. As we will see later, zeros and poles of the plant to be controlled impose restrictions on the behavior of these functions at particular complex frequencies in the ORHP. A powerful tool exists by means of which these restrictions in the ORHP can be transformed directly into equivalent constraints on the frequency response. Indeed, the imaginary axis can be looked upon as the boundary of the ORHP, which is the region where the special restrictions on the transfer functions occur, so that the broad mathematical problem is that of relating the behavior of a function inside a region to its behavior on the boundary of the region. A mechanism for this purpose is found in Cauchy's theory of analytic functions and its integrals around closed contours. The remainder of the book will make extensive use of this theory, and a self-contained review of this theory is given in Appendix A for convenience.

As a preliminary use of Cauchy's theory of complex integration, we will present two well-known applications. First we will establish the Nyquist stability theorem. We treat the continuous-time case; however, similar arguments apply to the discrete-time problem. As a second application, we will derive the Bode gain-phase relationship. The latter relationship will also serve as an introduction to the constraints introduced by nonminimum phase zeros on the achievable shape of the frequency response.

2.3.1 *Nyquist Stability Criterion*

The Nyquist criterion depends on a result known as the *Principle of the Argument*. This result uses the residue theorem (Theorem A.9.1 in Appendix A) to obtain information about the number of zeros of an analytic function (or about the number of zeros minus the number of poles of a *meromorphic* function.[10]

Theorem 2.3.1 (Principle of the Argument). Let C be a closed simple contour contained in a simply connected domain D. Let f be a meromorphic function in D and suppose that f has no zeros or poles on C. Let n_q be the number of zeros and n_p the number of poles of f in the interior of C, where a multiple zero or pole is counted according to its multiplicity. Then

$$\oint_C \frac{f'(s)}{f(s)}\, ds = j2\pi\,(n_q - n_p)\,, \tag{2.21}$$

where f' denotes the derivative of f, and integration in (2.21) is performed in the counter-clockwise direction.

Proof. The only possible singularities of the meromorphic function f'/f inside C are the zeros and poles of f. Suppose that s_0 represents a zero or pole of f with multiplicity n. We can write

$$f(s) = (s - s_0)^n \tilde{f}(s)\,, \tag{2.22}$$

where $\tilde{f}$ is analytic in a neighborhood of s_0 and $\tilde{f}(s_0) \neq 0$. Hence

$$\frac{f'(s)}{f(s)} = \frac{n}{s - s_0} + \frac{\tilde{f}'(s)}{\tilde{f}(s)} \tag{2.23}$$

in a neighborhood of s_0. Thus the residue of f'/f at s_0 is n (see §A.9.1 in Appendix A). Note that n is negative if s_0 is a zero of f, and n is positive if s_0 is a pole of f. Applying Theorem A.9.1 of Appendix A yields (2.21) and completes the proof. □

We will next see why Theorem 2.3.1 is called the principle of the argument. Let the equation for C be $s = s(\zeta)$, with ζ in $[a, b]$ and let $\xi = f(s)$.

[10] A function is *meromorphic* in a domain D if it is defined and analytic in D except for poles.

Consider now the image of C under f, $\tilde{C}$ say. Thus, the equation of $\tilde{C}$ in the ξ plane is

$$\xi = f(s(\zeta)) , \qquad \zeta \in [a, b] .$$

Since f is never zero on C, then the curve $\tilde{C}$ does not pass through the origin of the ξ plane. Hence, it is possible to define a continuous logarithm

$$F(\zeta) = \log f(s(\zeta))$$

on $\tilde{C}$. On any smooth portion of C, we have, by differentiation,

$$F'(\zeta) = \frac{f'(s(\zeta))}{f(s(\zeta))} s'(\zeta) ,$$

and hence, by the fundamental theorem of calculus (cf. (A.18) in Appendix A),

$$\int_a^b \frac{f'(s(\zeta))}{f(s(\zeta))} s'(\zeta)\, d\zeta = \log f(s(\zeta))\Big|_a^b .$$

This is equivalent to

$$\oint_C \frac{f'(s)}{f(s)}\, ds = \log f(s)\Big|_C = \log|f(s)|\Big|_C + j \arg f(s)\Big|_C .$$

The first term on the right is zero since C is closed. Then, dividing by 2π and using (2.21), we get

$$2\pi (n_q - n_p) = \arg f(s)\Big|_C . \tag{2.24}$$

That is, $n_q - n_p$ is the number of times the image curve $\tilde{C}$ winds around the origin in the ξ plane, when C is traversed in the counter-clockwise sense.

The principle of the argument finds immediate application in the Nyquist stability criterion. Specifically, let $L = GK$ be the open-loop transfer function of the feedback control system shown in Figure 2.1. Assume further that L is scalar. We have seen in §2.2.1 that the closed loop is internally stable if and only if there are no unstable cancelations in L, $L(\infty) \neq -1$, and all the zeros of the characteristic equation

$$1 + L(s) = 0 \tag{2.25}$$

have negative real part.

Consider next the contour C shown in Figure 2.3, consisting of the imaginary axis and a semicircle of infinite radius into the ORHP. If L has poles on the imaginary axis, then C must have small indentations to avoid them. Such a contour is called *Nyquist contour* and its image through $L(s)$ is called the *Nyquist plot* of L.

A necessary and sufficient condition for closed-loop stability is furnished by the following theorem.

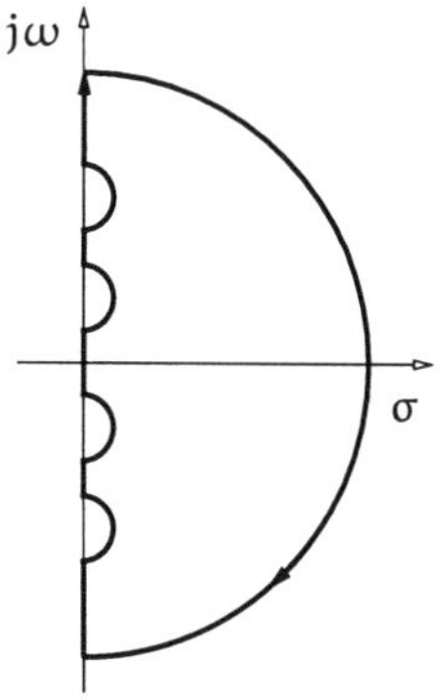

FIGURE 2.3. Nyquist contour.

Theorem 2.3.2 (Nyquist Stability Criterion). The feedback control system of Figure 2.1 is internally stable if and only if $L(\infty) \neq -1$ and the number of counter-clockwise revolutions made by the Nyquist plot of the open-loop transfer function $L = GK$ around the point $s = -1$, is equal to the number of unstable poles of G plus the number of unstable poles of K.

Proof. We apply Theorem 2.3.1 to the function $f = 1 + L$ using the Nyquist contour, C in Figure 2.3 (note that now integration is in the clockwise direction). Any pole or zero of $1 + L$ with positive real part will therefore lie within this contour. Thus, closed-loop stability is equivalent to $n_q = 0$ in (2.21) or (2.24).

It follows from (2.24) that the phase-shift increment of $1 + L(s)$ as s traverses the Nyquist contour is $2\pi(n_p - n_q)$, where n_p is the number of unstable poles of L. But, since L and $1+L$ have the same poles, it is equivalent to count revolutions of $L(s)$ around the point $(-1, 0)$. Thus the number of counter-clockwise revolutions of the Nyquist plot of $L = GK$ around the point $s = -1$ is different from the number of unstable poles of G plus the number of unstable poles of K if and only if there is an unstable cancelation in L, or $1 + L$ is nonminimum phase (i.e., $n_q \neq 0$). The result then follows. □

Example 2.3.1. Let L in Figure 2.1 be a strictly proper transfer function. Since $L(Re^{j\theta})$ vanishes as R becomes infinite, it follows that the phase-shift of $1 + L(s)$ as s traverses the Nyquist contour is given by its phase-shift when s moves on the imaginary axis. Moreover, since L has real coefficients, then it suffices to consider 2 times the phase-shift encountered as s moves along the positive imaginary axis. Also, note that $1 + L(j\omega)$ is the vector from the -1 point to the point on the Nyquist plot at frequency ω.

Now assume that L is strictly proper and has one unstable pole. According to the Nyquist stability criterion and the previous discussion, if the closed loop is stable, then the vector $1 + L(j\omega)$ must rotate an angle of π (in the counter-clockwise sense) as ω goes from 0 to ∞. But, since L is strictly proper, $1 + L(j\infty)$ is the vector from the -1 point to the origin of the complex plane. It thus follows that closed-loop stability requires $L(0) < -1$. ∘

2.3.2 *Bode Gain-Phase Relationships*

As a further application of analytic function theory, we will review the gain-phase relationships originally developed by (Bode 1945), which establish that, for a stable minimum phase transfer function, the phase of the frequency response is uniquely determined by the magnitude of the frequency response and vice versa.

We begin by showing that the real and imaginary parts of a proper stable rational function with real coefficients are dependent of each other. We consider this dependence at points of the imaginary axis.

Theorem 2.3.3 (Bode's Real-imaginary Parts Relationship). [11] Let H be a proper stable transfer function, and suppose that, at $s = j\omega$, $H(s)$ can be written as $H(j\omega) = U(\omega) + jV(\omega)$, where U and V are real valued. Then for any ω_0

$$V(\omega_0) = \frac{2\omega_0}{\pi} \int_0^\infty \frac{U(\omega) - U(\omega_0)}{\omega^2 - \omega_0^2} \, d\omega \, . \tag{2.26}$$

Proof. Let C be the clockwise oriented contour shown in Figure 2.4, and consisting of the imaginary axis, with infinitely small indentations at the points $+j\omega_0$ and $-j\omega_0$ (C_2 and C_3 in Figure 2.4), and the semicircle C_1, which has infinite radius in the ORHP[12].

Then the functions

$$\frac{H(s) - U(\omega_0)}{s - j\omega_0} \quad \text{and} \quad \frac{H(s) - U(\omega_0)}{s + j\omega_0}$$

are analytic on and inside C. Hence, applying Cauchy's integral theorem (see §A.5.2 in Appendix A) to both functions and subtracting yields

$$\oint_C \left(\frac{H(s) - U(\omega_0)}{s - j\omega_0} - \frac{H(s) - U(\omega_0)}{s + j\omega_0} \right) ds = 0 \, . \tag{2.27}$$

The integral above may be decomposed as

$$2\omega_0 \int_{-\infty}^\infty \frac{H(j\omega) - U(\omega_0)}{\omega^2 - \omega_0^2} \, d\omega + I_1 + I_2 + I_3 = 0 \, , \tag{2.28}$$

[11] This is, in fact, one of the many real-imaginary relationships derived by Bode .

[12] This should be understood as follows: the indentations on the imaginary axis have radii $\rho > 0$, and the large semicircle in the ORHP has finite radius R. These are then considered in the limit as $\rho \to 0$ and $R \to \infty$. In fact, the large semicircle is nothing but an indentation around ∞.

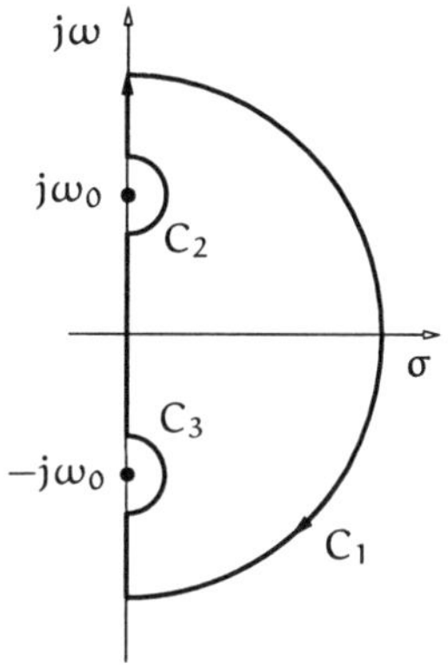

FIGURE 2.4. Contour used in Theorem 2.3.3.

where I_1, I_2, and I_3 are the integrals over C_1, C_2, and C_3 respectively. The integral on the imaginary axis can be written as

$$2\omega_0 \int_{-\infty}^{\infty} \frac{H(j\omega) - U(\omega_0)}{\omega^2 - \omega_0^2}\, d\omega = 4\omega_0 \int_0^{\infty} \frac{U(\omega) - U(\omega_0)}{\omega^2 - \omega_0^2}\, d\omega ,$$

since the real and imaginary parts of a transfer function evaluated at $s = j\omega$ are even and odd functions of ω respectively.

Next note that the integral I_1 vanishes because H is proper.

Now consider the integral I_2. Since the radius of C_2 is infinitely small, we can approximate $H(s)$ on C_2 by the constant $H(j\omega_0)$. We can also neglect the contribution of the fraction $1/(s+j\omega)$ in comparison with that of $1/(s-j\omega)$. Then

$$\begin{aligned} I_2 &= \int_{C_2} \frac{H(j\omega_0) - U(\omega_0)}{s - j\omega_0}\, ds \\ &= jV(\omega_0) \int_{C_2} \frac{1}{s - j\omega_0}\, ds \\ &= -\pi V(\omega_0) , \end{aligned}$$

where the last step follows from (A.32) in Appendix A.

Similarly, we can show that the integral I_3 equals $-\pi V(\omega_0)$. Finally, substituting into (2.28) and rearranging gives the desired expression. □

The above theorem shows that values on the $j\omega$-axis of the imaginary part of a stable and proper transfer function can be reconstructed from knowledge of the real part on the entire $j\omega$-axis. Conversely, under the assumption that the transfer function is *strictly* proper and stable, it can be shown that the real part can be obtained from the imaginary part, i.e.,

$$U(\omega_0) = -\frac{2}{\pi} \int_0^{\infty} \frac{\omega\, [V(\omega) - V(\omega_0)]}{\omega^2 - \omega_0^2}\, d\omega , \tag{2.29}$$

which follows similarly by adding instead of subtracting in (2.27).

It is shown in §A.6 of Appendix A that the values of a function analytic in a given region can be reconstructed from its values on the boundary. Combining this with relations (2.26) and (2.29), we deduce that the values of a stable and strictly proper transfer function on the CRHP are completely determined by the real or imaginary part of its frequency response.

We will next show that the gain and the phase of the frequency response of a stable, minimum phase transfer function are dependent on each other.

Theorem 2.3.4 (Bode's Gain-phase Relationship). Let H be a proper, stable, and minimum phase transfer function, such that $H(0) > 0$. Then, at any frequency ω_0, the phase $\phi(\omega_0) \triangleq \arg H(j\omega_0)$ satisfies

$$\phi(\omega_0) = \frac{1}{\pi}\int_{-\infty}^{\infty} \frac{d\log|H(j\omega_0 e^u)|}{du} \log\coth\left|\frac{u}{2}\right| du\,, \tag{2.30}$$

where $u = \log(\omega/\omega_0)$.

Proof. Consider

$$H(j\omega_0) = U(\omega_0) + jV(\omega_0) \triangleq |H(j\omega_0)|\, e^{j\phi(\omega_0)}\,. \tag{2.31}$$

Since H has no zeros in the CRHP, taking logarithms in (2.31) gives

$$\log H(j\omega_0) = \log|H(j\omega_0)| + j\phi(\omega_0) \triangleq m(\omega_0) + j\phi(\omega_0). \tag{2.32}$$

Comparing (2.31) and (2.32) shows that the magnitude characteristic $m(\omega_0)$ and the phase $\phi(\omega_0)$ are related to $\log H(j\omega_0)$ and to each other in the same way that $U(\omega_0)$ and $V(\omega_0)$ are related to $H(j\omega_0)$ and each other. Hence (2.29) and (2.26) immediately imply

$$m(\omega_0) = -\frac{2}{\pi}\int_0^{\infty} \frac{\omega\,[\phi(\omega) - \phi(\omega_0)]}{\omega^2 - \omega_0^2}\, d\omega\,, \tag{2.33}$$

$$\phi(\omega_0) = \frac{2\,\omega_0}{\pi}\int_0^{\infty} \frac{m(\omega) - m(\omega_0)}{\omega^2 - \omega_0^2}\, d\omega\,. \tag{2.34}$$

Note that the assumption that H has no zeros or poles in the CRHP guarantees the validity of the integrals above, since log H is analytic in the finite CRHP. The singularity at ∞ arising from a strictly proper H is ruled out in the chosen contour of integration (see footnote 12 on page 40). The fact that $|\log H(s)|/|s| \to 0$ when $|s| \to \infty$ eliminates the integral along the large semicircle in the ORHP.

Next, consider (2.34). Changing variables to $u = \log(\omega/\omega_0)$ and denoting $\tilde{m}(u) = m(\omega)$ gives

$$\begin{aligned}\phi(\omega_0) &= \frac{2}{\pi}\int_{-\infty}^{\infty} \frac{\tilde{m}(u) - m(\omega_0)}{e^u - e^{-u}}\, du\\ &= \frac{1}{\pi}\int_{-\infty}^{\infty} \frac{\tilde{m}(u) - m(\omega_0)}{\sinh(u)}\, du\,.\end{aligned}$$

Dividing the complete range of integration into separate ranges above and below $u = 0$ and integrating by parts yields

$$
\begin{aligned}
\phi(\omega_0) =& \frac{1}{\pi}\int_{-\infty}^{0} \frac{\tilde{m}(u) - m(\omega_0)}{\sinh(u)}\, du + \frac{1}{\pi}\int_{0}^{\infty} \frac{\tilde{m}(u) - m(\omega_0)}{\sinh(u)}\, du \\
=& \frac{1}{\pi}[\tilde{m}(u) - m(\omega_0)] \log\coth\left(\frac{-u}{2}\right)\Big|_{-\infty}^{0} \\
&+ \frac{1}{\pi}\int_{-\infty}^{0} \frac{d\tilde{m}(u)}{du} \log\coth\left(\frac{-u}{2}\right) du \\
&+ \frac{1}{\pi}[\tilde{m}(u) - m(\omega_0)] \log\coth\left(\frac{u}{2}\right)\Big|_{0}^{\infty} \\
&+ \frac{1}{\pi}\int_{0}^{\infty} \frac{d\tilde{m}(u)}{du} \log\coth\left(\frac{u}{2}\right) du\,.
\end{aligned}
\tag{2.35}
$$

Near $u = 0$, the quantity $\tilde{m}(u) - m(\omega_0)$ behaves proportionally to u, whilst $\log\coth(u/2)$ will vary as $-\log(u/2)$. Thus, at the limit $u \to 0$, the integrated portions of (2.35) behave as $u \log u$, which is known to vanish as $u \to 0$. As for the other limits, we have that $\lim_{u\to-\infty} \tilde{m}(u) = \lim_{\omega\to 0} m(\omega) = m(0)$, which is finite since H is stable and minimum phase; also $\lim_{u\to\infty} \tilde{m}(u)\log\coth(u/2) = \lim_{\omega\to\infty} m(\omega)2(\omega_0/\omega) = 0$. Hence both integrated portions in (2.35) are equal to zero. The result then follows on combining the remaining two integrals in (2.35). □

The implications of (2.30) can be easily appreciated using properties of the weighting function appearing in (2.30), namely

$$
\log\coth\left|\frac{u}{2}\right| = \log\left|\frac{\omega + \omega_0}{\omega - \omega_0}\right| . \tag{2.36}
$$

This function is plotted in Figure 2.5.

As we can see from this figure, the weighting function becomes logarithmically infinite at the point $\omega = \omega_0$. Thus, we conclude from (2.30) that the slope of the magnitude curve in the vicinity of ω_0, say c, determines the phase $\phi(\omega_0)$:

$$
\phi(\omega_0) \approx \frac{c}{\pi}\int_{-\infty}^{\infty} \log\coth\left|\frac{u}{2}\right| du = \frac{c}{\pi}\frac{\pi^2}{2} = c\frac{\pi}{2}\,.
$$

Hence, for stable and minimum phase transfer functions, a slope of $20c$ dB/decade in the gain in the vicinity of ω_0 implies a phase angle of approximately $c\,\pi/2$ rad sec^{-1}.

The above arguments lead classical designers to conclude that, to ensure closed-loop stability, the slope of the open-loop gain characteristic, $|L(j\omega)|$, should be in the range -20 to -30 dB/decade at the gain cross-over point

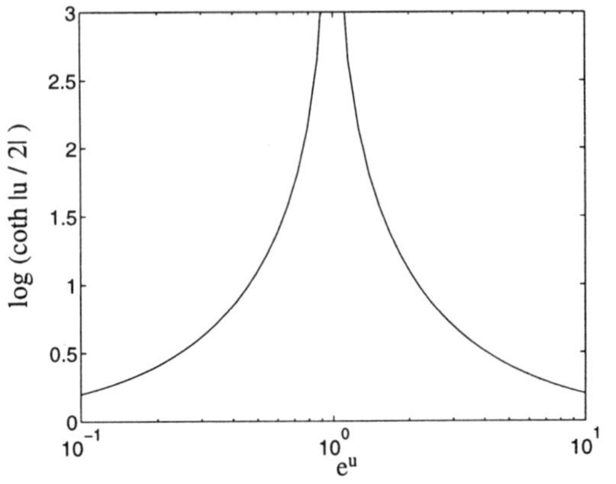

FIGURE 2.5. Weighting function of the Bode gain-phase relationship.

(i.e., at the frequency at which $|L| = 1$ or 0 dB), since this would imply the phase would be less than 180° i.e. the Nyquist plot would not encircle the '-1' point.

Example 2.3.2. Let us consider the X-29 aircraft design example discussed in §1.1 of Chapter 1. This system is (approximately) modelled by a strictly proper transfer function, G, which is unstable and nonminimum phase. For one flight condition, the unstable pole is at 6 and the nonminimum phase zero at 26. It is desired to use a stable, minimum phase controller in series with G, such that the closed loop is stable and has a phase margin of $\pi/4$. Consider the open-loop system formed by the cascade of plant and controller and modeled by a transfer function, L say. This system is neither stable nor minimum phase in open loop. However, we can associate with L another transfer function, $\tilde{L}$, defined as follows

$$L(s) = -\tilde{L}(s)\left(\frac{s-p}{s+p}\right)\left(\frac{s+q}{s-q}\right), \tag{2.37}$$

where $\tilde{L}$ is stable, minimum phase, and such that $\tilde{L}(0) > 0$, and where p and q correspond to the (real) ORHP pole and zero of G respectively.[13] The negative sign in (2.37) is necessary to guarantee a stable closed loop (see Example 2.3.1). We note that

$$|\tilde{L}(j\omega)| = |L(j\omega)| . \tag{2.38}$$

Now from (2.37) we have that $\phi(\omega_0) \triangleq \arg L(j\omega_0)$ is given by

$$\phi(\omega_0) = \arg \tilde{L}(j\omega_0) - \arg \frac{p - j\omega_0}{p + j\omega_0} - \arg \frac{j\omega_0 + q}{j\omega_0 - q} .$$

[13] Actually, the device used above to associate a stable, minimum phase transfer function with an unstable, nonminimum phase one will be used repeatedly in subsequent chapters.

Also, since $\tilde{L}$ is stable, minimum phase, and $\tilde{L}(0) > 0$, we can use (2.30) to obtain

$$\phi(\omega_0) = \frac{1}{\pi}\int_{-\infty}^{\infty} \frac{d\log|\tilde{L}(j\omega_0 e^u)|}{du} \log\coth\left|\frac{u}{2}\right| du - \arg\frac{(p-j\omega_0)(j\omega_0+q)}{(p+j\omega_0)(j\omega_0-q)} .$$

Finally, using (2.38) we obtain

$$\begin{aligned}\phi(\omega_0) &= \frac{1}{\pi}\int_{-\infty}^{\infty} \frac{d\log|L(j\omega_0 e^u)|}{du} \log\coth\left|\frac{u}{2}\right| du - \arg\frac{(p-j\omega_0)(j\omega_0+q)}{(p+j\omega_0)(j\omega_0-q)} \\ &= \frac{1}{\pi}\int_{-\infty}^{\infty} \frac{d\log|L(j\omega_0 e^u)|}{du} \log\coth\left|\frac{u}{2}\right| du - 2\arctan\frac{\omega_0/q+p/\omega_0}{1-p/q} .\end{aligned}$$

Say that we want a slope of -10 dB/decade at the gain cross-over frequency; then this means that the first term in the above expression is approximately $-\pi/4$. Now, the second term has its least negative value for $\omega_0 = \sqrt{pq}$, and for $p = 6$, $q = 26$, we obtain

$$\begin{aligned}\phi(\omega_0) &\leq (-\pi/4 - 1.79)\text{rad} \\ &= -147^\circ .\end{aligned}$$

Hence, the maximum possible phase margin in this case is 33°.

2.4 Summary

In this chapter we have provided an overview of system properties, with emphasis on the frequency response of LTI systems. We have considered stability, performance and robustness of feedback control loops. We have also shown that the sensitivity and complementary sensitivity functions can be used to quantify important feedback properties.

In addition, we have given two applications of Cauchy's integral theorems to systems theory, namely the Nyquist stability criterion and the Bode gain-phase relationship.

Notes and References

This chapter is mainly oriented towards people with some background in control and systems theory. For a more extensive introduction see e.g., Kailath (1980), Franklin, Powell & Emami-Naeini (1994) or Sontag (1990). The material included here was based on the references below.

LTI Systems

See e.g., Francis (1991), Maciejowski (1991), Middleton & Goodwin (1990), Kailath (1980) and Franklin et al. (1994).

Feedback Control Systems

See e.g., Freudenberg & Looze (1988), Doyle & Stein (1981), Maciejowski (1991) and Kwakernaak (1995).

A different approach to the concept of gains for multivariable systems is found in Postlethwaite & MacFarlane (1979), where emphasis is placed on the (frequency dependent) eigenvalues of the transfer matrix.

§2.3 is mainly based on Bode (1945).

Example 2.3.2 is taken from Åström (1995), where additional limitations on the achievable phase margin are also discussed.

3
SISO Control

In this chapter we present results on performance limitations in linear single-input single-output control systems. These results are the cornerstones of classical design. We focus on the sensitivity and complementary sensitivity functions, S and T. Although this chapter is based on contributions by many authors, mainly during the 80's, they may be seen with justice as direct descendants of the many ideas contained in Bode (1945).

3.1 Bode Integral Formulae

Bode's book (Bode 1945) mainly dealt with analysis and design of feedback amplifiers, but it settled the mathematical foundation for what is now known as classical control theory. Bode's methods had a deep and lasting impact in systems science, and many of his results still have repercussions. From his analysis, for example, it can be concluded (see §3.1.2) that, if the open-loop system is a stable rational function with relative degree at least two, then, provided the closed-loop system is stable, the sensitivity function must satisfy the following integral relation:

$$\int_0^\infty \log |S(j\omega)| d\omega = 0 \,. \tag{3.1}$$

This integral establishes that the net area subtended by the plot of $|S(j\omega)|$ on a logarithmic scale is zero. Hence, the contribution to this area of those frequency ranges where there is sensitivity amplification ($|S(j\omega)| > 1$)

must equal that of those ranges where there is sensitivity attenuation ($|S(j\omega)| < 1$). This trade-off is illustrated in Figure 3.1. This demonstrates that there is a compromise between sensitivity magnitudes in different frequency ranges. Moreover, since the sensitivity function quantifies (amongst other things) a system's ability to reject disturbances, the area constraint establishes that it is impossible to achieve arbitrary rejection at all frequencies.

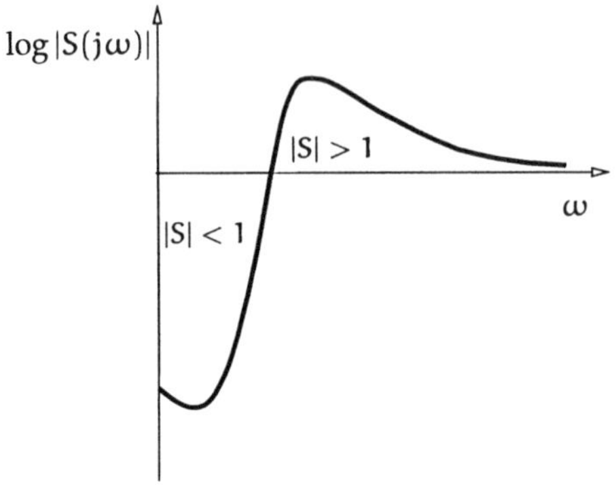

FIGURE 3.1. Area balance of the sensitivity integral.

Due to the aforementioned balance of areas, the relation (3.1) is sometimes called *the area formula*. As a matter of fact, (3.1) is a control system interpretation due to Horowitz (1963) of the original results of Bode (1945). These results dealt with the application of Cauchy's integral theorem (Theorem A.5.3 in Appendix A) to the real and imaginary parts of analytic functions (some of them have already been discussed in §2.3.2 of Chapter 2). In the following subsection we review the original formula of Bode that inspired (3.1).

3.1.1 Bode's Attenuation Integral Theorem

The theorem that we present in this section further develops the relationship between real and imaginary components of analytic functions. More specifically, it provides an expression for the behavior of the imaginary part at high frequencies, in terms of the integral of the real part, of a function H that satisfies the following conditions:

(i) $H(j\omega) = P(\omega) + jQ(\omega) = \overline{H(-j\omega)}$, where P and Q are real-valued functions of ω;

(ii) $H(s)$ is analytic at $s = \infty$ and in the CRHP except possibly for singularities $s_0 = j\omega_0$ on the finite imaginary axis that satisfy $\lim_{s \to s_0}(s - s_0)H(s) = 0$.

Bode called the real and imaginary parts, P and Q, *attenuation* and *phase*, respectively, since he studied functions of the form $H = \log \mu$, where μ represented characteristics of an electrical network. This suggested the name of the theorem.

Before presenting Bode's Attenuation Integral Theorem, we discuss a different interpretation of Theorem 2.3.3 in Chapter 2 that serves as motivation. Recall that Theorem 2.3.3[1] gave an expression of the imaginary part of H at any frequency ω_0 in terms of the integral of the complete frequency response of the real part of H. In the proof of Theorem 2.3.3, the function H was manipulated in order to create poles at $s = j\omega_0$ and $s = -j\omega_0$ in the integrand of (2.27). This led to the result that the integrand had residues of value $jQ(\omega_0)$ at each of these poles, and thus the real-imaginary parts relation (2.26) can be alternatively interpreted as an expression for the residue at a finite frequency. There is no reason why the same procedure cannot be applied to evaluate the residue at infinity of H, denoted by $\mathrm{Res}_{s=\infty} H(s)$ (see §A.9.1 in Chapter A). This is the essential format of Bode's Attenuation Integral Theorem.

Theorem 3.1.1 (Bode's Attenuation Integral Theorem). Let H be a function satisfying conditions (i) and (ii). Then, for $H(j\omega) = P(\omega) + jQ(\omega)$,

$$\int_0^\infty [P(\omega) - P(\infty)]\, d\omega = -\frac{\pi}{2} \operatorname*{Res}_{s=\infty} H(s)\,. \tag{3.2}$$

Proof. Since H is analytic at infinity, it follows from Example A.8.4 of Appendix A that it has a Laurent series expansion of the form

$$H(s) = \cdots + \frac{c_{-k}}{s^k} + \cdots + \frac{c_{-1}}{s} + c_0\,, \tag{3.3}$$

which is uniformly convergent in $R \le |s| < \infty$, for some $R > 0$. In view of the assumption (i) on H, it follows that the coefficients $\{c_k\}$ are real and that $c_0 = P(\infty)$.

Consider next the contour of integration shown in Figure 2.3 of Chapter 2, where the small indentations correspond to possible singularities of H on the imaginary axis, and where the large semicircle has radius larger than R. Denote this semicircle by C_R. Then, according to Cauchy's integral theorem

$$\oint [H(s) - P(\infty)]\, ds = 0\,. \tag{3.4}$$

The contribution of the integrals on the small indentations on the imaginary axis can be shown (see the proof of Lemma A.6.2 in Appendix A) to

[1] Theorem 2.3.3 was proven for proper and stable transfer functions, but the result also holds for any function H satisfying conditions (i) and (ii).

tend to zero as the indentations vanish, due to the particular type of singularities characterized in (ii). The integral on the large semicircle of the terms in s^k, $k \le 2$, of (3.3) tends to zero as the radius becomes infinite (see Example A.4.1 in Appendix A). Hence, taking limits as $R \to \infty$, (3.4) reduces to

$$\begin{aligned} 0 &= \int_{-\infty}^{\infty} [P(\omega) + jQ(\omega) - P(\infty)]\, j\, d\omega + \lim_{R\to\infty} \int_{C_R} \frac{c_{-1}}{s}\, ds \\ &= 2j \int_{0}^{\infty} [P(\omega) - P(\infty)]\, d\omega - j\pi c_{-1} \,, \end{aligned}$$

where the second line is obtained using the symmetry properties of P and Q, and Example A.5.1 of Appendix A. Equation (3.2) then follows using $\text{Res}_{s=\infty} H(s) = -c_{-1}$ (see (A.80) in Appendix A). □

Note that, from (3.3), $Q(\omega) = \text{Im}\, H(j\omega) = -c_{-1}/\omega + c_{-3}/\omega^3 + \cdots$. Hence (3.2) can also be written as

$$\int_{0}^{\infty} [P(\omega) - P(\infty)]\, d\omega = -\frac{\pi}{2} \lim_{\omega\to\infty} \omega\, Q(\omega) \,.$$

This gives an expression for the behavior at infinity of the imaginary component of H.

A formula parallel to (3.2) but applicable at zero frequency is easily obtained from the above result using the fact that H is analytic at zero. This is established in the following corollary.

Corollary 3.1.2. Let H be a function such that $H(1/\xi)$ satisfies conditions (i) and (ii). Then

$$\int_{0}^{\infty} [P(\omega) - P(0)]\, \frac{d\omega}{\omega^2} = \frac{\pi}{2} \lim_{s\to 0} \frac{dH(s)}{ds} \,. \tag{3.5}$$

Proof. Since H is analytic at zero it has a power series expansion of the form

$$H(s) = a_0 + a_1 s + \cdots + a_k s^k + \cdots \,, \tag{3.6}$$

which is uniformly convergent in $|s| \le r$, for some $r > 0$. Consider the function $H(1/\xi)$, which is analytic at infinity and admits — from (3.6) — a Laurent series expansion of the form

$$H(1/\xi) = \cdots + \frac{a_k}{\xi^k} + \cdots + \frac{a_1}{\xi} + a_0 \,,$$

which is uniformly convergent in $|\xi| \ge 1/r$. Then, using (3.2), we have

$$\int_{0}^{\infty} [P(1/\nu) - P(0)]\, d\nu = \frac{\pi}{2} a_1 \,.$$

The proof is completed on making the change of variable of integration $\omega = 1/\nu$ and noting that $a_1 = \lim_{s\to 0}[dH(s)/ds]$. □

The theorem and corollary given above, as well as Theorem 2.3.3 in Chapter 2, give examples of the great variety of formulae that can be obtained via a judicious choice of contour of integration and integrand in Cauchy's theorems. Bode collected a number of such formulae in his book; see, for example, the table at the end of Chapter 13 of Bode (1945). Our choice of the relations presented so far is justified by their immediate application in systems theory. Otherwise, to quote Bode,

> " ... it is extremely difficult to organize all the possible relations in any very coherent way. In a purely mathematical sense most of the formulae are related to one another by such obvious transformations and changes of variable that there is no good reason for picking out any particular set as independent. Basically they are merely reflections of Cauchy's theorem. Thus the expressions which one chooses to regard as distinctive must be selected for their physical meaning for the particular problem in hand."

3.1.2 *Bode Integrals for* S *and* T

This subsection starts this book's main journey through results concerning *complementary mappings*, by which we mean mappings relating signals of the loop and such that their sum is a constant mapping. We consider here the sensitivity function, S, and the complementary sensitivity function, T, of the classical unity feedback configuration of linear feedback control theory, represented in Figure 3.2.

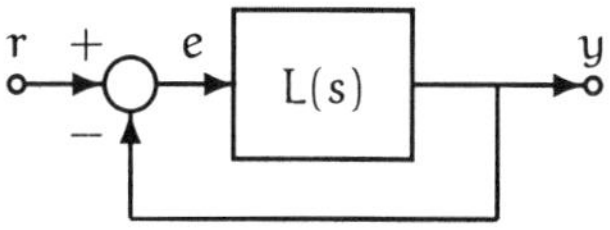

FIGURE 3.2. Feedback control system.

These two mappings satisfy the complementarity constraint

$$S(s) + T(s) = 1 . \tag{3.7}$$

This represents an algebraic trade-off (Freudenberg & Looze 1985), since it constrains the properties of the closed-loop system at each frequency. It implies, for example, that the magnitudes of S and T cannot both be smaller than $1/2$ at the same frequency (Doyle, Francis & Tannenbaum 1992).

Assuming minimality, S has zeros at the unstable open-loop plant poles and T has zeros at the nonminimum phase open-loop plant zeros. These are generally known as *interpolation constraints*. More precisely, let the open-loop system be L and assume that it is free of unstable hidden modes[2]. Then, the sensitivity and complementary sensitivity functions have the forms

$$S(s) = \frac{1}{1 + L(s)}, \qquad \text{and} \qquad T(s) = \frac{L(s)}{1 + L(s)}. \tag{3.8}$$

The following lemma formalizes the interpolation constraints that S and T must satisfy at the CRHP poles and zeros of L.

Lemma 3.1.3 (Interpolation Constraints). Assume that the open-loop system L is free of unstable hidden modes. Then S and T must satisfy the following conditions.

(i) If $p \in \overline{\mathbb{C}^+}$ is a pole of L, then

$$S(p) = 0, \qquad \text{and} \qquad T(p) = 1. \tag{3.9}$$

(ii) If $q \in \overline{\mathbb{C}^+}$ is a zero of L, then

$$S(q) = 1, \qquad \text{and} \qquad T(q) = 0. \tag{3.10}$$

Proof. Follows immediately from (3.8). □

This result states that the CRHP zeros of S and T are determined by those of L^{-1} and L. We introduce for convenience the following notation for the ORHP zeros of S and T.[3]

$$\begin{aligned} \mathcal{Z}_S &\triangleq \{s \in \mathbb{C}^+ : S(s) = 0\}, \\ \mathcal{Z}_T &\triangleq \{s \in \mathbb{C}^+ : T(s) = 0\}. \end{aligned} \tag{3.11}$$

Lemma 3.1.3 then establishes the fact that $\mathcal{Z}_S$ is the set of ORHP poles of L and $\mathcal{Z}_T$ is the set of ORHP zeros of L. In other words, we have translated the open-loop characteristics of instability and "nonminimum phaseness" into properties that the functions S and T must satisfy in the CRHP. In particular, note that if the open-loop system is formed by the cascade of the plant, G, and the controller, K, i.e., $L = GK$, then the constraints imposed by CRHP poles and zeros of G hold *irrespective* of the choice of K, provided that L has no unstable zero-pole cancelation.

As a first extension of Bode's results to integrals on sensitivity functions, we will derive Horowitz's formula (3.1) and a complementary result for T using Theorem 3.1.1 and Corollary 3.1.2.

[2]Recall from Chapter 2 that L is free of unstable hidden modes if there are no cancelations of CRHP zeros and poles between the plant and controller whose cascade connection forms L.

[3]In the sequel, when defining a set of zeros of a transfer function, the zeros are repeated according to their multiplicities.

Example 3.1.1 (Horowitz's formula for S). Suppose that L is a *proper rational function without poles in the ORHP*. Assume that the closed-loop system is stable (i.e., the numerator of $1 + L$ is Hurwitz and $L(\infty) \neq -1$) and consider the corresponding sensitivity function S. It follows that $\mathcal{Z}_S$ in (3.11) is empty and that the function $H = \log S$ satisfies assumptions (i) and (ii) of §3.1.1. Thus, we can apply Theorem 3.1.1 to the real part, $\log|S(j\omega)|$, to obtain

$$\int_0^\infty [\log|S(j\omega)| - \log|S(j\infty)|]\, d\omega = -\frac{\pi}{2} \operatorname*{Res}_{s=\infty} \log S(s) \, .$$

If we further assume that the open-loop plant has *relative degree two or more*, then[4] $S(j\infty) = 1$ and $\text{Res}_{s=\infty} \log S(s) = 0$. This recovers Horowitz's area formula given in (3.1). ◦

Example 3.1.2 (Horowitz's formula for T). Assume that $L(s)$ is a *minimum phase rational function such that* $L(0) \neq 0$. Then, if we consider the complementary sensitivity function T, we have that $\mathcal{Z}_T$ in (3.11) is empty. Next, notice that the function $T(1/\xi)$ can be written as

$$T(1/\xi) = \frac{1}{1 + 1/L(1/\xi)} \, . \tag{3.12}$$

Then, under the assumption of closed-loop stability (which implies $L(0) \neq -1$), the function $H(1/\xi) = \log T(1/\xi)$ satisfies conditions (i) and (ii) of §3.1.1. We thus obtain, from Corollary 3.1.2,

$$\int_0^\infty [\log|T(j\omega)| - \log|T(0)|] \frac{d\omega}{\omega^2} = \frac{\pi}{2} \frac{1}{T(0)} \lim_{s\to 0} \frac{dT(s)}{ds} \, .$$

If we further assume that L *has at least two integrators* then $T(0) = 1$ and $\lim_{s\to 0} dT(s)/ds = 0$, which yields

$$\int_0^\infty \log|T(j\omega)| \frac{d\omega}{\omega^2} = 0 \, .$$

◦

It is instructive at this point to reflect on the complementarity of these formulae for S and T, as well as the hypotheses required on the open-loop system L to derive them. Under appropriate conditions, S and T exhibit symmetry with respect to frequency inversion (Kwakernaak 1995). For example, the condition that L be proper and $L(j\infty) \neq -1$ (or alternatively, that L be strictly proper) imply the analyticity of S and $\log S$ at infinity. On the other hand, the requirement that $L(0) \neq \{0, -1\}$ (or that L has poles at zero) gives the analyticity of T and $\log T$ at zero.

[4] See Example A.9.2 in Appendix A.

The extensions to unstable open-loop plants and to plants with time delay were derived by Freudenberg & Looze (1985, 1987). The complementary result for T was obtained by Middleton & Goodwin (1990). These two area formulae will be given next under the name of the *Bode Integrals* for S and T. We choose these name since the results are natural extensions of Bode's integral (3.2) to the sensitivity and complementary sensitivity functions.

Theorem 3.1.4 (Bode Integral for S). Let S be the sensitivity function defined by (3.8). Let $\{p_i : i = 1, \ldots, n_p\}$ be the set of poles in the ORHP of the open-loop system L. Then, assuming closed-loop stability,

(i) if L is a proper rational function,

$$\int_0^\infty \log \left| \frac{S(j\omega)}{S(j\infty)} \right| d\omega = \frac{\pi}{2} \lim_{s \to \infty} \frac{s[S(s) - S(\infty)]}{S(\infty)} + \pi \sum_{i=1}^{n_p} p_i \, ; \qquad (3.13)$$

(ii) if $L(s) = L_0(s)e^{-s\tau}$, where $L_0(s)$ is a strictly proper rational function and $\tau > 0$,

$$\int_0^\infty \log |S(j\omega)| \, d\omega = \pi \sum_{i=1}^{n_p} p_i \, . \qquad (3.14)$$

Proof. Consider the contour of Figure 3.3, where the indentations $C_1, C_2, \ldots, C_{n_p}$ into the right half plane avoid the branch cuts of log S corresponding to the zeros of S (see Figure A.19 in Appendix A). Let C_0 consist of the remaining portions of the imaginary axis and let C_R be the semicircle of radius R.

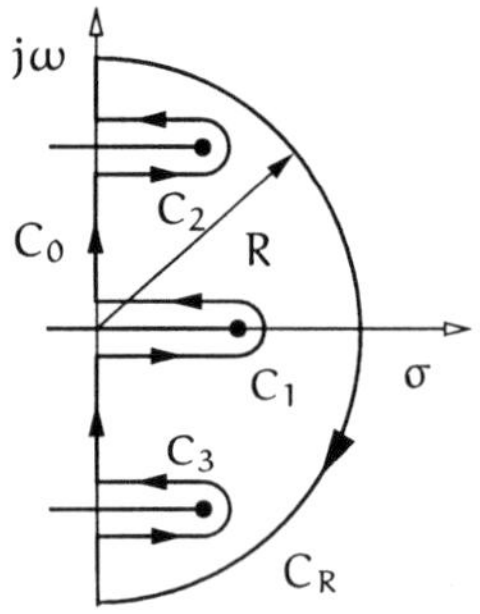

FIGURE 3.3. Contour for Bode Sensitivity Integral.

Since $\log[S(s)/S(\infty)]$ is analytic on and inside the total contour, denoted by C, then, by Cauchy's integral theorem,

$$\lim_{\substack{\epsilon \to 0 \\ R \to \infty}} \int_C \log \frac{S(s)}{S(\infty)} \, ds = 0 \, .$$

Note that the computation of the portions of the integral on the imaginary axis and on the indentations is the same for both cases (i) and (ii) (the only distinction being that in (ii) we have that $S(\infty) = 1$). However, the integral on the large semicircle requires a different analysis in each case.

The portion of the integral on C_0 gives[5]

$$\lim_{\substack{\epsilon\to 0\\ R\to\infty}} \int_{C_0} \log\frac{S(s)}{S(\infty)}\, ds = 2j \int_0^{\infty} \log\left|\frac{S(j\omega)}{S(j\infty)}\right| d\omega\,.$$

For the integral on the branch cut indentations we have, using (A.85) of Appendix A,

$$\sum_{i=1}^{n_p} \int_{C_i} \log\frac{S(s)}{S(\infty)}\, ds = -j2\pi \sum_{i=1}^{n_p} \operatorname{Re} p_i = -j2\pi \sum_{i=1}^{n_p} p_i\,,$$

where the last equality follows since complex zeros must appear in conjugate pairs.

It remains to compute the integral on C_R for each of the cases (i) and (ii).

(i) If L is a proper rational function, the assumption of closed-loop stability guarantees that the function $\log S(s)$ is analytic at infinity. We can then use Example A.9.2 of Appendix A to compute the integral on C_R in the limit when $R \to \infty$ as

$$\begin{aligned}\lim_{R\to\infty} \int_{C_R} \log\frac{S(s)}{S(\infty)}\, ds &= j\pi \operatorname*{Res}_{s=\infty} \log\frac{S(s)}{S(\infty)}\\ &= j\pi \frac{1}{S(\infty)} \lim_{s\to\infty} s[S(\infty) - S(s)]\,,\end{aligned} \tag{3.15}$$

where we have used (A.84) from Appendix A.

(ii) If $L = L_0 e^{-s\tau}$, with $\tau > 0$ and L_0 strictly proper, we have that there is an $r > 0$ such that, for all s with $\operatorname{Re} s \geq 0$ and $|s| > r$, the modulus of $L(s)$ satisfies $|L(s)| < 1$. Then, a similar use of the expansion of $\log(1 + s)$ as in Example A.9.2 of Appendix A yields

$$\log S(s) = -L_0(s)e^{-s\tau} + \frac{L_0^2(s)e^{-2s\tau}}{2} + \cdots\,,$$

in $\{s : |s| > r \text{ and } \operatorname{Re} s \geq 0\}$. Since L_0 has relative degree at least one, then the above expansion has the form

$$\log S(s) = \frac{c_{-1}}{s} e^{-s\tau} + \cdots\,, \qquad \text{in} \quad \{s : |s| > r \text{ and } \operatorname{Re} s \geq 0\}\,.$$

[5]Under the convention that the phase of a negative real number is equal to π for $\omega \geq 0$ and $-\pi$ for $\omega < 0$.

Then, using the result in Example A.4.3 of Appendix A, we finally have that

$$\lim_{R\to\infty} \int_{C_R} \log S(s)\, ds = 0 \, .$$

The result then follows by combining the integrals over all portions of the total contour for each of the cases (i) and (ii). □

Theorem 3.1.4 shows that the presence of open-loop unstable poles further restricts the compromise between areas of sensitivity attenuation and amplification imposed by the integral. Specifically, since the term $\sum_{i=1}^{n_p} p_i$ is nonnegative, the contribution to the integral of those frequencies where $|S(j\omega)| < 1$ is clearly reduced if L has unstable poles. Moreover, the farther from the $j\omega$-axis the poles are, the worse will be their effect.

The first term on the RHS of (3.13) appears only if the plant has relative degree 0 or 1 and no time delay, and admits an interpretation in terms of the time response of L. Indeed, if $l(t)$ denotes the response of the plant to a unitary step input, we can alternatively write

$$\begin{aligned} \lim_{s\to\infty} \frac{s[S(s) - S(\infty)]}{S(\infty)} &= \lim_{s\to\infty} s \left[\frac{L(\infty) - L(s)}{1 + L(s)}\right] \\ &= -\frac{\dot{l}(0^+)}{1 + L(\infty)} \, , \end{aligned}$$

where the last step follows from the Initial Value Theorem of the Laplace transform (e.g., Franklin et al. 1994). Thus, $\dot{l}(0^+)$ is the initial slope of $l(t)$, as illustrated in Figure 3.4.

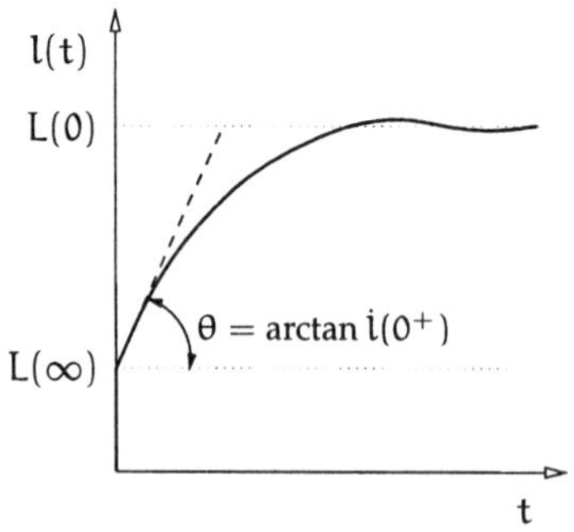

FIGURE 3.4. Step response of a plant of relative degree 0.

If the plant has relative degree 0 or 1, a large positive value for $\dot{l}(0^+)$ will ameliorate the integral constraint for S. This, as we will see in more detail in §3.1.3, is justified by the fact that a larger value for $\dot{l}(0^+)$ is associated with larger bandwidth of the system.

If the plant has relative degree greater than 1, then $\dot{l}(0^+)$ vanishes, and (3.13) reduces to (3.14). We will see in §3.1.3, however, that larger bandwidth still alleviates the trade-offs induced by this integral constraint.

Naturally, an equivalent relation holds for the complementary sensitivity function, as stated in the following theorem.

Theorem 3.1.5 (Bode Integral for T). Let T be the complementary sensitivity function defined by (3.8). Let $\{q_i : i = 1, \dots, n_q\}$ be the set of zeros in the ORHP of the open-loop system L, and suppose that $L(0) \neq 0$. Then, assuming closed-loop stability,

(i) if L is a proper rational function

$$\int_0^\infty \log\left|\frac{T(j\omega)}{T(0)}\right| \frac{d\omega}{\omega^2} = \frac{\pi}{2}\frac{1}{T(0)} \lim_{s\to 0} \frac{dT(s)}{ds} + \pi \sum_{i=1}^{n_q} \frac{1}{q_i} ; \qquad (3.16)$$

(ii) if $L(s) = L_0(s)e^{-s\tau}$, where $L_0(s)$ is a strictly proper rational function and $\tau > 0$,

$$\int_0^\infty \log\left|\frac{T(j\omega)}{T(0)}\right| \frac{d\omega}{\omega^2} = \frac{\pi}{2}\frac{1}{T(0)} \lim_{s\to 0} \frac{dT(s)}{ds} + \pi \sum_{i=1}^{n_q} \frac{1}{q_i} + \frac{\pi}{2}\tau . \qquad (3.17)$$

Proof. The proof follows along the same lines as that of Theorem 3.1.4 with inverted symmetry in the roles played by the points $s = 0$ and $s = \infty$. The interested reader may find the details in Middleton (1991). □

It is interesting to note that the integral for T is in fact the same as that for S under frequency inversion. Indeed, by letting $\nu = 1/\omega$, we can alternatively express

$$\int_0^\infty \log\left|\frac{T(j\omega)}{T(0)}\right| \frac{d\omega}{\omega^2} = \int_0^\infty \log\left|\frac{T(1/j\nu)}{T(0)}\right| d\nu .$$

Accordingly, as seen in (3.16) and (3.17), nonminimum phase zeros of L play for T an entirely equivalent role to that of unstable poles for S; i.e., they worsen the integral constraint. In this case, zeros in the ORHP that are *closer* to the $j\omega$-axis will pose a greater difficulty in shaping T.

The first term on the RHSs of (3.16) and (3.17) has an interpretation in terms of steady-state properties of the plant. Consider the feedback loop shown in Figure 3.5. The Laplace transform of the error signal $e = r - y$ is given by

$$E(s) = \left[1 - \frac{T(s)}{T(0)}\right] R(s) .$$

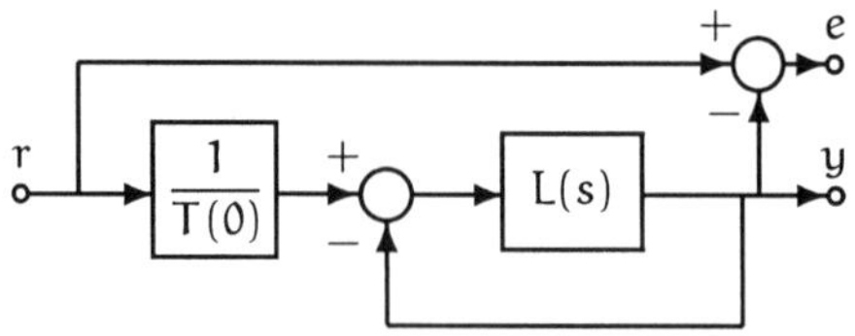

FIGURE 3.5. Type-1 feedback system.

Hence, if r is a unitary ramp, i.e., $R(s) = 1/s^2$, then the corresponding steady-state value of $e(t)$ (see Figure 3.6) can be computed using the Final Value Theorem (e.g., Franklin et al. 1994) and L'Hospital's rule (e.g., Widder 1961, p. 260) as

$$\begin{aligned} e_{ss} &= \lim_{t\to\infty} e(t) \\ &= \lim_{s\to 0} sE(s) \\ &= \lim_{s\to 0} \frac{1 - \dfrac{T(s)}{T(0)}}{s} \\ &= -\frac{1}{T(0)} \lim_{s\to 0} \frac{dT(s)}{ds} . \end{aligned}$$

Thus the constant $1/T(0)dT/ds|_{s=0}$ plays a role similar to that played by the reciprocal of the *velocity constant* in a Type-1 feedback system (e.g., Truxal 1955, p. 286). Consequently, the corresponding term on the RHSs of (3.16) and (3.17) can ameliorate the severity of the design trade-off *only* if the steady-state error to a ramp input is large and positive, so that the output lags the reference input significantly.

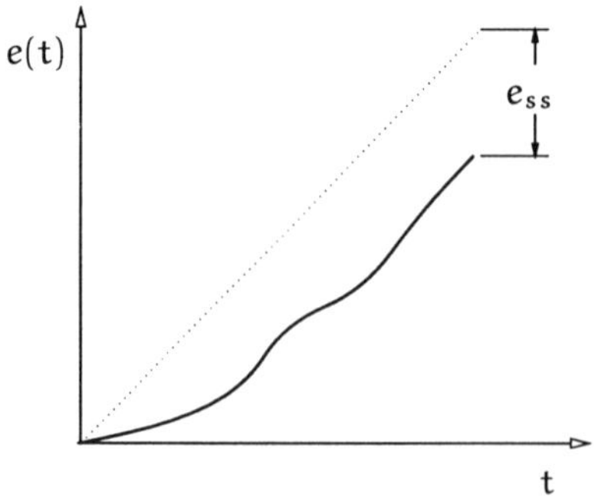

FIGURE 3.6. Steady-state error to ramp.

Finally, the last term on the RHS of (3.17) shows that the trade-off also worsens if the plant has a time delay. In a sense this is not surprising, since a time delay may be approximated by nonminimum phase zeros[6], which already appear in connection with the Bode integral for T.

In the following section we will discuss more detailed design implications of the integral constraints given by Theorem 3.1.4 and Theorem 3.1.5.

3.1.3 *Design Interpretations*

As discussed in §3.1, the integral constraint (3.1) tells us that, if the feedback loop is designed to hold $|S(j\omega)|$ smaller than one over a given frequency range, then it will be larger than one in another range. In the case of open-loop unstable plants, the constraint (3.14) shows that the trade-off is worsened since the integral must be positive in this case.

Heuristically, this effect can be seen from the Nyquist plot of the open-loop transfer function, L, as follows. Consider first that L is a stable rational function of relative degree two or more; then $L(j\omega)$ will ultimately have phase lag of at least $-\pi$. This situation is depicted on the left of Figure 3.7 for $L(s) = 1.5/(1+s)^2$. Note that $S^{-1}(j\omega) = 1+L(j\omega)$ is the vector from the -1 point to the point on the Nyquist plot corresponding to the frequency ω. Thus, we see that there is a portion of the plot where $|S(j\omega)|$ is less than one and another portion where $|S(j\omega)|$ increases above one. A similar situation occurs for strictly proper plants with time delay, as shown on the right of Figure 3.7 for $L(s) = 1.5e^{-3s}/(s+2)$. In this case, however, the Nyquist plot alternates between areas of sensitivity attenuation and amplification infinitely many times.

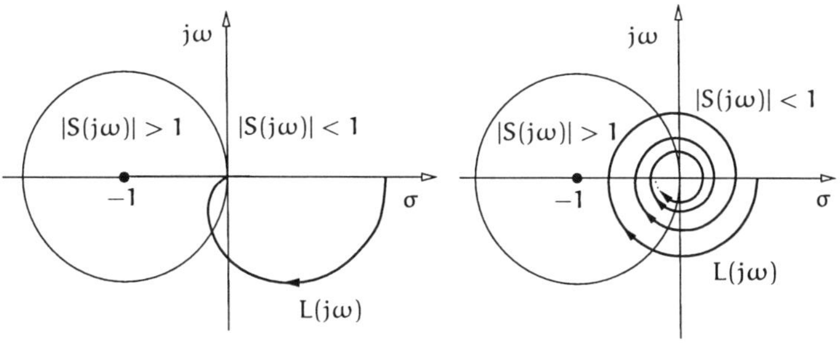

FIGURE 3.7. Graphical interpretation of the area formula.

The possibility of achieving a negative value of the integral in (3.13) for stable open-loop systems having relative degree one can also be deduced from Figure 3.7. Indeed, these systems will exhibit an asymptotic phase lag of $-\pi/2$, and hence their Nyquist plot can be kept largely outside the region of sensitivity increase.

[6]This is evident in the Padé approximations of $e^{-s\tau}$ (e.g., Middleton & Goodwin 1990).

Example 3.1.3. Let $L = K/(s+1)$, $K > 0$, i.e., a first order plant cascaded with a proportional controller. Then

$$S(s) = \frac{s+1}{s+1+K}\,.$$

The value of the integral in (3.13) is $-K$ (which is minus the initial slope of the step response of L). The Nyquist plot of L lies entirely in the forth quadrant, and therefore there is no area of sensitivity amplification. ∘

Despite the area balance effect, one cannot immediately conclude from either (3.1) or (3.14) that a peak that is significantly greater than one will occur outside the frequency range over which sensitivity reduction has been achieved. Actually, it is possible that area of sensitivity reduction over some finite range of frequencies may be compensated by an area where $|S|$ is allowed to be slightly greater than one over an arbitrarily large range of frequencies.

Any practical design, however, is affected by bandwidth constraints. Indeed, several factors such as undermodeling, sensor noise, plant bandwidth, etc., lead to the desirability of decreasing the open-loop gain at high frequencies, thus putting a limit on the bandwidth of the closed loop. It is reasonable to assume, then, that the open-loop gain satisfies a design specification of the type

$$|L(j\omega)| \le \delta \left(\frac{\omega_c}{\omega}\right)^{1+k}, \qquad \forall \omega \ge \omega_c\,, \tag{3.18}$$

where $\delta < 1/2$ and $k > 0$. Note that k, ω_c and δ can be adjusted to provide upper limits to the slope of magnitude roll-off, the frequency where roll-off starts and the gain at that frequency, respectively.

By convention, the *bandwidth* of the closed-loop system in Figure 3.2 is defined (e.g., Franklin et al. 1994) as the frequency of a sinusoidal input, r, at which the output y is attenuated to a factor of $1/\sqrt{2}$. From the definition of T in (3.8), the bandwidth ω_b is the frequency at which $|T(j\omega_b)| = 1/\sqrt{2}$. It is possible to approximate ω_b by a factor (generally between 1 and 2) of the *crossover frequency*, ω_0, defined as the lowest frequency at which the magnitude of the *open-loop system* is one, i.e., $|L(j\omega_0)| = 1$ (e.g., Franklin et al. 1994). Since ω_c in (3.18) is such that $|L(j\omega_c)| < 1/2$, it follows that $\omega_0 < \omega_c$ and hence the bandwidth ω_b can be roughly taken as $\omega_b \approx \omega_c$.

The following result shows that a condition such as (3.18) on the open-loop gain puts an upper limit on the area of sensitivity increase that can be present at frequencies greater than ω_c.

Corollary 3.1.6. Suppose that L is a rational function of relative degree two or more and satisfying the bandwidth restriction (3.18). Then

$$\left| \int_{\omega_c}^{\infty} \log|S(j\omega)|\, d\omega \right| \le \frac{3\delta\omega_c}{2k}\,. \tag{3.19}$$

Proof. We use the fact that if $|s| < 1/2$, then $|\log(1+s)| \le 3|s|/2$ (See Example A.8.6 in Appendix A). Then

$$\begin{aligned}\left|\int_{\omega_c}^{\infty} \log|S(j\omega)|\, d\omega\right| &\le \int_{\omega_c}^{\infty} |\log|S(j\omega)||\, d\omega \\ &\le \int_{\omega_c}^{\infty} |\log S(j\omega)|\, d\omega \\ &= \int_{\omega_c}^{\infty} |\log[1+L(j\omega)]|\, d\omega \\ &\le \frac{3\delta\omega_c^{1+k}}{2}\int_{\omega_c}^{\infty} \frac{1}{\omega^{1+k}}\, d\omega \\ &= \frac{3\delta\omega_c}{2k}\,.\end{aligned}$$

□

The above result shows that the area of the tail of the Bode sensitivity integral over the infinite frequency range $[\omega_c, \infty)$ is *limited*. Hence, any area of sensitivity reduction must be compensated by a *finite* area of sensitivity increase, which will necessarily lead to a large peak in the sensitivity frequency response before ω_c. To see this, suppose that S is required to satisfy the following reduction condition

$$|S(j\omega)| \le \alpha < 1\,, \qquad \forall \omega \le \omega_1 < \omega_c\,. \tag{3.20}$$

Now, using (3.14) and the bounds (3.18) and (3.20), it is easy to show that

$$\sup_{\omega\in(\omega_1,\omega_c)} \log|S(j\omega)| \ge \frac{1}{\omega_c - \omega_1}\left[\pi \sum_{p\in\mathcal{Z}_s} p + \omega_1 \log\frac{1}{\alpha} - \frac{3\delta\omega_c}{2k}\right]. \tag{3.21}$$

The bound (3.21) shows that any attempt to increase the area of sensitivity reduction by requiring α to be small and/or ω_1 to be close to ω_c will necessarily result in a large sensitivity peak in the range (ω_1, ω_c). Hence, the Bode sensitivity integral (3.14) imposes a clear design trade-off when natural bandwidth constraints are assumed for the closed-loop system. Notice, however, that this trade-off is alleviated if the closed-loop bandwidth is large, i.e., large ω_c in (3.18).

Example 3.1.4. The inequality (3.21) can be used to derive a lower bound on the closed-loop bandwidth in terms of the sum of open-loop unstable poles. Indeed, suppose that the frequency ω_1 in (3.20) is taken to be a fraction of the bandwidth, say

$$\omega_1 = k_1\omega_c\,,$$

and assume that we desire that the peak sensitivity on the LHS of (3.21) be less than or equal to a number $S_m > 1$. Then, necessarily, the lower bound on the RHS of (3.21) must be less than or equal to S_m. Imposing this condition yields the following lower bound on the bandwidth, which we take as $\omega_b = \omega_c$,

$$\omega_b \geq B(S_m) \sum_{p \in \mathcal{Z}_s} p \,, \tag{3.22}$$

where $B(S_m)$ is

$$B(S_m) \triangleq \frac{\pi}{(1 - k_1)(S_m + 1.5\,\delta/k + k_1 \log \alpha)} \,.$$

The factor $B(S_m)$ is plotted in Figure 3.8 as a function of the desired peak sensitivity S_m, for $\delta = 0.45$, $k = 1$, $k_1 = 0.7$ and $\alpha = 0.5$.

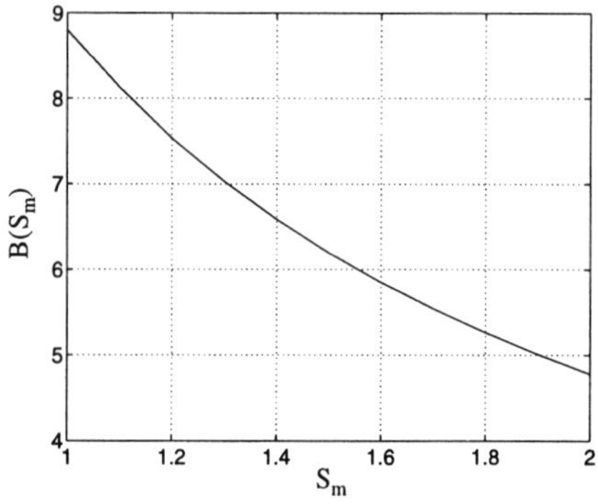

FIGURE 3.8. Lower bound on the bandwidth as a function of the peak sensitivity.

We may conclude from this figure, for example, that an open-loop unstable system having relative degree two requires a bandwidth of at least 6.5 times the sum of its ORHP poles if it is desired that $|S|$ be smaller than $1/2$ over 70% of the closed-loop bandwidth while keeping the lower bound on the peak sensitivity smaller than $S_m = \sqrt{2}$. ∘

3.2 The Water-Bed Effect

The Bode sensitivity and complementary sensitivity integrals represent a constraint imposed by nonminimum phase zeros and unstable poles of the open-loop system. We have just seen in §3.1.3 that these integral constraints, in conjunction with a further requirement on the closed-loop bandwidth, lead to a quantifiable trade-off in design. Indeed, from the comment at the end of §3.1.3, we see that an arbitrary reduction of sensitivity over a range of frequencies necessarily implies an arbitrarily large increase at other frequencies. This kind of phenomenon has been referred to as a *push-pop* or *water-bed effect* (e.g., Shamma 1991, Doyle et al. 1992).

The water-bed effect was first recognized by Francis & Zames (1984), who showed that a similar trade-off exists with respect to sensitivity minimization over a frequency interval when the open-loop system has a zero in the ORHP. This will be established below. As opposed to the previous results, which were based on the Cauchy integral theorem, the result that follows relies on an application of the maximum modulus principle (Theorem A.10.2 in Appendix A).

Theorem 3.2.1 (Water-Bed Effect). Let S be the sensitivity function defined by (3.8). Assume that S is proper and stable. Then, if the open-loop plant L has a zero in the ORHP, there exists a positive number m such that

$$\sup_{\omega\in[\omega_1,\omega_2]} |S(j\omega)| \geq 1/\|S\|_\infty^m ,$$

where

$$\|S\|_\infty = \sup_\omega |S(j\omega)|$$

is the infinity norm of S.

Proof. Let q be a zero of L in the ORHP. It follows from (3.7) that $S(q) = 1$. Consider the mapping from the CRHP onto the unit disk given by

$$z = \frac{q-s}{\overline{q}+s}, \qquad s = \frac{q-\overline{q}z}{1+z} .$$

Then the interval $\{s = j\omega : \omega \in [\omega_1, \omega_2]\}$ is mapped onto the arc

$$\{z = e^{j\theta} : \theta \in [\theta_1, \theta_2]\} . \tag{3.23}$$

Let R be the following function,

$$R(z) \triangleq S\left(\frac{q-\overline{q}z}{1+z}\right) .$$

It follows that R is analytic in $|z| < 1$, $R(0) = 1$, and moreover

$$\sup_{\theta\in[\theta_1,\theta_2]} \left|R\left(e^{j\theta}\right)\right| = \sup_{\omega\in[\omega_1,\omega_2]} |S(j\omega)| ,$$

and

$$\sup_\theta \left|R\left(e^{j\theta}\right)\right| = \|S\|_\infty .$$

Now let ϕ be the angle $\phi = |\theta_2 - \theta_1|$ and let n be any integer greater than $2\pi/\phi$. Define the auxiliary function P as

$$P(z) \triangleq \prod_{k=0}^{n-1} R\left(z\, e^{j2\pi k/n}\right) . \tag{3.24}$$

Then P is also analytic in $|z| < 1$, and $P(0) = 1$. Since the angle $2\pi/n$ is less than ϕ, at least one of the points $z\, e^{j2\pi k/n}$, $k = 0, \cdots, n-1$, lies on the arc (3.23) for each z on the unit circle. Thus, from (3.24) and the maximum modulus principle

$$\begin{aligned}
1 &= |P(0)| \\
&\leq \sup_{\theta} \left|P\left(e^{j\theta}\right)\right| \\
&\leq \left[\sup_{\theta} \left|R\left(e^{j\theta}\right)\right|\right]^{n-1} \sup_{\theta \in [\theta_1, \theta_2]} \left|R\left(e^{j\theta}\right)\right| \\
&= \|S\|_{\infty}^{n-1} \sup_{\omega \in [\omega_1, \omega_2]} |S(j\omega)| .
\end{aligned}$$

The result then follows by assigning $m = n - 1$. □

This result shows the water-bed effect for linear nonminimum phase systems: the magnitude of S can be made arbitrarily small over a frequency interval only at the expense of having the magnitude arbitrarily large off this interval (Francis & Zames 1984). We will next obtain alternative expressions for the sensitivity trade-offs using Poisson integral formula. In section §3.3.2, these results will be used to give a more explicit form for the water-bed trade-off.

3.3 Poisson Integral Formulae

Consider again the feedback control configuration of Figure 3.2 and the sensitivity and complementary sensitivity functions defined in (3.8). In §3.1.2 we discussed the algebraic trade-off imposed by the complementarity constraint (3.7), which, for example, precludes the possibility of having both small $|S(j\omega)|$ and $|T(j\omega)|$ at the same frequency. We will see in this section that it is particularly revealing to evaluate this algebraic constraint at the nonminimum phase zeros and unstable poles of the open-loop transfer function L. Indeed, with the aid of the Poisson integral formula for the half plane (see §A.6.1 in Appendix A), we will show that this algebraic constraint actually leads to integral constraints on the overall frequency responses of S and T.

3.3.1 Poisson Integrals for S and T

Before deriving the Poisson integral theorems for S and T, we need some preliminary notation. Recall that, if L is free of unstable hidden modes, the ORHP poles and zeros of L become zeros of S and T, as stated in Lemma 3.1.3. Let q_i, $i = 1, \ldots, n_q$, be the zeros of L in the ORHP, and p_i, $i = 1, \ldots, n_p$, be the poles of L in the ORHP, all counted with multiplicities. Following the notation introduced in (3.11), we then have that the ORHP zeros of S and T are given by

$$\begin{aligned} \mathcal{Z}_S &= \{p_i : i = 1, \ldots, n_p\} , \\ \mathcal{Z}_T &= \{q_i : i = 1, \ldots, n_q\} . \end{aligned} \tag{3.25}$$

We next introduce the *Blaschke products* of the ORHP zeros of S and T, given by (see also (2.37) in Chapter 2)

$$B_S(s) = \prod_{i=1}^{n_p} \frac{p_i - s}{\overline{p}_i + s} , \qquad \text{and} \qquad B_T(s) = \prod_{i=1}^{n_q} \frac{q_i - s}{\overline{q}_i + s} . \tag{3.26}$$

Blaschke products are "all-pass" functions, since their magnitude is constant and equal to one on the $j\omega$-axis. Using B_S and B_T, we factorize L as

$$L(s) = \tilde{L}(s) B_S^{-1}(s) B_T(s) e^{-s\tau} , \tag{3.27}$$

where $\tau \geq 0$ is a possible time delay in the open-loop system, and where the factor $\tilde{L}(s)$ is a proper rational function having no poles or zeros in the ORHP. Then S and T in (3.8) can be factored as

$$\begin{aligned} S(s) &= \tilde{S}(s) B_S(s) , \\ T(s) &= \tilde{T}(s) B_T(s) e^{-s\tau} . \end{aligned} \tag{3.28}$$

Note that $\tilde{S}$ and $\tilde{T}$ have no zeros in the finite ORHP since the factors B_S and B_T remove the finite zeros of S and T. Also, if the closed-loop system is stable, both $\tilde{S}$ and $\tilde{T}$ have no poles in the finite CRHP. The point $s = \infty$ requires some more care. If $\tau = 0$ and $\tilde{L}$ is proper, then both $\tilde{S}$ and $\tilde{T}$ are analytic at infinity. On the other hand, if $\tau > 0$ and $\tilde{L}$ is proper, S and T have an essential singularity at infinity (see Example A.8.3 in Appendix A). However, both functions $\log \tilde{S}$ and $\log \tilde{T}$ are of class $\mathcal{R}$, i.e., their growth at infinity in the CRHP is restricted.[7]

The above discussion suggests that we can apply the Poisson integral formula — more specifically Corollary A.6.3 in Appendix A — to the functions $\log \tilde{S}$ and $\log \tilde{T}$, as we show next.

---

[7]Recall that a function f is of class $\mathcal{R}$ if $\lim_{R\to\infty} \sup_{\theta \in [-\pi/2, \pi/2]} |f(Re^{j\theta})| = 0$ (§A.6.1 in Appendix A). Since L is proper and the closed-loop system is stable, it is not difficult to see that $\tilde{S}$ is bounded on the CRHP by a constant, S_m say. Then $\log |\tilde{S}| \leq \log S_m$ on the CRHP and hence $\log |\tilde{S}|/R \to 0$ as $R \to \infty$. A similar argument shows that $\tilde{T}$ is of class $\mathcal{R}$.

Theorem 3.3.1 (Poisson Integral for S). Let S be the sensitivity function defined by (3.8). Assume that the open-loop system L can be factored as in (3.27) and let $q = \sigma_q + j\omega_q$ be an ORHP zero of L. Then, if the closed-loop system is stable,

$$\int_{-\infty}^{\infty} \log|S(j\omega)| \frac{\sigma_q}{\sigma_q^2 + (\omega_q - \omega)^2}\, d\omega = \pi \log\left|B_S^{-1}(q)\right| . \tag{3.29}$$

Proof. We will apply Corollary A.6.3 of Appendix A to the function $\log \tilde{S}$. Since the closed-loop system is stable and in view of the factorization (3.28), then $\tilde{S}$ has no zeros or poles in the ORHP (see comments after (3.28)) and hence $\log \tilde{S}$ is analytic there. If $\tau > 0$ then $\log \tilde{S}$ has an essential singularity at infinity, but these are allowed by Corollary A.6.3. Then, since $\log \tilde{S}$ is of class $\mathcal{R}$ (see footnote on page 65), (A.48) in Corollary A.6.3 can be used with with $f = \log \tilde{S}$ and $s_0 = q$. The constraint (3.29) then follows on noting that $|\tilde{S}(j\omega)| = |S(j\omega)|$ (since $|B_S(j\omega)| = 1, \forall\omega$) and that $\tilde{S}(q) = B_S^{-1}(q)$ (since $S(q) = 1$). □

The corresponding result for T is as follows.

Theorem 3.3.2 (Poisson Integral for T). Let T be the complementary sensitivity function defined by (3.8). Assume that the open-loop system L can be factored as in (3.27) and let $p = \sigma_p + j\omega_p$ be an ORHP pole of L. Then, if the closed-loop system is stable,

$$\int_{-\infty}^{\infty} \log|T(j\omega)| \frac{\sigma_p}{\sigma_p^2 + (\omega_p - \omega)^2}\, d\omega = \pi \log\left|B_T^{-1}(p)\right| + \pi\sigma_p\tau . \tag{3.30}$$

Proof. We will apply Corollary A.6.3 of Appendix A to the function $\log \tilde{T}$. Since the closed-loop system is stable and in view of the factorization (3.28), $\tilde{T}$ has no zeros or poles in the ORHP (see comments after (3.28)). If $\tilde{L}$ is strictly proper, and/or if $\tau > 0$, then $\log \tilde{T}$ has a singularity at infinity, but these are allowed by Corollary A.6.3. Then, since $\log \tilde{T}$ is of class $\mathcal{R}$ (see footnote on page 65), (A.48) in Corollary A.6.3 can be used with with $f = \log \tilde{T}$ and $s_0 = p$. The constraint (3.29) then follows on noting that $|\tilde{T}(j\omega)| = |T(j\omega)|$ (since $\left|B_S(j\omega)e^{-j\omega\tau}\right| = 1, \forall\omega$) and that $\tilde{T}(p) = B_T^{-1}(p)$ (since $T(p) = 1$). □

Note that, as was the case for Corollary A.6.3 in Appendix A, Theorems 3.3.1 and 3.3.2 still hold when L has poles or zeros on the imaginary axis. Also, if the ORHP zero q in Theorem 3.3.1 has multiplicity $m > 1$, additional integral constraints on the first $m - 1$ derivatives of $\log \tilde{S}$ can be obtained. These, however, do not seem to be as insightful as (3.29). A similar comment is in order in the case of multiple ORHP poles in Theorem 3.3.2.

Similar to the Bode integral formulae of §3.1.2, the Poisson integral constraints (3.29) and (3.30) represent a balance between areas of sensitivity or complementary sensitivity attenuation and amplification. To see this notice: first that, for $s_0 = \sigma_0 + j\omega_0$, $\sigma_0 > 0$, the weighting function

$$W_{s_0}(\omega) \triangleq \frac{\sigma_0}{\sigma_0^2 + (\omega_0 - \omega)^2} \tag{3.31}$$

is positive for all ω, and second that the RHSs of both equations are nonnegative since, for a Blaschke product B, we have that $\left|B^{-1}(s)\right| \geq 1$ for s in the ORHP.

The area balance implied by the Poisson integrals, however, differs from that of the Bode integrals. Indeed, the presence of the weighting function (3.31) in the Poisson integrals precludes the possibility of compensating an area of sensitivity reduction over a finite range of frequencies by an area where $|S|$ (or $|T|$) is allowed to be slightly greater than one over an arbitrarily large range of frequencies. This is because the weighted area of the imaginary axis is *finite* and equal to π, as seen from integrating (3.31). More insights into the properties of the weighting function will be given in the following section.

The case where the open-loop system is both unstable and nonminimum phase is even more problematic. Indeed, note that both RHSs are positive if the open-loop system has at least one ORHP zero and one ORHP pole. If this is the case the weighted area of sensitivity increase must be larger than the area of sensitivity reduction. Moreover, since $B_T^{-1}(p)$ in (3.29) has poles at $p = q_i$, $i = 1, \dots, n_q$ (similarly, $B_S^{-1}(q)$ in (3.30) has poles at $q = p_i$, $i = 1, \dots, n_p$), the constraints are aggravated if there is an approximate pole-zero cancelation in the ORHP.

In the following section we will discuss design trade-offs implied by Theorems 3.3.1 and 3.3.2 with respect to sensitivity and complementary sensitivity minimization over a frequency interval.

3.3.2 *Design Interpretations*

In §3.1.3 we gave a graphical interpretation of the area balance implied by (3.1) using the Nyquist plot of the open-loop transfer function. It was seen in Figure 3.7 that, as long as the phase of the open-loop system surpasses $-\pi$, the plot must enter the area of sensitivity increase. It is well known that ORHP zeros introduce additional phase lag[8] to the open-loop system, which is evident from the fact that each factor of B_T in (3.26) satisfies

$$\arg \frac{q_i - j\omega}{\overline{q}_i + j\omega} \to -\pi \quad \text{as} \quad \omega \to \infty. \tag{3.32}$$

[8]By *additional phase lag* introduced by an ORHP zero we mean the difference between the phase of the system with the ORHP zero and that of a system where the ORHP zero is replaced by its reflection in the OLHP.

Hence the trade-off between areas of sensitivity attenuation and amplification can also be observed in the Nyquist plot of a nonminimum phase open-loop system. As an example, Figure 3.9 shows the Nyquist plot of the function $L(s) = (1-s)/(1+s)^2$ together with its minimum phase counterpart $\tilde{L}(s) = 1/(1+s)$. It can be seen from this figure that the area balance is apparent for the nonminimum phase system whilst the Nyquist plot of the minimum phase system can be kept inside the area of sensitivity reduction save, of course, at infinite frequency.

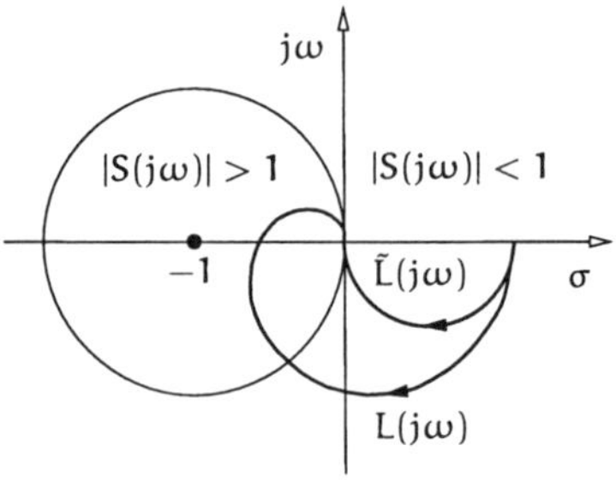

FIGURE 3.9. Area balance for a nonminimum phase system.

From the expression $T = 1/(1 + 1/L)$, it is evident that a similar graphic analysis can be done for the complementary sensitivity function by studying the Nyquist plot of $1/L$. Hence, the additional "phase lag" introduced by ORHP poles (of L) to the function $1/L$ are now seen to be the cause of the need to balance the weighted areas where $|T(j\omega)| < 1$ and $|T(j\omega)| > 1$.

The weighting function (3.31) — or Poisson kernel for the right half plane — in the integral constraints (3.29) and (3.30) takes explicit account of the effect of the additional phase lag introduced by ORHP zeros or poles.[9] To see this, given an interval $[-\omega_1, \omega_1]$, $\omega_1 > 0$, consider the integral of the weighting function (3.31), i.e.,

$$\begin{aligned}\Theta_{s_0}(\omega_1) &\triangleq \int_{-\omega_1}^{\omega_1} W_{s_0}(\omega)\, d\omega \\ &= \int_{-\omega_1}^{\omega_1} \frac{\sigma_0}{\sigma_0^2 + (\omega_0 - \omega)^2}\, d\omega \\ &= \arctan\frac{\omega_1 - \omega_0}{\sigma_0} + \arctan\frac{\omega_1 + \omega_0}{\sigma_0}\,.\end{aligned} \tag{3.33}$$

It is then easy to see that, for a real zero $s_0 = q = \sigma_q$, we have

$$\Theta_q(\omega_1) = -\arg\frac{\sigma_q - j\omega_1}{\sigma_q + j\omega_1}\,,$$

[9]Since the poles are here seen as zeros of $1/L$, we will use the word "zeros" in the discussion about the weighting function.

whereas for a pair of complex conjugate zeros $s_0 = q$ and $\overline{q}$, we have

$$\Theta_q(\omega_1) = -\frac{1}{2}\left[\arg\frac{q - j\omega_1}{\overline{q} + j\omega_1} + \arg\frac{\overline{q} - j\omega_1}{q + j\omega_1}\right] . \tag{3.34}$$

Note that $\Theta_{s_0}(\omega_1)$ represents the weighted length (by the ORHP zero s_0) of the frequency interval $[-\omega_1, \omega_1]$; thus, the weighted length of a such an interval is equal to minus the additional phase lag introduced by the real ORHP zero (or to minus half the additional phase lag introduced by the pair of complex conjugate zeros) at the upper endpoint of the interval.

The above interpretation of the weighted length of the frequency interval is useful in quantifying trade-offs implied by the Poisson integrals relative to the location of the ORHP zero. As an example, we will next revisit the water-bed effect of §3.2. We view it here in the light of the integral constraint (3.29).

Suppose, as in §3.1.3, that the feedback loop has been designed to achieve

$$|S(j\omega)| \leq \alpha_1 < 1 , \qquad \forall\omega \in \Omega_1 \triangleq [-\omega_1, \omega_1] . \tag{3.35}$$

Let q be an ORHP zero of L and consider the weighted length of the interval Ω_1, i.e., $\Theta_q(\omega_1)$ as in (3.33). Then the infinity norm of the sensitivity function has a lower bound as shown in the following result.

Corollary 3.3.3. Let S be the sensitivity function defined by (3.8) and suppose that the open-loop system L can be factored as in (3.27). Assume that the closed-loop system is stable and that the goal (3.35) has been achieved. Then, for each ORHP zero of L, q, we have

$$\|S\|_\infty \geq \left(\frac{1}{\alpha_1}\right)^{\frac{\Theta_q(\omega_1)}{\pi - \Theta_q(\omega_1)}} \left|B_S^{-1}(q)\right|^{\frac{\pi}{\pi - \Theta_q(\omega_1)}} , \tag{3.36}$$

where $\Theta_q(\omega_1)$ is as in (3.33) with $s_0 = q$.

Proof. Dividing the range of integration in (3.29), and using the inequalities (3.35) and $|S(j\omega)| \leq \|S\|_\infty, \forall\omega$, we have

$$\log\alpha_1\ \Theta_q(\omega_1) + \log\|S\|_\infty\,[\pi - \Theta_q(\omega_1)] \geq \pi\log\left|B_S^{-1}(q)\right| . \tag{3.37}$$

The result then follows by exponentiating both sides. □

It is immediate from (3.36) that the lower bound on the sensitivity peak is strictly greater than one. Indeed, this follows from the fact that $\left|B_S^{-1}(q)\right| \geq 1$, $\alpha_1 < 1$ and $\Theta_q(\omega_1) < \pi$. Note also that if the open-loop system is both nonminimum phase and unstable, then the sensitivity peak is greater than one even if there is no region of sensitivity reduction.

Remark 3.3.1. Corollary 3.3.3 provides another interpretation for the water-bed effect of Theorem 3.2.1: if the open-loop system is nonminimum phase, then requiring $|S(j\omega)|$ to be small over a range of frequencies will necessarily lead to a large sensitivity peak outside that range. In this new form, however, the lower bound on the sensitivity peak exhibits an explicit dependence on the location of the ORHP zero of the open-loop system. Indeed, recall from our previous discussion that $\Theta_q(\omega_1)$ is equal to minus the additional phase lag introduced by the real ORHP zero q (or to minus half the additional phase lag introduced by the pair of complex conjugate zeros q and $\overline{q}$) at the frequency ω_1. Thus, if at $\omega = \omega_1$ the ORHP zero contributes with a significant amount of lag ($\Theta_q(\omega_1)$ close to π, say) the first term on the LHS of (3.37) will be large (and negative) and hence will require a large value of $\|S\|_\infty$ to satisfy the lower bound on the RHS. On the other hand, if the lag introduced by the zero at ω_1 is small then the trade-off implied by (3.36) will not be severe.

Observe that, since $|S(j\omega)| \leq \alpha_1 < 1$ implies $|L(j\omega)| \geq 1/\alpha_1 - 1$, a desirable feedback design objective would be to roll-off the open-loop gain well before the phase lag introduced by the ORHP zero becomes significant. This confirms a well-known rule of thumb used in classical feedback control design. ∘

A result parallel to Corollary 3.3.3 for the complementary sensitivity function can be easily obtained if we assume that the following goal has been achieved (e.g., for robustness purposes)

$$|T(j\omega)| \leq \alpha_2 < 1\,, \qquad \forall \omega \in \Omega_2 \triangleq [-\infty, -\omega_2] \cup [\omega_2, \infty]\,. \tag{3.38}$$

We then have the following result.

Corollary 3.3.4. Let T be the sensitivity function defined by (3.8) and suppose that the open-loop system L can be factored as in (3.27). Assume that the closed-loop system is stable and that the goal (3.38) has been achieved. Then, for each ORHP pole of L, p, we have

$$\|T\|_\infty \geq \left(\frac{1}{\alpha_2}\right)^{\frac{\pi - \Theta_p(\omega_2)}{\Theta_p(\omega_2)}} \left|B_T^{-1}(p) e^{\sigma_p \tau}\right|^{\frac{\pi}{\Theta_p(\omega_2)}}\,, \tag{3.39}$$

where $\Theta_p(\omega_1)$ is as in (3.33) with $s_0 = p$.

Proof. Similar to the proof of Corollary 3.3.3, using the bound (3.38). □

Notice that, according to the definition of closed-loop bandwidth in §3.1.3, ω_2 in (3.38) is equal to the bandwidth ω_b if α_2 is taken to be $1/\sqrt{2}$. In view of this observation, Corollary 3.3.4 can be used to obtain bounds on the closed-loop bandwidth, as shown in the following example.

Example 3.3.1. Consider the feedback system of Figure 3.2, and assume, for simplicity, that the open-loop system L has a real ORHP pole $p = \sigma_p$, is minimum phase and has no time delay, i.e., $q = 0$ and $\tau = 0$ in (3.27). Assume that the closed-loop system is stable and satisfies (3.38) for $\alpha_2 = 1/\sqrt{2}$ (and hence $\omega_2 = \omega_b$). Then

$$\|T\|_\infty \geq \sqrt{2}^{\frac{\pi - \Theta_p(\omega_b)}{\Theta_p(\omega_b)}} .$$

Suppose further that we impose the condition that the RHS of the above expression be less than or equal to T_m, which is necessary if we require that the peak complementary sensitivity (i.e., the LHS of the above expression) be less than or equal to T_m. Using this additional requirement we obtain the following lower bound on the closed-loop bandwidth

$$\omega_b \geq p \tan\left(\frac{\pi}{2 + 2\log T_m / \log\sqrt{2}}\right) . \tag{3.40}$$

We conclude from (3.40), for example, that an open-loop system having a real ORHP pole, p, requires a bandwidth at least equal to p if it is desired to keep the lower bound on the peak complementary sensitivity smaller than $T_m = \sqrt{2}$.

Notice that the lower bounds on the bandwidth given by (3.22) and (3.40) are not directly comparable, since (3.22) was obtained from requirements on S while (3.40) follows from specifications on T. ∘

A more realistic design will require both (3.35) and (3.38) to be satisfied. Note that (3.38) implies $|S(j\omega)| \leq 1 + \alpha_2, \forall \omega \in \Omega_2$. Figure 3.10 represents the combined shape specifications on $|S(j\omega)|$.

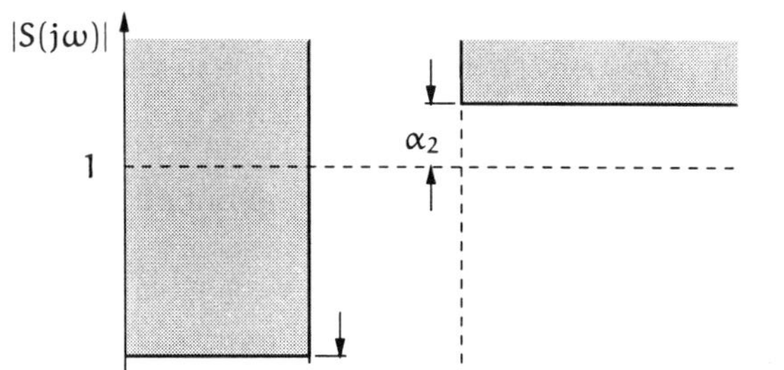

FIGURE 3.10. Frequency specifications for S.

Using this additional information as in the proof of Corollary 3.3.3 we obtain

$$\|S\|_\infty \geq \left(\frac{1}{\alpha_1}\right)^{\frac{\Theta_q(\omega_1)}{\Theta_q(\omega_2)-\Theta_q(\omega_1)}} \left(\frac{1}{1+\alpha_2}\right)^{\frac{\pi-\Theta_q(\omega_2)}{\Theta_q(\omega_2)-\Theta_q(\omega_1)}} \times \left|B_S^{-1}(q)\right|^{\frac{\pi}{\Theta_q(\omega_2)-\Theta_q(\omega_1)}} . \tag{3.41}$$

The following example examines the severity of constraint (3.41).

Example 3.3.2. Suppose that we desire to control a plant with a single nonminimum phase zero $q > 0$. We assume for simplicity that the plant is open-loop stable. We desire for the closed-loop system a sensitivity reduction of at least α_1 within the interval of frequencies $[0, \omega_1]$, where ω_1 is chosen to be three quarters of the desired closed-loop bandwidth frequency ω_b, i.e., the specification (3.35) holds with $\omega_1 = 0.75\omega_b$.

Following the definition of closed-loop bandwidth on page 60, we can write the requirement of bandwidth by setting $\omega_2 = \omega_b$ and $\alpha_2 = 1/\sqrt{2}$ in (3.38) (see also Example 3.3.1).

Figure 3.11 plots the lower bound (3.41) on $\|S\|_\infty$ versus the position of the nonminimum phase zero (relative to the desired bandwidth), and for different values of sensitivity reduction.

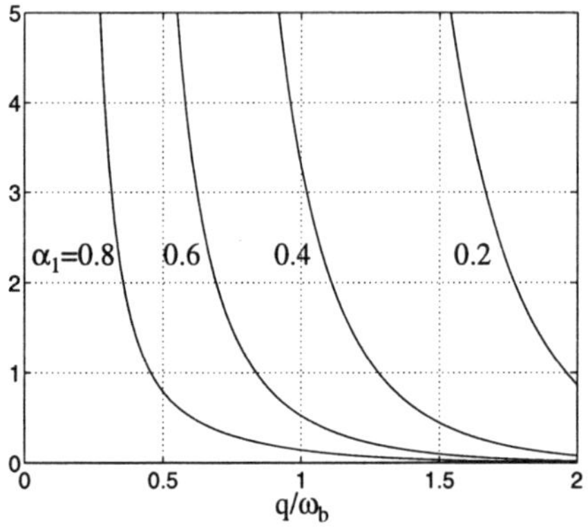

FIGURE 3.11. Lower bound (3.41) on $\|S\|_\infty$.

As anticipated in Remark 3.3.1, the picture shows that the constraints imposed by a nonminimum phase zero worsen the more the zero is within the desired closed-loop bandwidth, which manifests as higher peaks in $|S(j\omega)|$. Since, as seen in Chapter 2, large peaks in $|S(j\omega)|$ are associated with poor sensitivity and robustness properties, then a nonminimum phase zero imposes a trade-off in design that limits the achievable closed-loop bandwidth of the system. The same conclusion holds for complex nonminimum phase zeros. ◦

Notice that the bound (3.41) is satisfied with equality for an "ideal" sensitivity function that is sectionally constant and equal to α_1 in Ω_1, to $\|S\|_\infty$ in $[-\omega_2, -\omega_1] \cup [\omega_1, \omega_2]$, and to $1 + \alpha_2$ in Ω_2. Hence, the lower bound given by (3.41) is in fact *conservative*. The closer the actual sensitivity is to the "ideal" function, the tighter the bound (3.41) is.[10] Although, in theory, this sectionally constant function can be arbitrarily approximated by a rational function, there exist a number of practical limitations in achieving this result. For example, a restriction on the complexity of the controller would mean that the ideal case cannot be achieved.

Finally, if the plant has more than one ORHP zero, bounds similar to (3.36) or (3.41) hold for each of them, which, in general, will provide different information. Thus, since each of these bounds may not be tight if considered alone, it will be necessary to analyze them in combination to obtain the overall trade-offs imposed by the set of ORHP zeros. However, this will not be as insightful and simple as in the case of a single zero, which shows a limitation of the approach. A different technique, based on Nevanlinna-Pick theory, has been described by O'Young & Francis (1985). This technique provides an iterative procedure to compute "hard" bounds on the sensitivity function for multivariable plants with several ORHP zeros.

3.3.3 *Example: Inverted Pendulum*

Consider again the inverted pendulum shown in Figure 3.12, studied in §1.3.3 of Chapter 1 in the time domain.

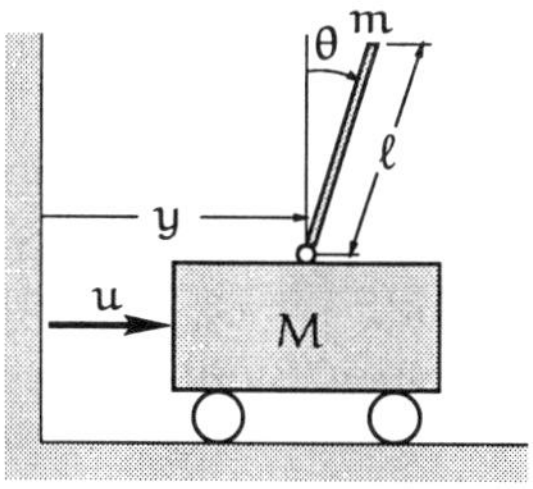

FIGURE 3.12. Inverted pendulum.

It was seen in §1.3.3 that the linearized model for this system has a transfer function from u to y of the form

$$\frac{Y(s)}{U(s)} = \frac{(s-q)(s+q)}{M\, s^2 (s-p)(s+p)},$$

[10]Less conservative bounds can be also derived from these integral constraints by considering more realistic shape specifications; see for example Middleton (1991) or Middleton & Goodwin (1990, Chapter 13).

i.e., it has one nonminimum phase zero and one unstable pole. Assume that the parameters in the model are chosen as in §1.3.3, i.e., so that the ORHP zero is $q = 1$ and the pole is moved according to four different values of the mass ratio m/M. For example, for $m/M = 1$, the pole is $p = \sqrt{2}$. Using Corollary 3.3.3 and Corollary 3.3.4 with the latter values of p and q, and assuming no particular range of sensitivity reduction (i.e., taking $\omega_1 = 0$ in (3.35) and $\omega_2 = \infty$ in (3.38)), then the peak sensitivity and complementary sensitivity satisfy

$$\|S\|_\infty \geq 5.8284, \qquad \text{and} \qquad \|T\|_\infty \geq 5.8284. \tag{3.42}$$

Figure 3.13 shows $\log|S(j\omega)|$ and $\log|T(j\omega)|$ achieved by the same LQG-LQR design used in §1.3.3, for $m/M = 0.1, 0.2, 0.4, 1$. These figures corre-

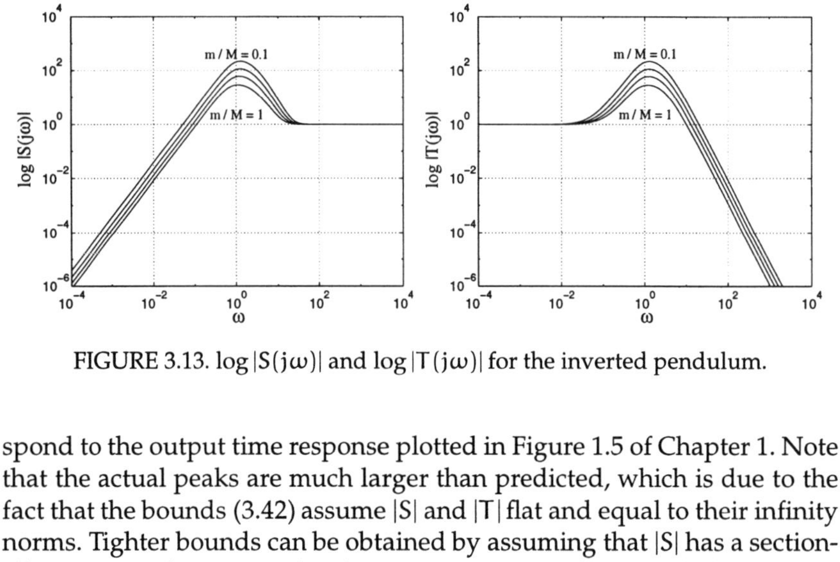

FIGURE 3.13. $\log|S(j\omega)|$ and $\log|T(j\omega)|$ for the inverted pendulum.

spond to the output time response plotted in Figure 1.5 of Chapter 1. Note that the actual peaks are much larger than predicted, which is due to the fact that the bounds (3.42) assume $|S|$ and $|T|$ flat and equal to their infinity norms. Tighter bounds can be obtained by assuming that $|S|$ has a sectionally constant shape as in (3.41), or using more sophisticated shapes as in Middleton (1991). Note, however, that the bounds derived in this chapter are valid for any design methodology, and hence it is expected that they will be conservative for some particular design choices.

3.4 Discrete Systems

This section describes the extension of the Poisson and Bode integral relations to discrete-time feedback control systems. We again focus on sensitivity and complementary sensitivity functions of feedback control systems, but the transfer functions are here complex functions of the variable z, i.e., they represent the $\mathcal{Z}$-transform of impulse responses of discrete-time systems. We will see that the continuous-time results extend in a straightforward fashion to discrete time by means of the Poisson integral formula for the unit disk, treated in §A.6.2 of Appendix A. We will also see, however, that in general, the discrete constraints are more demanding since the sensitivity trade-offs must be achieved on the finite frequency interval $[0, \pi]$.

The results in this section apply to systems (plant and controller) whose original description is in discrete time; if the original description of the plant is in continuous time and the controller is digital, we call the overall system a *sampled-data system*, and the analysis is more adequately performed in continuous time. Sampled-data control systems are discussed in Chapter 6.

3.4.1 *Poisson Integrals for* S *and* T

Consider the feedback control loop of Figure 3.14, where the open-loop system, L, is a proper transfer function of the complex variable z.

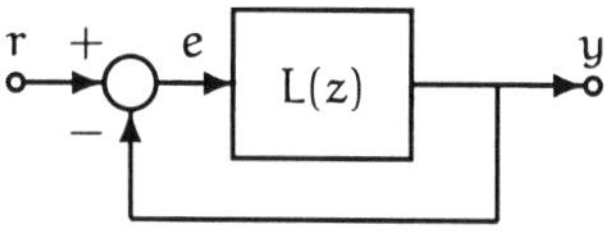

FIGURE 3.14. Discrete-time feedback control system.

As in (3.8), the sensitivity and complementary sensitivity functions of the loop in Figure 3.14 are defined as

$$S(z) = \frac{1}{1 + L(z)} \quad \text{and} \quad T(z) = \frac{L(z)}{1 + L(z)}, \tag{3.43}$$

respectively. In common with the continuous-time case, the sensitivity and complementary sensitivity functions for the discrete case satisfy interpolations constraints imposed by zeros and poles of the open-loop system. Indeed, if L is free of unstable pole-zero cancelations,[11] then S has zeros at the unstable open-loop plant poles and T has zeros at the nonminimum phase open-loop plant zeros; i.e., Lemma 3.1.3 holds with $\overline{\mathbb{C}^+}$ replaced by the region outside the open unit disk, which we denote by $\mathbb{D}^c$.

For the purpose of deriving Poisson integral constraints, we need to extract from S and T all the zeros outside the *closed*[12] unit disk — denoted by

[11] Recall from Chapter 2 that, for discrete-time systems, a transfer function is nonminimum phase if it has zeros outside the open unit disk, $\mathbb{D}$, and it is unstable if it has poles outside $\mathbb{D}$. Thus, L is free of unstable pole-zero cancelations if there are no cancelations of zeros and poles outside $\mathbb{D}$ between the plant and controller whose cascade connection form L.

[12] Recall that, in the continuous-time case, only the zeros and poles of the plant in the ORHP impose integral constraints. Those on the imaginary axis do not contribute to the value of the integrals (see Lemma A.6.2 and Corollary A.6.3 in Appendix A).

$\overline{\mathbb{D}}^c$. Let q_i, $i = 1, \ldots, n_q$, be the zeros of L in $\overline{\mathbb{D}}^c$, and p_i, $i = 1, \ldots, n_p$, be the poles of L in $\overline{\mathbb{D}}^c$, all counted with multiplicities. It follows from our previous discussion that the q_i's are the zeros of T in $\overline{\mathbb{D}}^c$, and the p_i's are the zeros of S in $\overline{\mathbb{D}}^c$. We introduce the *discrete Blaschke products* of the zeros of S and T, given by

$$B_S(z) = \prod_{i=1}^{n_p} \frac{\overline{p_i}}{|p_i|} \frac{z - p_i}{1 - \overline{p_i} z} , \qquad \text{and} \qquad B_T(z) = \prod_{i=1}^{n_q} \frac{\overline{q_i}}{|q_i|} \frac{z - q_i}{1 - \overline{q_i} z} . \tag{3.44}$$

We will define $B_S(z) = 1$ if L is stable, and $B_T(z) = 1$ if L is minimum phase. Similar to their continuous-time counterpart, the discrete Blaschke products are "all-pass" functions, since their magnitude is constant and equal to one on the unit circle $|z| = 1$. Using (3.44), and extracting the zeros at infinity in a similar fashion as was done in (2.7) in Chapter 2, we can factor the open-loop transfer function, L, as

$$L(z) = B_S^{-1}(z)\, B_T(z)\, \tilde{L}(z)\, z^{-\delta} , \tag{3.45}$$

where $\tilde{L}$ is a stable, minimum-phase transfer function, having relative degree zero. It is then easy to see that S and T in (3.43) can be factored as

$$\begin{aligned} S(z) &= \tilde{S}(z) B_S(z) , \\ T(z) &= \tilde{T}(z) B_T(z) z^{-\delta} , \end{aligned} \tag{3.46}$$

where $\tilde{S}$ and $\tilde{T}$ have no zeros in the region outside the closed unit disk (including infinity). Note that here the zeros at infinity of T (which are branch points of $\log T$, see §A.9.2 in Appendix A) have to be factored explicitly since they will contribute to the value of the Poisson integral of $\log T$ on the unit circle. This is because the point at infinity is an interior point of the region "encircled" by the unit circle, $\overline{\mathbb{D}}^c$, and thus it should be treated as any other finite singularity of $\log T$ in $\overline{\mathbb{D}}^c$. Note that this is not the case for continuous-time systems, i.e., the singularities at infinity of $\log T$ arising from a strictly proper L do not add to the value of the Poisson integral of $\log T$ on the imaginary axis (see Corollary A.6.3 in Appendix A, and the proof of Theorem 3.3.2).

Assuming stability of the closed-loop system, and since, as just seen, $\tilde{S}$ and $\tilde{T}$ in (3.46) have no zeros in $\overline{\mathbb{D}}^c$, then the functions $\log \tilde{S}$ and $\log \tilde{T}$ are analytic in $\overline{\mathbb{D}}^c$. The real parts of these functions are harmonic in $\overline{\mathbb{D}}^c$, and they thus satisfy the conditions of Corollary A.6.5 in Appendix A. We then obtain the following results.

Theorem 3.4.1 (Poisson Integral for S). Let S be the discrete sensitivity function defined by (3.43). Assume that the open-loop system L can be factored as in (3.45) and let $q = r_q e^{j\theta_q}$ be a zero of L in $\overline{\mathbb{D}}^c$. Then, if the closed-loop system is stable,

$$\int_{-\pi}^{\pi} \log|S(e^{j\theta})| \frac{r_q^2 - 1}{1 - 2r_q \cos(\theta - \theta_q) + r_q^2} \, d\theta = 2\pi \log\left|B_S^{-1}(q)\right| . \qquad (3.47)$$

Proof. Under the assumptions of the theorem, S can be factored as in (3.46), where $\tilde{S}$ is a stable, minimum-phase transfer function, having relative degree zero. It follows that the function $\log|\tilde{S}|$ is harmonic in $\overline{\mathbb{D}}^c$. Using Corollary A.6.5 in Appendix A with $u = \log|\tilde{S}|$ and $s_0 = q$, and noting that $|\tilde{S}(e^{j\theta})| = |S(e^{j\theta})|$ (since $|B_S(e^{j\theta})| = 1, \forall\theta$) and that $\tilde{S}(q) = B_S^{-1}(q)$ (since $S(q) = 1$), yields the desired result. □

The corresponding result for T is as follows.

Theorem 3.4.2 (Poisson Integral for T). Let T be the complementary sensitivity function defined by (3.43). Assume that the open-loop system L can be factored as in (3.45) and let $p = r_p e^{j\theta_p}$ be a pole of L in $\overline{\mathbb{D}}^c$. Then, if the closed-loop system is stable,

$$\int_{-\pi}^{\pi} \log|T(e^{j\theta})| \frac{r_p^2 - 1}{1 - 2r_p \cos(\theta - \theta_p) + r_p^2} \, d\theta = 2\pi \log\left|B_T^{-1}(p)\right| + 2\pi\delta \log|p| . \qquad (3.48)$$

Proof. The proof uses Corollary A.6.5 in Appendix A with $u = \log|\tilde{T}|$, where $\tilde{T}$ is given in (3.46), and $s_0 = p$. Then (3.48) follows by similar arguments to those used in the proof of Theorem 3.4.1. □

As in the continuous-time case, two technical comments are in order. We first note that Theorems 3.4.1 and 3.4.2 still hold when L has poles or zeros on the unit circle. Second, if the zeros and/or poles of L outside the unit disk have multiplicities greater than one, then additional integral constraints on the derivatives of $\log\tilde{S}$ and of $\log\tilde{T}$ can be derived.

Also similar to the Poisson integral constraints for continuous-time systems, the relations (3.47) and (3.48) represent a balance between weighted areas of sensitivity or complementary sensitivity attenuation and amplification. This is because: (i) for $z_0 = r_0 e^{j\theta_0}$ with $r_0 > 1$, the weighting function (or Poisson kernel for the unit disk)

$$W_{z_0}(\theta) \triangleq \frac{r_0^2 - 1}{1 - 2r_0 \cos(\theta - \theta_0) + r_0^2} \qquad (3.49)$$

is positive for all θ, and (ii) the RHSs of both equations are nonnegative since, for a discrete Blaschke product B, we have that $\left|B^{-1}(z)\right| \geq 1$ for $|z| > 1$. These facts imply that the weighted area of sensitivity increase must be, at least, as large as the weighted area of sensitivity reduction. This situation is aggravated if the open-loop system is both unstable and nonminimum phase, since in this case both $|B_S^{-1}(q)|$ and $|B_T^{-1}(p)|$ on the RHSs of (3.47) and (3.48) are strictly greater than one. Moreover, the closer the unstable zeros and poles are to each other, the larger the RHSs of both equations become.

Note also the similarities between Theorem 3.4.2 and its counterpart for continuous-time systems, Theorem 3.3.2. In particular, the term due to the relative degree of the plant in the discrete case in (3.48) is analogous to that due to a pure time delay in (3.30). Hence, the constraints imposed on T by open loop time delays and unstable poles are also present in the discrete case, and again, the "more unstable" the pole, and the larger the time delay, the worse these constraints will be.

In the following section we will discuss some of the design trade-offs implied by Theorems 3.4.1 and 3.4.2.

3.4.2 *Design Interpretations*

The Poisson integrals of Theorem 3.4.1 and Theorem 3.4.2 can be used to derive lower bounds on the infinity norms of the discrete sensitivity functions. As in the continuous-time case, these lower bounds exhibit the water-bed effect experienced by nonminimum phase plants when sensitivity reduction is required over some frequency range.

To analyze the water-bed effect, we require the weighted length of an interval on the unit disk by the kernel (3.49) (cf. the corresponding for the weighting function for the half plane, given by (3.33)), i.e.,

$$
\begin{aligned}
\Theta_{z_0}(\theta_1) &\triangleq \int_{-\theta_1}^{\theta_1} W_{z_0}(\theta)\, d\theta \\
&= \arctan\left[\frac{r_0-1}{r_0+1}\tan\frac{\theta_1-\theta_0}{2}\right] + \arctan\left[\frac{r_0-1}{r_0+1}\tan\frac{\theta_1+\theta_0}{2}\right],
\end{aligned}
\tag{3.50}
$$

where the interval of interest is $\Theta_1 = [-\theta_1, \theta_1]$, $\theta_1 < \pi$. It is possible to show that $\Theta_{z_0}(\theta_1)$ is equal to minus half the sum of the phases of the discrete Blaschke products corresponding to z_0 and its conjugate, evaluated at $z = e^{j\theta_1}$, i.e.,

$$
\Theta_{z_0}(\theta_1) = -\frac{1}{2}\left[\arg\frac{\overline{z_0}}{r_0}\,\frac{e^{j\theta_1}-z_0}{1-\overline{z_0}e^{j\theta_1}} + \arg\frac{z_0}{r_0}\,\frac{e^{j\theta_1}-\overline{z_0}}{1-z_0 e^{j\theta_1}}\right].
$$

This was also the case for the weighted length of an interval on the imaginary axis by the Poisson kernel for the right half plane, as can be seen from equation (3.34).

Suppose next that the discrete feedback control loop has been designed to achieve the goal

$$
|S(e^{j\theta})| \le \alpha < 1, \qquad \forall \theta \in \Theta_1 \triangleq [-\theta_1, \theta_1], \quad \theta_1 < \pi. \tag{3.51}
$$

Let q be a zero of L in $\overline{\mathbb{D}}^c$ and consider the weighted length of the interval Θ_1, i.e., $\Theta_q(\theta_1)$ as in (3.50). Then the infinity norm of the sensitivity function has a lower bound as shown in the following result.

Corollary 3.4.3. Let S be the sensitivity function defined by (3.43) and suppose that the open-loop system L can be factored as in (3.45). Assume that the closed-loop system is stable and that the goal (3.51) has been achieved. Then, for each zero of L in $\overline{\mathbb{D}}^c$, $q = r_q e^{j\theta_q}$,

$$\|S\|_\infty \geq \left(\frac{1}{\alpha}\right)^{\frac{\Theta_q(\theta_1)}{2\pi-\Theta_q(\theta_1)}} \left|B_S^{-1}(q)\right|^{\frac{2\pi}{2\pi-\Theta_q(\theta_1)}} , \tag{3.52}$$

where Θ_q is the function defined in (3.50).

Proof. The proof follows that of Corollary 3.3.3. □

Similar comments to those following Corollary 3.3.3 are also relevant here. In particular, we note that the lower bound on the RHS of (3.52) is strictly greater than one and it becomes larger when the open-loop system has both zeros and poles in $\overline{\mathbb{D}}^c$. Also, parallel lower bounds can be derived for the complementary sensitivity function, as well as bounds arising from combined specifications for S and T. The interested reader is encouraged to restate the results in §3.3.2 for the discrete-time case.

3.4.3 Bode Integrals for S and T

In this section we obtain Bode integral constraints for the discrete-time sensitivity and complementary sensitivity functions. Recall that, in Theorems 3.1.4 and 3.1.5, we established the corresponding results for continuous-time systems by direct contour integration. We will not follow the same procedure here, but instead we will derive the Bode integrals from the Poisson integral formula for the disk. This is done in the following results.

Theorem 3.4.4 (Bode Integral for S). Let S be the sensitivity function defined by (3.43). Assume that the open-loop system, L, is strictly proper and let $\{p_i : i = 1, \dots, n_p\}$ be the set of poles of L in $\overline{\mathbb{D}}^c$. Then, assuming closed-loop stability,

$$\int_0^\pi \log|S(e^{j\theta})|\, d\theta = \pi \sum_{i=1}^{n_p} \log|p_i| . \tag{3.53}$$

Proof. As in the proof of Theorem 3.4.1, we use the factorization (3.46) and apply Corollary A.6.5 in Appendix A with $u = \log|\tilde{S}|$ and $s_0 = r, r > 1$.

This yields

$$\frac{1}{2\pi}\int_{-\pi}^{\pi}\log|S(e^{j\theta})|\frac{r^2-1}{1-2r\cos\theta+r^2}\,d\theta = \log|\tilde{S}(r)|$$
$$= \log\left|B_S^{-1}(r)\right| + \log|S(r)|\,. \quad (3.54)$$

Next, we take limits as $r \to \infty$ on both sides of (3.54). Using the uniform convergence theorem to take limits inside the integral (Levinson & Redheffer 1970, p. 335), and the fact that $|S(e^{-j\theta})| = |S(e^{j\theta})|$, we have that the LHS in (3.54) tends to

$$\frac{1}{\pi}\int_0^{\pi}\log|S(e^{j\theta})|\,d\theta\,.$$

As for the RHS, note that $\lim_{r\to\infty}\log|S(r)| = 0$, since L is strictly proper, and

$$\lim_{r\to\infty}\log\left|B_S^{-1}(r)\right| = \lim_{r\to\infty}\log\prod_{i=1}^{n_p}\left|\frac{1-\overline{p_i}r}{r-p_i}\right|$$
$$= \sum_{i=1}^{n_p}\log|p_i|\,.$$

The result (3.53) then follows. □

The parallel result for T is given in the following theorem.

Theorem 3.4.5 (Bode Integral for T). Let T be the complementary sensitivity function defined by (3.43). Assume that the open-loop system L can be factored as in (3.45) and let $\{q_i : i = 1, \dots, n_q\}$ be the set of zeros of L in $\overline{\mathbb{D}}^c$. Suppose further that $L(1) \neq 0$. Then, assuming closed-loop stability,

$$\int_0^{\pi}\log\left|\frac{T(e^{-j\theta})}{T(1)}\right|\frac{d\theta}{1-\cos\theta} = \frac{\pi}{T(1)}\lim_{z\to 1}\frac{dT(z)}{dz} + \pi\sum_{i=1}^{n_q}\frac{|q_i|^2-1}{|q_i-1|^2} + \pi\delta\,, \quad (3.55)$$

where δ is the relative degree of L.

Proof. From the factorization (3.46), we write the Poisson Integral formula (Corollary A.6.5) for $\log(\tilde{T}/T(1))$ evaluated at a real point $r > 1$ (note that $T(1) \neq 0$ since $L(1) \neq 0$ by assumption). This gives

$$\int_0^{\pi}\log\left|\frac{\tilde{T}(e^{j\theta})}{T(1)}\right|\frac{r^2-1}{1-2r\cos(\theta)+r^2}\,d\theta = \pi\log\left|\frac{T(r)}{T(1)}\right|$$
$$+ \pi\log\left|B_T^{-1}(r)\right| + \pi\delta\log r\,. \quad (3.56)$$

Next, we divide both sides of (3.56) by $(r-1)$ and take limits as $r \to 1$. We compute each term in turn. First, by the uniform convergence theorem, the limit on the LHS of (3.56) can be brought inside the integral, and then, using the fact that $|\tilde{T}(e^{j\theta})| = |T(e^{j\theta})|$, it is easy to check that the LHS of (3.55) is obtained.

Then consider the first term on the RHS of (3.56). We have

$$\pi \lim_{r\to 1} \frac{1}{(r-1)} \log\left|\frac{T(r)}{T(1)}\right| = \frac{\pi}{2} \lim_{r\to 1} \frac{1}{(r-1)} \log\left(\frac{T(r)}{T(1)}\right)^2 \tag{3.57}$$

$$= \frac{\pi}{T(1)} \lim_{z\to 1} \frac{dT(z)}{dz}, \tag{3.58}$$

which follows by applying L'Hospital's rule. This is the first term on the RHS of (3.55). The second term on the RHS of (3.55) is also obtained from a straightforward application of L'Hospital's rule to the following limit

$$\lim_{r\to 1} \frac{1}{(r-1)} \log\left|B_T^{-1}(r)\right| = \lim_{r\to 1} \frac{1}{2(r-1)} \sum_{i=1}^{n_q} \log \frac{(1-r\bar{q}_i)(1-rq_i)}{(r-q_i)(r-\bar{q}_i)} .$$

Finally, the proof is concluded by noting that

$$\lim_{r\to 1} \frac{\pi\delta \log r}{r-1} = \pi\delta .$$

□

We see in (3.55) that Theorem 3.4.5 is entirely analogous to its continuous-time counterpart, Theorem 3.1.5. Indeed, the first term on the RHS of (3.55) is precisely minus the reciprocal of the velocity constant of a *discrete-time* system equivalent to that in Figure 3.5. The effect of nonminimum phase zeros of L is also equivalent to that in continuous-time, and so is the effect of the relative degree of L, which corresponds to a time delay of δ discrete units. Hence, the same interpretations that were given for Theorem 3.1.5 apply to Theorem 3.4.5 for discrete-time systems.

Nevertheless, an important difference between these integrals in the continuous- and discrete-time cases is that the latter involves restrictions over a *finite* interval. We see this in more detail in the following subsection, where we study the design trade-offs induced by Bode's discrete sensitivity integral.

3.4.4 Design Interpretations

Bode integral formulae, both in the discrete and continuous-time case, indicate that there is a balance of areas of sensitivity reduction and increase. In the continuous-time case, however, the Bode integral formula does not imply a direct trade-off in design if additional bandwidth constraints are not imposed (see §3.1.3). Indeed, since integration is performed over an infinite range, the area of sensitivity reduction over a finite range of frequencies may be compensated by an area where $|S|$ is allowed to be slightly greater than one over an arbitrarily large range of frequencies.

On the other hand, in the discrete-time case, the Bode integral implies a nontrivial sensitivity trade-off even if no bandwidth constraint is assumed. This is seen from the following immediate corollary.

Corollary 3.4.6. Suppose that the conditions of Theorem 3.4.4 hold. Assume further that the goal (3.51) has been achieved. Then necessarily

$$\|S\|_\infty \geq \left(\frac{1}{\alpha}\right)^{\frac{\theta_1}{2\pi-\theta_1}} \left|\prod_{i=1}^{n_p} p_i\right|^{\frac{2\pi}{2\pi-\theta_1}}. \tag{3.59}$$

Proof. The proof follows by splitting the range of integration in (3.53) and using the bounds (3.51) and $|S(e^{j\theta})| \leq \|S\|_\infty$, $\forall\theta$. □

Similar results can be derived from Theorem 3.4.5 for the discrete complementary sensitivity function, showing that discrete nonminimum phase zeros and time delays induce design trade-offs even if no bandwidth constraints are imposed on the discrete system.

However, the above conclusions might be irrelevant to the real design problem if the discrete system corresponds to the discretization of a continuous-time system where the controller is implemented digitally. This is discussed in the following remark.

Remark 3.4.1. Consider the discrete system shown in Figure 3.15. In this system, K_d is a digital controller and $(GH)_d$ denotes the discrete equivalent of the cascade of a hold, H, and a continuous-time plant, G. This *discretized plant* is depicted in Figure 3.16, and usually arises in the analysis and design of continuous-time systems where the controller is implemented digitally.[13]

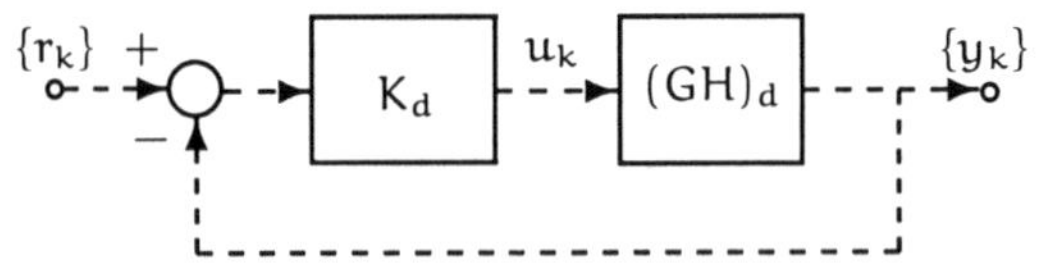

FIGURE 3.15. Discrete-time representation of a sampled-data control system.

The signals in Figures 3.15 and 3.16 represent discrete-time signals, where $\{u_k\}$ is the discrete output of the digital controller, and $\{r_k\}$, $\{y_k\}$ are

[13]We give a more thorough treatment of these systems in Chapter 6, to which we refer for further details.

the *sampled* values of the continuous-time reference and output signals, respectively.

If we define L as the product of discrete controller and discretized plant, $L = (GH)_d K_d$, then the system may be studied by analyzing the corresponding discrete sensitivity and complementary sensitivity functions S and T defined in (3.43). The model of the system obtained by discretization is LTI, due to the periodicity of the sampling process, which greatly simplifies the analysis. However, a limitation of this method is that such models fail to represent the full response of the system, since *intersample behavior* is inherently lost or hidden.[14]

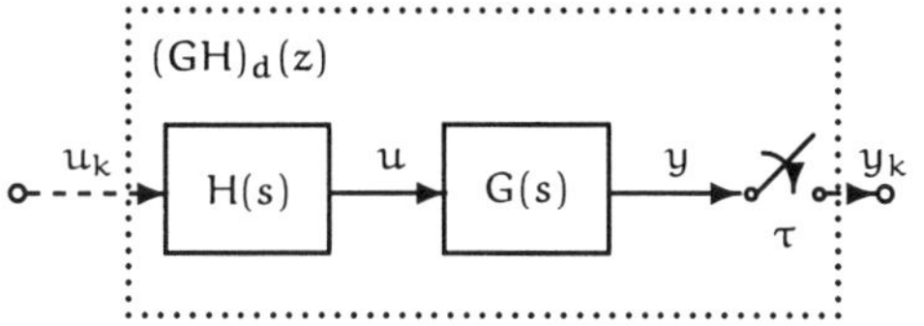

FIGURE 3.16. Discrete equivalent of the cascade of hold and plant.

It is well known that the poles of the discretized plant are determined by those of the continuous-time plant via the mapping $z = e^{s\tau}$, where τ is the sampling period (e.g., Middleton & Goodwin 1990). The discrete zeros of $(GH)_d$, however, bear no simple relation to those of G and, in fact, *may be arbitrarily assigned by suitable choice of the hold device* (Åström & Wittenmark 1990, p. 74).

From this observation, it may be tempting to conclude that design limitations due to nonminimum phase zeros of the analogue plant may be circumvented by assigning the zeros of the discretized plant to be minimum phase.

Unfortunately, as we will see in Chapter 6, the difficulties imposed by nonminimum phase zeros of the analogue plant remain when the controller is implemented digitally and, moreover, are independent of the type of hold used. It is important that the intersample behavior be examined if the problems are to be detected, since analyzing the system response only at the sampling instants may be misleading. ○

3.5 Summary

In this chapter, we have considered both continuous and discrete-time scalar systems in unity feedback control loops. For these systems, we

[14]Some intersample information can still be handled in a discrete model by using the *modified $\mathcal{Z}$-transform*. However, this line of work will not be pursued here.

have presented integral relations on the frequency response of the sensitivity and complementary sensitivity functions. These integrals — of the Bode and Poisson type — follow from applications of Cauchy's integral relations, and reveal constraints on the closed-loop system imposed by ORHP open-loop poles and zeros. One can use these constraints to study fundamental limitations on the achievable performance of the closed loop. For example, it is possible to show that, if the open-loop system is nonminimum phase, then requiring $|S(j\omega)|$ to be small over a range of frequencies will necessarily lead to a large sensitivity peak outside that range. Moreover, the relations allow one to quantify this peak in terms of parameters of the open-loop plant. Thus, it is possible to use the results presented here to establish bounds on the achievable frequency responses for S and T, which hold for all possible controller designs.

Notes and References

Bode Integral Formulae

§3.1.1 is taken mainly from Bode (1945); §3.1.2 is based on results of Horowitz (1963), Freudenberg & Looze (1985, 1987), Middleton & Goodwin (1990) and Middleton (1991). §3.1.3 follows Freudenberg & Looze (1988), but it is also possible to obtain similar results assuming functional bounds, as was done in Middleton (1991).

The Water-Bed Effect

This result was first discussed by Francis & Zames (1984), and then revisited by Freudenberg & Looze (1987) and Doyle et al. (1992). The result presented here is the one given in Francis & Zames (1984).

Poisson Integral Formulae

This section is largely taken from Freudenberg & Looze (1985, 1988).

For an interesting discussion on the complementarity of S and T see Kwakernaak (1995).

Discrete Systems

§3.4 is based on Sung & Hara (1988). A unification of both continuous and discrete-time results has been made in Middleton (1991), where the Bode integral for T is also derived.

4
MIMO Control

This chapter investigates sensitivity limitations in multivariable linear control. There are different ways of extending the scalar results to a multivariable setting. We follow here two approaches, namely, one that considers *integral constraints on the singular values of the sensitivity functions*, and a second that develops *integral constraints on sensitivity vectors*. These approaches complement each other, in the sense that they find application in different problems, and hence both are needed to obtain a general view of multivariable design limitations imposed by ORHP zeros and poles. In order to avoid repetition, we use the first approach to derive the multivariable version of Bode's integral theorems, whilst the second approach is taken to obtain the multivariable extension of the Poisson integrals. Both approaches emphasize the multivariable aspects of the problem by taking into account, in addition to location, the *directions* of zeros and poles.

4.1 Interpolation Constraints

Consider the unity feedback configuration of Figure 4.1, where the open-loop system, L, is a square (i.e., $n \times n$), proper transfer matrix. As in Chapter 3, let the sensitivity and complementary sensitivity functions be given by

$$S(s) = [I + L(s)]^{-1}, \qquad \text{and} \qquad T(s) = L(s)\,[I + L(s)]^{-1}. \tag{4.1}$$

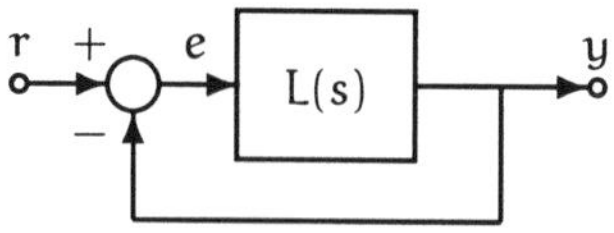

FIGURE 4.1. Feedback control system.

We will assume that the open-loop system is formed by the cascade of the plant, G, and the controller, K, i.e.,

$$L = GK .$$

The sensitivities in (4.1) are defined at the plant output. A similar pair of sensitivities can be defined at the plant input. In this chapter we will focus on the sensitivities given in (4.1), but we point out that parallel results can be derived for the sensitivities defined at the plant input.

In the sequel, we will impose the following assumption, which precludes the possibility of hidden pole-zero cancelations in the open-loop system.[1]

Assumption 4.1. The sets of frequency locations of CRHP zeros and poles of the open-loop system L are disjoint. ∘

Let the plant and controller have the following coprime factorizations over the ring of proper and stable transfer matrices:

$$\begin{aligned} G &= \tilde{D}_G^{-1}\tilde{N}_G = N_G D_G^{-1} , \\ K &= \tilde{D}_K^{-1}\tilde{N}_K = N_K D_K^{-1} . \end{aligned} \tag{4.2}$$

It then follows from Chapter 2 that q is a zero of G if and only if there exist vectors $\Psi_i, \Psi_o \in \mathbb{C}^n$ such that $\tilde{N}_G(q)\Psi_i = 0$ and $\Psi_o^* N_G(q) = 0$. Ψ_i and Ψ_o are the input and output zero directions associated with q, and can be normalized to be unitary vectors, i.e., such that $\Psi_i^*\Psi_i = 1$ and $\Psi_o^*\Psi_o = 1$. Similarly, p is a pole of G if and only if there exist vectors $\Phi_i, \Phi_o \in \mathbb{C}^n$ such that $\tilde{D}_G(p)\Phi_i = 0$ and $\Phi_o^* D_G(p) = 0$; Φ_i and Φ_o are the input and output pole directions associated with p.[2] Recall that a zero (pole) direction is said to be canonical if it has only one nonzero component.

Using the factorizations of plant and controller given in (4.2), the sensitivity and complementary sensitivity functions (4.1) can be expressed as

$$\begin{aligned} S &= D_K(\tilde{D}_G D_K + \tilde{N}_G N_K)^{-1}\tilde{D}_G , \\ T &= N_G(\tilde{N}_K N_G + \tilde{D}_K D_G)^{-1}\tilde{N}_K . \end{aligned} \tag{4.3}$$

Note that the zeros of S and T are easily identified from (4.3). This is formalized in the following lemma.

[1]A relaxed form of this requirement will be considered in §4.3.5.

[2]Note that if G is invertible, then p is a zero of G^{-1}.

Lemma 4.1.1 (Interpolation Constraints). Under Assumption 4.1, the sensitivity and complementary sensitivity functions must satisfy the following conditions.

(i) If $p \in \overline{\mathbb{C}^+}$ is a pole of G with input direction $\Phi \in \mathbb{C}^n$, then

$$S(p)\Phi = 0, \qquad \text{and} \qquad T(p)\Phi = \Phi. \tag{4.4}$$

(ii) If $q \in \overline{\mathbb{C}^+}$ is a zero of G with output direction $\Psi \in \mathbb{C}^n$, then

$$\Psi^* S(q) = \Psi^*, \qquad \text{and} \qquad \Psi^* T(q) = 0. \tag{4.5}$$

(iii) If $p \in \overline{\mathbb{C}^+}$ is a pole of K with output direction $\Phi \in \mathbb{C}^n$, then

$$\Phi^* S(p) = 0, \qquad \text{and} \qquad \Phi^* T(p) = \Phi^*.$$

(iv) If $q \in \overline{\mathbb{C}^+}$ is a zero of K with input direction $\Psi \in \mathbb{C}^n$, then

$$S(q)\Psi = \Psi, \qquad \text{and} \qquad T(q)\Psi = 0.$$

Proof. Note first that Assumption 4.1 guarantees that there is no pole-zero cancelation in S and T. Next, let $p \in \overline{\mathbb{C}^+}$ be a pole of G with input direction $\Phi \in \mathbb{C}^n$. Then $\tilde{D}_G(p)\Phi = 0$ and (4.4) thus holds from (4.3). The proof of the other cases if similar. □

Lemma 4.1.1 is the multivariable generalization of Lemma 3.1.3 in Chapter 3 and, as was the case with the latter lemma, it translates the open-loop characteristics of instability and nonminimum phaseness into properties that the functions S and T must satisfy in the CRHP. In the multivariable version, these properties also involve directions of zeros and poles. Note that the constraints given in (4.4) and (4.5) depend only on poles and zeros of the plant and hence hold *irrespective* of the controller design.

In §4.2 and §4.3 we will use Lemma 4.1.1 to derive integral constraints of the Bode and Poisson type, respectively.

4.2 Bode Integral Formulae

In this section we will derive Bode integral formulae for the singular values of the sensitivity function. An important obstacle in extending the scalar integral constraints to the multivariable case lies in the fact that the singular values of an analytic transfer matrix are not themselves analytic. A notable step towards the multivariable results was taken by Boyd & Desoer (1985), who established inequality versions of the Bode and Poisson integral formulae. These inequalities use the fact that the logarithm of the largest singular value of an analytic transfer matrix is a subharmonic function.[3] By restricting the class of systems under consideration — to those for which the multiplicity of the singular values of S is constant in the CRHP — Chen (1995) established *equality* versions of the Bode and Poisson integral formulae. These relations were derived using Green's formula for functions of two real variables (see §A.5.1 in Appendix A), which holds for arbitrary functions having continuous second derivatives over a region. Bode's integral formulae for the singular values of S will be given in §4.2.2, after some preliminary definitions and results.

4.2.1 Preliminaries

Given a unitary vector $\Phi \in \mathbb{C}^n$, we call the one-dimensional subspace spanned by Φ the *direction* of Φ. The *angle* between the directions of two unitary vectors Φ_1 and Φ_2 is defined to be the *principal angle* (Björck & Golub 1973) between the two corresponding subspaces spanned by the two vectors. This angle, denoted by $\angle(\Phi_1, \Phi_2)$, is given by

$$\cos\angle(\Phi_1, \Phi_2) \triangleq |\Phi_1^*\Phi_2| .$$

As discussed in Björck & Golub (1973) and Golub & Van Loan (1983), the principal angle between two subspaces serves as a distance measure, and it quantifies how well the two subspaces are aligned.

Our aim in the following subsection is to apply Green's formula to the logarithm of the singular values of S. It is thus necessary to extract from S all its ORHP zeros. It is well known that a nonminimum phase transfer function H admits a factorization that consists of a minimum phase part and an all-pass factor.[4] This factorization can be obtained by the following sequential procedure, which amounts to repeated use of a formula developed in Wall, Jr., Doyle & Harvey (1980). Let $q_i \in \mathbb{C}^+, i = 1, \cdots, n_q$, be the ORHP zeros of H. Then, H can be factorized as $H = H^1 B_1$, where

$$B_1(s) = I - \frac{2\,\mathrm{Re}\, q_1}{(s + \overline{q}_1)} \Phi_1\Phi_1^* ,$$

and Φ_1 is the input direction of q_1. Note that after this factorization, q_1 is no longer a zero of the transformed transfer matrix H^1. This procedure can be continued to obtain $H^{i-1} = H^i B_i$, where $B_i(s) = I - 2\,\mathrm{Re}\, q_i/(s + \overline{q}_i)\, \Phi_i\Phi_i^*$, and Φ_i, satisfying $\Phi_i^*\Phi_i = 1$, is obtained as if it were the direction of q_i but is computed from H^{i-1} rather than H. As such, Φ_i need not coincide with the input direction of q_i; however, it is easy to see that it is a

[3]Recall (cf. §A.3.1 in Appendix A) that a continuous function f is subharmonic if its Laplacian $\nabla^2 f \triangleq \partial^2 f/\partial\sigma^2 + \partial^2 f/\partial\omega^2$ is nonnegative.

[4]An *all-pass transfer function* is a stable transfer function such that the magnitude of its largest singular value equals one at all points on the imaginary axis.

linear combination of the zero directions. By repeating this procedure, we obtain a factorization of the form

$$H = \tilde{H} \prod_{i=1}^{n_q} B_i \,,$$

where $\tilde{H}$ is minimum phase and B_i, obtained as described above, is an all-pass factor corresponding to q_i.

Let p_i, $i = 1, \ldots, n_p$, be the poles of the open-loop system L in the ORHP, repeated according to their geometric multiplicities.[5] Under Assumption 4.1, it follows from Lemma 4.1.1 and the factorization described above that S can be expressed as

$$S = \tilde{S} \prod_{i=1}^{n_p} B_i \,, \tag{4.6}$$

where $\tilde{S}$ is minimum phase and B_i is the all-pass factor corresponding to p_i, given by

$$B_i(s) = I - \frac{2 \operatorname{Re} p_i}{(s + \overline{p}_i)} \Psi_i \Psi_i^* \,. \tag{4.7}$$

We note that the all-pass factors in the factorization (4.6) can be constructed *independently of the controller if the controller is stable.* In particular, suppose that Assumption 4.1 holds and that the plant and controller are expressed as in (4.2). Assume further that the controller denominator D_K has no ORHP zeros. Then, it is clear from the expression for S in (4.3) that the all-pass factors of the factorization (4.6) can be computed from $\tilde{D}_G$ alone. This will be relevant in the following section when the interest is to obtain performance limitations that can be computed before selecting any particular controller design.

In the remaining of §4.2, we make the following assumption.

Assumption 4.2.

(i) The closed-loop system of Figure 4.1 is stable.

(ii) $\displaystyle\lim_{R\to\infty} \sup_{\substack{s \in \overline{\mathbb{C}^+} \\ |s| \geq R}} R\,\overline{\sigma}(L(s)) = 0.$

(iii) The singular values of $\tilde{S}$ in (4.6), i.e., $\sigma_i(\tilde{S}(s))$, $i = 1, \cdots, n$, have continuous second order derivatives for all $s \in \overline{\mathbb{C}^+}$.

○

[5] Recall that the geometric multiplicity of a pole of $H = \tilde{D}^{-1}\tilde{N}$ is equal to the deficiency in rank of $\tilde{D}$ at the frequency of the pole.

Assumption 4.2-(i) is the same as for the SISO case, and amounts to the analyticity of S in the CRHP. Assumption 4.2-(ii) states that the largest singular value of the open-loop transfer function has a roll-off rate of more than one pole-zero excess. It will hold, for example, if each entry of L has relative degree larger than one.[6] Assumption 4.2-(iii) is necessary for the use of Green's theorem. A sufficient condition for this assumption to hold is that the multiplicity of $\sigma_i(\tilde{S}(s))$, $i = 1, \cdots, n$, be constant for all $s \in \overline{\mathbb{C}^+}$ (Chen 1995) — which in fact guarantees that $\sigma_i(\tilde{S})$ has continuous derivatives of all orders in the CRHP.

It is important to note that, even if S is analytic in the CRHP and the singular values of $\tilde{S}$ have continuous derivatives of all orders in the CRHP, the function $\sigma_i(\tilde{S}(s))$ is not, in general, a harmonic function (Boyd & Desoer 1985, Freudenberg & Looze 1987). This precludes the use of Cauchy's and Poisson's formulae, which apply to analytic and harmonic functions.

We end this subsection with a technical lemma that will prove useful in the subsequent analysis.

Lemma 4.2.1. For any $s \in \overline{\mathbb{C}^+}$ such that $s \neq q_i$, we have that

$$\overline{\sigma}(B_i^{-1}(s)) = \left|\frac{s+\overline{q}_i}{s-q_i}\right| , \tag{4.8}$$

$$\sigma_j(B_i^{-1}(s)) = 1 , \qquad j = 2, \cdots, n , \tag{4.9}$$

where B_i is given by (4.7).

Proof. Using the matrix inversion lemma (Golub & Van Loan 1983), we have, for $s \neq q_i$,

$$\begin{aligned} B_i^{-1}(s) &= \left(I - \frac{2\,\mathrm{Re}\, q_i}{(s+\overline{q}_i)}\,\Phi_i\Phi_i^*\right)^{-1} \\ &= I + \frac{2\,\mathrm{Re}\, q_i}{(s+\overline{q}_i)}\,\Phi_i\Phi_i^*\,\frac{1}{1 - 2\,\mathrm{Re}\, q_i \Phi_i^*\Phi_i/(s+\overline{q}_i)} \\ &= I + \frac{2\,\mathrm{Re}\, q_i}{s-q_i}\Phi_i\Phi_i^* . \end{aligned}$$

Consider next the matrix Υ_i, such that $[\Phi_i, \Upsilon_i]$ form an orthonormal basis of $\mathbb{C}^n$. Then $I = \Phi_i\Phi_i^* + \Upsilon_i\Upsilon_i^*$, and

$$\begin{aligned} B_i^{-1}(s) &= \Phi_i\Phi_i^* + \Upsilon_i\Upsilon_i^* + \frac{2\,\mathrm{Re}\, q_i}{s-q_i}\Phi_i\Phi_i^* \\ &= [\Phi_i \quad \Upsilon_i] \begin{bmatrix} \dfrac{s+\overline{q}_i}{s-q_i} & 0 \\ 0 & I \end{bmatrix} \begin{bmatrix} \Phi_i^* \\ \Upsilon_i^* \end{bmatrix} . \end{aligned}$$

[6]This follows from the inequality $\overline{\sigma}(A) \leq n \max_{1\leq i,j\leq n} |a_{ij}|$, which holds for any $A = [a_{ij}] \in \mathbb{C}^{n\times n}$ (see e.g., Golub & Van Loan 1983).

Thus

$$\sigma_j(B_i^{-1}(s)) = \sigma_j\left(\begin{bmatrix} \frac{s+\overline{q}_i}{s-q_i} & 0 \\ 0 & I \end{bmatrix}\right),$$

and the result follows. □

4.2.2 Bode Integrals for S

In this section we will state the Bode integral theorem for the logarithm of the singular values of the sensitivity function. The proof involves the use of Green's formula and a careful computation of contour integrals. Since it is rather long and tedious, we defer it to §B.1 of Appendix B.

The main result of the section is the following.

Theorem 4.2.2 (Bode Integral for S). Let S be factored as in (4.6). Then, under Assumption 4.2,

$$\int_0^{\infty} \log \sigma_j(S(j\omega))\, d\omega = F_j + K_j\,, \tag{4.10}$$

where

$$F_j = \frac{1}{2}\iint\limits_{\overline{\mathbb{C}^+}} \sigma\, \nabla^2 \log \sigma_j(\tilde{S}(\sigma + j\omega))\, d\sigma\, d\omega\,, \qquad \text{and}$$

$$K_j = \lim_{R\to\infty} \int_{-\pi/2}^{\pi/2} R \log \sigma_j \left(\prod_{i=1}^{n_p} B_i^{-1}(Re^{j\theta})\right) \cos\theta\, d\theta\,.$$

Proof. The proof follows by applying Green's formula (A.30) in Appendix A for Ω equal to the typical semicircular domain of radius R into the ORHP, and the choices of functions $f(s) = \log \sigma_j(\tilde{S}(s))$ and $g(s) = \log(|\eta + s|/|\eta - s|)$, where $\eta > R$. The details are given in §B.1 of Appendix B. □

Note that F_j and K_j are functions of the particular factorization of S given in (4.6). As discussed before, this factorization will, in general, depend on the choice of controller. It is possible, however, to derive an inequality that is independent of the controller as long as the controller is stable. This is stated in the following corollary.

Corollary 4.2.3. Let the plant be given by $G = \tilde{D}_G^{-1}\tilde{N}_G$, and assume that the controller K is stable. Let S be factorized as in (4.6), where the all-pass factors are computed from $\tilde{D}_G$ alone. Then, under Assumption 4.2,

$$\int_0^{\infty} \log \overline{\sigma}(S(j\omega))\, d\omega \geq \lim_{R\to\infty} \int_{-\pi/2}^{\pi/2} R \log \overline{\sigma} \left(\prod_{i=1}^{n_p} B_i^{-1}(Re^{j\theta})\right) \cos\theta\, d\theta\,. \tag{4.11}$$

Proof. If the controller is stable, then it is possible to factorize S as in (4.6), where the all-pass factors are computed from $\tilde{D}_G$ alone. Consider next (4.10) for $j = 1$, i.e., focusing on the largest singular value of S. Then the function $\overline{\sigma}(\tilde{S})$ is subharmonic in $\overline{\mathbb{C}^+}$ (Boyd & Desoer 1985, Freudenberg & Looze 1987). This implies that $\nabla^2 \log \overline{\sigma}(\tilde{S}) \geq 0$ and hence $F_1 \geq 0$. Inequality (4.11) then follows. □

The integral constraint given in (4.10) is an extension of Bode's integral given previously for scalar systems in Theorem 3.1.4 of Chapter 3. Note that for SISO systems, the singular values, $\sigma_j(\tilde{S})$, all collapse into $|\tilde{S}|$. Since, $\tilde{S}$ is analytic in the CRHP it follows that $\log|\tilde{S}|$ is harmonic in the CRHP. Hence $\nabla^2 \log|\tilde{S}| = 0$ in $\overline{\mathbb{C}^+}$, giving $F_j = 0$ in (4.10). We next focus on K_j in (4.10). Note that for SISO systems, the all-pass factor in (4.7) have the form $B_i(s) = (s - p_i)/(s + \overline{p}_i)$. Using the uniform convergence theorem (Levinson & Redheffer 1970, p. 335), and Example A.8.7 of Appendix A, K_j in (4.10) can be written as

$$
\begin{aligned}
K_j &= \sum_{i=1}^{n_p} \int_{-\pi/2}^{\pi/2} \lim_{R\to\infty} R \log \left| \frac{Re^{j\theta} + \overline{p}_i}{Re^{j\theta} - p_i} \right| \cos\theta \, d\theta \\
&= \sum_{i=1}^{n_p} \int_{-\pi/2}^{\pi/2} \operatorname{Re} \lim_{R\to\infty} \left[R \log\left(1 + \frac{\overline{p}_i e^{-j\theta}}{R}\right) - R \log\left(1 - \frac{p_i e^{-j\theta}}{R}\right) \right] \cos\theta \, d\theta \\
&= \sum_{i=1}^{n_p} \int_{-\pi/2}^{\pi/2} \operatorname{Re}(\overline{p}_i e^{-j\theta} + p_i e^{-j\theta}) \cos\theta \, d\theta \\
&= 2 \sum_{i=1}^{n_p} \operatorname{Re} p_i \int_{-\pi/2}^{\pi/2} \cos^2\theta \, d\theta \\
&= \pi \sum_{i=1}^{n_p} \operatorname{Re} p_i \, .
\end{aligned}
$$

Hence, for scalar systems, (4.10) reduces to (3.13) (particularized to the case $S(\infty) = 1$).

We next discuss the terms F_j and K_j in (4.10) for the general multivariable case. It is easy to see from the proof of Theorem 4.2.2 (see §B.1 in Appendix B) that F_j will be present for both stable and unstable open-loop systems. In particular, for $j = 1$, i.e., focusing on the largest singular value of S, we have seen in the proof of Corollary 4.2.3 that $F_1 \geq 0$. We thus conclude that, even for open-loop stable systems (satisfying Assumption 4.2), the Bode integral of the largest singular value of S has a nonnegative value.

It is also clear from the proof of Theorem 4.2.2 that the terms K_j completely quantify the effect of open-loop unstable systems on the Bode integrals for S. It is easy to see that $K_j \geq 0$. Indeed, this follows from the inequality

$$\sigma_j\left(\prod_{i=1}^{n_p} B_i^{-1}(Re^{j\theta})\right) \geq \prod_{i=1}^{n_p} \underline{\sigma}\left(B_i^{-1}(Re^{j\theta})\right) ,$$

and the fact that the RHS above equals 1 by Lemma 4.2.1. Thus, open-loop unstable poles contribute an additional nonnegative term, which has the potential for additional limitations upon sensitivity properties.

In order to obtain a better understanding of the multivariable nature of the result in Theorem 4.2.2, we will focus on some particular cases, for which the term K_j has a simpler explicit expression and bound. It is instructive to first examine two extreme cases. The first case corresponds to $\Psi_i^*\Psi_j = 0$ for all $i, j = 1, \cdots, n_p$, $i \neq j$, and the second case corresponds to $\Psi_1 = \Psi_2 = \cdots = \Psi_{n_p}$. For simplicity, we assume that $n_p \leq n$. We then have the following result.

Proposition 4.2.4. Let S be factorized as in (4.6) and assume, without loss of generality, that $\operatorname{Re} p_1 \geq \cdots \geq \operatorname{Re} p_{n_p}$. Also, let $n_p \leq n$. Then, under Assumption 4.2,

(i) if $\Psi_i^*\Psi_j = 0$ for all $i, j = 1, \cdots, n_p$, $i \neq j$, we have that

$$\int_0^\infty \log \sigma_j(S(j\omega))\, d\omega = \pi \operatorname{Re} p_j + F_j , \qquad j = 1, \cdots, n_p , \tag{4.12}$$

and

$$\int_0^\infty \log \sigma_j(S(j\omega))\, d\omega = F_j , \qquad j = n_p + 1, \cdots, n ; \tag{4.13}$$

(ii) if $\Psi_i = \Psi$ for all $i = 1, \cdots, n_p$, we have that

$$\int_0^\infty \log \overline{\sigma}(S(j\omega))\, d\omega = \pi \sum_{i=1}^{n_p} \operatorname{Re} p_i + F_1 , \tag{4.14}$$

and

$$\int_0^\infty \log \sigma_j(S(j\omega))\, d\omega = F_j , \qquad j = 2, \cdots, n . \tag{4.15}$$

Proof. In case (i), it is easy to show that

$$\prod_{i=1}^{n_p} B_i^{-1}(s) = I + \sum_{i=1}^{n_p} \frac{2 \operatorname{Re} p_i}{s - p_i} \Psi_i \Psi_i^* .$$

Consider next the matrix Υ, such that $[\Psi_1, \cdots, \Psi_{n_p}, \Upsilon]$ form an orthonormal basis of $\mathbb{C}^n$. Then $I = \sum_{i=1}^{n_p} \Psi_i\Psi_i^* + \Upsilon\Upsilon^*$, and by a similar argument to that used in the proof of Lemma 4.2.1, it follows that

$$\sigma_j\left(\prod_{i=1}^{n_p} B_i^{-1}(s)\right) = \sigma_j\left(\begin{bmatrix} \frac{s+\overline{p}_1}{s-p_1} & & & \\ & \ddots & & \\ & & \frac{s+\overline{p}_{n_p}}{s-p_{n_p}} & \\ & & & I_{n-n_p}\end{bmatrix}\right).$$

It is then clear from (4.10) and the above relation that (4.13) holds. As for (4.12), a straightforward manipulation shows that for $s \in \overline{\mathbb{C}^+}$ and $|s| \geq R$ with R being sufficiently large, the inequality

$$\left|\frac{s+\overline{p}_1}{s-p_1}\right| \geq \left|\frac{s+\overline{p}_2}{s-p_2}\right|$$

holds if $\operatorname{Re} p_1 \geq \operatorname{Re} p_2$, as was assumed. Then, as $R \to \infty$

$$\sigma_j\left(\prod_{i=1}^{n_p} B_i^{-1}(s)\right) = \left|\frac{s+\overline{p}_j}{s-p_j}\right|, \qquad j = 1, \cdots, n_p .$$

This gives K_j in (4.10) equal to $\pi \operatorname{Re} p_j$, thus showing (4.12).

For case (ii), one can construct a matrix Υ whose columns form an orthonormal basis of $\mathbb{C}^n$ together with Ψ. This leads to the expression

$$\prod_{i=1}^{n_p} B_i^{-1}(s) = \begin{bmatrix}\Psi & \Upsilon\end{bmatrix} \begin{bmatrix} \prod_{i=1}^{n_p} \frac{s+\overline{p}_i}{s-p_i} & \\ & I_{n-1}\end{bmatrix} \begin{bmatrix}\Psi^* \\ \Upsilon^*\end{bmatrix},$$

and the rest of the proof is similar to that of case (i). □

Proposition 4.2.4 suggests that the limitations imposed by open-loop unstable poles in multivariable systems are related not only to the location in frequency, but also to the *directions* of poles (or rather a linear combination of such directions) and further to the *relative geometric configuration* of these directions. Case (i) corresponds to the situation where the pole directions are *mutually orthonormal*, for which the integral pertaining to each singular value is related solely to one unstable pole (with a corresponding distance to the imaginary axis), as if each "channel" of the system is decoupled from the others. Case (ii) corresponds to the situation where all the pole directions are *parallel*, for which the unstable poles affect only the integral of the largest singular value, as if the channel corresponding to this singular value contains all unstable poles.

The following result — stated without proof — specializes Theorem 4.2.2 to the case of two open-loop unstable poles, and it further shows that the relative geometry of the directions of these poles is indeed crucial in the study of sensitivity limitations.

Theorem 4.2.5 (Chen, 1995). Let $n_p = 2$ and let the conditions of Theorem 4.2.2 hold. Then

$$\int_0^\infty \log \overline{\sigma}(S(j\omega))\, d\omega = a(1,2) + F_1 \,, \tag{4.16}$$

$$\int_0^\infty \log \sigma_2(S(j\omega))\, d\omega = b(1,2) + F_2 \,, \tag{4.17}$$

and

$$\int_0^\infty \log \sigma_j(S(j\omega))\, d\omega = F_j \,, \qquad j = 3, \cdots, n \,, \tag{4.18}$$

where

$$a(i,j) \triangleq \frac{\pi}{2}\left(\mathrm{Re}(p_i+p_j)+\sqrt{[\mathrm{Re}(p_i-p_j)]^2+4\,\mathrm{Re}\,p_i\,\mathrm{Re}\,p_j\cos^2\angle(\Psi_i,\Psi_j)}\right),$$

$$b(i,j) \triangleq \frac{\pi}{2}\left(\mathrm{Re}(p_i+p_j)-\sqrt{[\mathrm{Re}(p_i-p_j)]^2+4\,\mathrm{Re}\,p_i\,\mathrm{Re}\,p_j\cos^2\angle(\Psi_i,\Psi_j)}\right).$$

∘

The above result fully characterizes the limitation imposed by a pair of open-loop unstable poles on the sensitivity reduction properties. This limitation depends, not only on the distances of the poles to the imaginary axis, but also on the *principal angle* between the pole directions.

Using similar arguments as in the proof of Theorem 4.2.5, Chen (1995) obtained the following lower bounds for the Bode integral of the largest singular value of the sensitivity function.

Corollary 4.2.6 (Chen, 1995). Let the conditions of Theorem 4.2.2 hold. Then, for any $\Phi \in \mathbb{C}^n$ satisfying $\Phi^*\Phi = 1$, we have

$$\int_0^\infty \log \overline{\sigma}(S(j\omega))\, d\omega \geq \pi \sum_{i=1}^{n_p} \mathrm{Re}\,p_i \cos^2 \angle(\Phi, \Psi_i) + F_1 \,, \tag{4.19}$$

and, in particular, for any $j = 1, \cdots, n_p$,

$$\int_0^\infty \log \overline{\sigma}(S(j\omega))\, d\omega \geq \pi \sum_{i=1}^{n_p} \mathrm{Re}\,p_i \cos^2 \angle(\Psi_i, \Psi_j) + F_1 \,. \tag{4.20}$$

∘

From Theorem 4.2.2 and Corollary 4.2.6, it can be inferred that there will exist a frequency range over which the largest singular value of the sensitivity function exceeds one if it is to be kept below one at other frequencies. We will further see in §4.2.3 below that, in the presence of bandwidth constraints, this will result in a sensitivity trade-off in different frequency ranges. The above results also show unique features of multivariable feedback systems. First, owing to the fact that $\log \overline{\sigma}(\tilde{S}(j\omega))$ is a subharmonic function in the CRHP, the results show (via the terms F_j) that design limitations due to bandwidth constraints are, in a sense, more stringent than in scalar systems.[7] Moreover, how stringent these limitations are depends on the Laplacian of $\log \overline{\sigma}(\tilde{S}(j\omega))$. The second multivariable feature highlighted by the results, is that the additional cost associated with open-loop unstable poles (captured in the terms K_j) is a function not only of the position of the poles but also of the pole directions. Moreover, Proposition 4.2.4 and Theorem 4.2.5 show that the relative geometry of the pole directions is also important.

A multivariable version of the Poisson integral for the largest singular value of the sensitivity function can be used to study the effect of open-loop nonminimum phase zeros upon sensitivity limitations. This result was given in Chen (1995), who developed an extension of the Poisson integral formula for real functions $f : \mathbb{C} \to \mathbb{R}$ with continuous derivatives of all orders in $\overline{\mathbb{C}^+}$, but that are not necessarily harmonic functions. As expected, this new formula has an additional term involving the Laplacian of f. Accordingly, the Poisson integral theorem for the largest singular value of the sensitivity function displays a term depending on the Laplacian of $\log \overline{\sigma}(\tilde{S}(j\omega))$. This Poisson integral also shows that the sensitivity reduction ability of the system may be severely limited if the system has both open-loop unstable poles and nonminimum phase zeros, especially when these poles and zeros are close to each other and the principal angle between their directions are small. More discussion on this version of the Poisson integral for multivariable systems can be found in Chen (1995).

As a final comment, we remark that using standard modifications to the proof of Theorem 4.2.2, similar to the ones used in the proof of Theorem 3.1.4 of Chapter 3, it is possible to show that open-loop poles on the imaginary axis do not contribute to the values of the Bode and Poisson integrals. Also, parallel results can be obtained for the complementary sensitivity function.

4.2.3 *Design Interpretations*

In this section we briefly show the utility of the integral constraint (4.10) to study design trade-offs imposed on sensitivity reduction by open-loop unstable poles in conjunction with bandwidth constraints. These trade-offs are similar to those obtained in §3.1.3 of Chapter 3 for scalar systems. For the purpose of illustration, we will consider limitations related to the

[7]This conclusion holds if the multivariable problem is approached from a singular value perspective. Yet we will see in §4.3 that, when studying integral constraints on sensitivity vectors, the trade-offs appear to be alleviated with respect to the scalar case.

largest singular value of S, which we have seen in Chapter 2 has relevance in quantifying feedback properties of the closed-loop system of Figure 4.1. Following §3.1.3, we will assume that the largest singular value of the open-loop system satisfies the bound

$$\overline{\sigma}(L(j\omega)) \leq \frac{\delta}{\omega^{k+1}} \leq \epsilon < 1 \,, \qquad \forall \omega \in [\omega_c, \infty] \,, \tag{4.21}$$

where $\delta > 0$ and $k > 0$ are given constants. A bandwidth constraint like (4.21) may be necessary to ensure robust stability against uncertainty in the plant model (see Chapter 2). Note that a different upper bound of the form (4.21) may hold for each singular value.

Next, assume that the feedback loop of Figure 4.1 has been designed to achieve the following reduction specification

$$\overline{\sigma}(S(j\omega)) \leq \alpha \,, \qquad \forall \omega \in [-\omega_1, \omega_1] \,, \tag{4.22}$$

where $\omega_1 < \omega_c$ and $\alpha > 0$ is a given small constant. The following result shows that the reduction specification (4.22), together with the bandwidth constraint (4.21), can be achieved only at the expense of sensitivity increase over the range $[\omega_1, \omega_c]$.

Corollary 4.2.7. Let the conditions of Theorem 4.2.2 hold. In addition, suppose that (4.21) and (4.22) are satisfied for some ω_1 and ω_c such that $\omega_1 < \omega_c$. Then,

$$\sup_{\omega \in (\omega_1, \omega_c)} \log \overline{\sigma}(S(j\omega)) \geq \frac{1}{\omega_c - \omega_1} \left[F_1 + K_1 + \omega_1 \log \frac{1}{\alpha} + \omega_c \log(1-\epsilon) \right] . \tag{4.23}$$

Proof. The proof is similar to that of (3.21) in §3.1.3, and follows using (4.10) together with the bounds (4.21) and (4.22), and the observation that for $\omega \geq \omega_c$

$$\begin{aligned} \overline{\sigma}(S(j\omega)) &= \frac{1}{\underline{\sigma}(I + L(j\omega))} \\ &\leq \frac{1}{1 - \overline{\sigma}(L(j\omega))} \\ &\leq \frac{1}{1 - \delta/\omega^{k+1}} \,. \end{aligned}$$

□

Similar to its scalar counterpart, given in (3.21) in Chapter 3, the bound (4.23) shows that any attempt to increase the area of sensitivity reduction, by requiring α to be small and/or ω_1 to be close to ω_c, will necessarily result in a large sensitivity peak in the range (ω_1, ω_c). Note that the constant

K_1 — which is the multivariable version of the term $\pi \sum \operatorname{Re} p_i$ in (3.21) — can be precisely evaluated for some particular cases as shown in Proposition 4.2.4 and Theorem 4.2.5. In the general case K_1 is nonnegative, thus potentially increasing the sensitivity peak. The main multivariable feature appears in the term F_1. As we have seen, the constant F_1 is also nonnegative, leading to an additional possibility of an increase in the sensitivity peak.

We remark that similar trade-offs can be analyzed by means of the Poisson integral formula for the largest singular value of the sensitivity function (Chen 1995).

4.3 Poisson Integral Formulae

In this section, we use the approach of Gómez & Goodwin (1995) to obtain integral constraints — of the Poisson type — on *columns* of the sensitivity function. These constraints are useful in the study of performance limitations in multivariable feedback problems where *structural* features — such as diagonalization, triangularization, etc.— are of interest. The result uses the Poisson integral formula developed in §A.6.1 of Appendix A. Similar integral constraints can be obtained on rows of the complementary sensitivity function. The interested reader is referred to Gómez & Goodwin (1995) for this latter analysis.

4.3.1 *Preliminaries*

In this subsection we introduce some preliminary notions. We denote by S_{ik}, the element in the i-row and k-column of the $n \times n$ square transfer matrix S. If Ψ is a vector in $\mathbb{C}^n$, we denote its elements by ψ_i, $i = 1, \dots, n$, i.e., $\Psi = [\psi_1^*, \psi_2^*, \dots, \psi_n^*]^*$. Also, we introduce the index set $\mathcal{I}_\Psi \triangleq \{i \in \mathbb{N} : \psi_i \neq 0\}$ as the set of indices of the nonzero elements of Ψ.

As in Chapter 3, given a set $\mathcal{Z}_k = \{s_i, i = 1, \dots, n_k\}$ of complex numbers in the ORHP, we define its Blaschke product as the function

$$B_k(s) = \prod_{i=1}^{n_k} \frac{s - s_i}{s + \bar{s}_i} .$$

If $\mathcal{Z}_k$ is empty, we define $B_k(s) = 1, \forall s$.

In the remainder of §4.3, we make the following assumption.

Assumption 4.3.

(i) The closed-loop system of Figure 4.1 is stable.

(ii) The open-loop system L is a proper transfer matrix.

○

Assumption 4.3-(i) is the same as for the Bode integral given in §4.2. Assumption 4.3-(ii) is necessary to restrict the behavior at infinity of the sensitivity function, and is more relaxed than Assumption 4.2-(ii) used to derive the Bode integral.

Next, let $\Psi \in \mathbb{C}^n$, $\Psi \neq 0$, be the output direction of an ORHP zero of the plant G, as described in Lemma 4.1.1 (ii). Consider the vector function $\Psi^* S : \mathbb{C} \to \mathbb{C}^n$, whose n elements are proper, stable, scalar rational functions. Pick one of these elements, say

$$\rho_k(s) \triangleq \sum_{i=1}^{n} \psi_i^* S_{ik}(s) , \tag{4.24}$$

where k is in $\mathcal{J}_\Psi$, and let B_k be the Blaschke product of the zeros of ρ_k in the ORHP. Then, the function $\tilde{\rho}_k(s) \triangleq B_k^{-1}(s)\rho_k(s)$ is proper, stable, and minimum phase, which implies that $\log \tilde{\rho}_k(s)$ is analytic in the ORHP and satisfies the conditions of the Poisson integral formula (see Theorem A.6.1 in Appendix A). In the following subsection we will use this information to derive Poisson integrals for S.

4.3.2 *Poisson Integrals for S*

In the following theorem, we translate the interpolation constraint given in (4.5) into Poisson integrals on the columns of S.

Theorem 4.3.1 (Poisson Integral for S). Let $q = \sigma_q + j\omega_q$, $\sigma_q > 0$, be a zero of the plant G, and let $\Psi \in \mathbb{C}^n$, $\Psi \neq 0$, be its output direction, as described in Lemma 4.1.1 (ii). Then, under Assumption 4.3, for each index k in $\mathcal{J}_\Psi$,

$$\frac{1}{\pi}\int_{-\infty}^{\infty} \log\left|\sum_{i=1}^{n} \frac{\psi_i^*}{\psi_k^*} S_{ik}(j\omega)\right| \frac{\sigma_q}{\sigma_q^2 + (\omega_q - \omega)^2}\, d\omega = \log|B_k^{-1}(q)| , \tag{4.25}$$

where B_k is the Blaschke product of the ORHP zeros of ρ_k in (4.24).

Proof. The proof is an application of the Poisson integral formula, as in Theorem 3.3.1 in Chapter 3, to the scalar elements ρ_k of the vector function $\Psi^* S : \mathbb{C} \to \mathbb{C}^n$, which, under Assumption 4.3, are proper, stable, rational functions.

Let ρ_k in (4.24), where k is in $\mathcal{J}_\Psi$, be one of these elements, and let B_k be the Blaschke product of the zeros of ρ_k in the ORHP. As noted before, we can then apply the Poisson integral formula to the function $\log \tilde{\rho}_k(s)$, where $\tilde{\rho}_k(s) \triangleq B_k^{-1}(s)\rho_k(s)$. Evaluating this integral at $s = q$, and using the fact that $|\tilde{\rho}_k(j\omega)| = |\rho_k(j\omega)|$ yields

$$\begin{aligned}\frac{1}{\pi}\int_{-\infty}^{\infty} \log\left|\sum_{i=1}^{n} \psi_i^* S_{ik}(j\omega)\right| \frac{\sigma_q}{\sigma_q^2+(\omega_q-\omega)^2}\,d\omega &= \log\left|B_k^{-1}(q)\sum_{i=1}^{n} \psi_i^* S_{ik}(q)\right| \\ &= \log|B_k^{-1}(q)| + \log|\psi_k^*| ,\end{aligned} \tag{4.26}$$

where the last step follows from the interpolation condition $\Psi^* S(q) = \Psi^*$ given in (4.5) (notice that $\psi_k \neq 0$ by assumption). Since

$$\frac{1}{\pi}\int_{-\infty}^{\infty} \frac{\sigma_q}{\sigma_q^2 + (\omega_q - \omega)^2}\,d\omega = 1 ,$$

subtracting $\log|\psi_k^*|$ from both sides of (4.26) yields (4.25), thus completing the proof. □

Note that the result in Theorem 4.3.1 depends on the particular choice of controller. This is due to the presence of the Blaschke product B_k of ORHP zeros of ρ_k in the RHS of (4.25). Indeed, ρ_k in (4.24) is formed by combining entries of S, which clearly are functions of both plant and controller. However, the following corollary establishes a constraint that holds for any controller, and can therefore be used to identify, a priori, design limitations imposed by plant characteristics.

Corollary 4.3.2. Under the conditions of Theorem 4.3.1, the sensitivity function S satisfies, for each index k in $\mathcal{I}_\Psi$,

$$\int_{-\infty}^{\infty} \log\left|\sum_{i=1}^{n} \frac{\psi_i^*}{\psi_k^*} S_{ik}(j\omega)\right| \frac{\sigma_q}{\sigma_q^2 + (\omega_q - \omega)^2}\,d\omega \geq 0. \tag{4.27}$$

Proof. The proof follows from (4.25), on noting that $|B_k^{-1}(s)| \geq 1$ at any point s in $\mathbb{C}^+$, and so $\log|B_k^{-1}(q)| \geq 0$. □

Theorem 4.3.1 and Corollary 4.3.2 establish that, for each nonminimum phase zero of the plant, q, there is a set of integral constraints that limit the values of S on the $j\omega$-axis in an intrinsically vectorial fashion. Observe that, depending on the number of nonzero elements of the output direction Ψ associated with the zero of the plant — i.e., the number of nonzero elements in the index set $\mathcal{I}_\Psi$ — up to n integral constraints of the form (4.25) can be stated, each corresponding to a column of S. Moreover, if ν is the geometric multiplicity of the ORHP zero of the plant, i.e., the dimension of the (left) null space of the matrix $N(q)$, then up to ν integral constraints of the form (4.25) arise for each column of S. Since the components of the null space are linearly independent, it is expected that each integral constraint will give different information. Hence, it is reasonable to expect that, the greater the drop in rank caused by a particular zero, the more restrictive the constraint becomes.[8]

Note that, if the zero direction happens to be canonical, i.e., the index set $\mathcal{I}_\psi$ has only one element, k say, then the constraints reduce to those of the SISO case for the element S_{kk}. This is seen more clearly from the following corollary.

Corollary 4.3.3. Under the conditions of Theorem 4.3.1, the diagonal elements, S_{kk}, of the sensitivity function satisfy, for each index k in $\mathcal{I}_\psi$,

$$\begin{aligned}&\int_{-\infty}^{\infty} \log|S_{kk}(j\omega)| \frac{\sigma_q}{\sigma_q^2 + (\omega_q - \omega)^2}\, d\omega \geq \\ \int_{-\infty}^{\infty} \log&\left|\frac{\psi_k^* S_{kk}(j\omega)}{\sum_{i=1}^{n} \psi_i^* S_{ik}(j\omega)}\right| \frac{\sigma_q}{\sigma_q^2 + (\omega_q - \omega)^2}\, d\omega\,.\end{aligned} \tag{4.28}$$

Proof. The LHS of (4.27) can be alternatively written as

$$\int_{-\infty}^{\infty} \log\left|S_{kk}\left(1 + \sum_{\substack{i=1\\ i\neq k}}^{n} \frac{\psi_i^* S_{ik}}{\psi_k^* S_{kk}}\right)\right| \frac{\sigma_q}{\sigma_q^2 + (\omega_q - \omega)^2}\, d\omega =$$

$$\int_{-\infty}^{\infty} \log|S_{kk}| \frac{\sigma_q}{\sigma_q^2 + (\omega_q - \omega)^2}\, d\omega + \int_{-\infty}^{\infty} \log \frac{\left|\sum_{i=1}^{n} \psi_i^* S_{ik}\right|}{|\psi_k^* S_{kk}|} \frac{\sigma_q}{\sigma_q^2 + (\omega_q - \omega)^2}\, d\omega\,,$$

from which (4.28) follows. □

As discussed before, if the zero direction is canonical, then the index set $\mathcal{I}_\psi$ has only one element, k say. In this particular case, Corollary 4.3.3 gives,

$$\int_{-\infty}^{\infty} \log|S_{kk}(j\omega)| \frac{\sigma_q}{\sigma_q^2 + (\omega_q - \omega)^2}\, d\omega \geq 0\,,$$

which is the same constraint that would hold for a scalar system having a nonminimum phase zero at $s = q$ and a sensitivity function equal to S_{kk}. Corollary 4.3.3 also shows that, if the zero direction is not canonical, i.e., the set $\mathcal{I}_\psi$ has more than one element, then additional degrees of freedom arise in multivariable systems, which can potentially be exploited to reduce the cost associated with nonminimum phase zeros. This will be analyzed further in the following two sections.

---

[8] If the geometric multiplicity of a zero is greater than one, then there is no unique way of choosing a basis for the corresponding null space. Each choice of basis, in principle, generates alternative constraints. However, these constraints are interrelated and do not provide independent information regarding performance limitations.

It is clear from the previous discussion that the results given above display the essential multivariable nature of the problem, in the sense that the constraints stress the importance of directions of ORHP open-loop zeros and poles in connection with spatial — i.e., related to the structure of the transfer matrix — sensitivity allocation. In the following section we analyze design implications and trade-offs induced by the integral constraints just given.

4.3.3 *Design Interpretations*

As in the case of scalar systems, Theorem 4.3.1 and Corollary 4.3.2 can be used to give insights into the frequency domain trade-offs in sensitivity. A straightforward corollary of these results emphasizes the vectorial nature of the associated trade-offs. Let $\Omega_1 \triangleq [0, \omega_1]$ denote a given range of frequencies of interest, and assume that the k-column of S satisfies the following design specifications:

$$|S_{ik}(j\omega)| \leq \alpha_{ik} \quad \text{for } \omega \text{ in } [-\omega_1, \omega_1], \quad i = 1, \ldots, n, \tag{4.29}$$

where α_{ik}, $i = 1, \ldots, n$, are small positive numbers. Let $\Theta_q(\omega_1)$ be the weighted length of the interval $[-\omega_1, \omega_1]$, as defined in (3.33) in Chapter 3, which can be expressed as

$$\Theta_q(\omega_1) = \int_0^{\omega_1} \frac{\sigma_q}{\sigma_q^2 + (\omega_q - \omega)^2} + \frac{\sigma_q}{\sigma_q^2 + (\omega_q + \omega)^2} \, d\omega.$$

We then have the following corollary.

Corollary 4.3.4. Assume that all the conditions of Theorem 4.3.1 hold. Then, if the k-column of S, where k is in $\mathcal{J}_\psi$, achieves the specifications given in (4.29), the following inequality must be satisfied,

$$\|S_{kk}\|_\infty + \sum_{\substack{i=1 \\ i \neq k}}^{n} \left| \frac{\psi_i^*}{\psi_k^*} \right| \|S_{ik}\|_\infty \geq \left(\frac{1}{\alpha_{kk} + \sum_{\substack{i=1 \\ i \neq k}}^{n} \left| \frac{\psi_i^*}{\psi_k^*} \right| \alpha_{ik}} \right)^{\frac{\pi - \Theta_q(\omega_1)}{\Theta_q(\omega_1)}} . \tag{4.30}$$

Proof. Dividing the range of integration and using the weighted length of the interval $[-\omega_1, \omega_1]$, the inequality (4.27) implies

$$\log \max_{\omega \in [-\omega_1, \omega_1]} \left| \sum_{i=1}^{n} \frac{\psi_i^*}{\psi_k^*} S_{ik}(j\omega) \right| \Theta_q(\omega_1) +$$
$$\log \left\| \sum_{i=1}^{n} \frac{\psi_i^*}{\psi_k^*} S_{ik}(j\omega) \right\|_\infty [\pi - \Theta_q(\omega_1)] \geq 0 .$$

Exponentiating both sides above yields

$$\left[\max_{\omega\in[-\omega_1,\omega_1]}\left|\sum_{i=1}^{n}\frac{\psi_i^*}{\psi_k^*}S_{ik}(j\omega)\right|\right]^{\Theta_q(\omega_1)}\left\|\sum_{i=1}^{n}\frac{\psi_i^*}{\psi_k^*}S_{ik}(j\omega)\right\|_\infty^{[\pi-\Theta_q(\omega_1)]}\geq 0\,.$$

Using the specifications (4.29) and the triangular inequality, we have

$$\left(\alpha_{kk}+\sum_{\substack{i=1\\ i\neq k}}^{n}\left|\frac{\psi_i^*}{\psi_k^*}\right|\alpha_{ik}\right)^{\Theta_q(\omega_1)}\left(\|S_{kk}\|_\infty+\sum_{\substack{i=1\\ i\neq k}}^{n}\left|\frac{\psi_i^*}{\psi_k^*}\right|\|S_{ik}\|_\infty\right)^{[\pi-\Theta_q(\omega_1)]}\geq 0\,,$$

from which (4.30) follows immediately. □

As in the SISO case, the above corollary shows that integral relation (4.25) implies lower bounds on the infinity norm of elements of S. Notice that the exponent on the RHS of (4.30) is a positive number and its base is likely to be larger than one if the coefficients α_{ik} are small enough; hence, the more demanding the specifications, the larger these lower bounds are.

If the zero direction is not canonical, an important difference in the MIMO case is that the lower bounds apply to a combination of norms of elements, which somehow relaxes the constraint over the SISO case (where there is only one element). Also, the lower bounds are smaller than in the SISO case due to the presence of extra positive terms in the denominator of the RHS of (4.30). As a consequence, when the zero direction is not canonical, there is an *alleviation* of the cost associated with the corresponding zero as compared with the scalar case. To gain further insight into these ideas, we will next consider the special case of a diagonally decoupled design.

4.3.4 The Cost of Decoupling

Multivariable systems are, by their intrinsic nature, subject to coupling between different outputs and inputs. This means that, in general, one input affects more than one output and, conversely, one output is affected by more than one input. A natural approach to tackle the additional design difficulty arising from this interaction is to devise methodologies that translate — or approximate — the MIMO problem into a collection of SISO problems. There are alternative ways to do this, but they all involve achieving special structures for the closed-loop transfer matrices in such a way that the coupling is eliminated or is easier to handle. Examples of these decoupling methodologies include diagonalization, triangularization, diagonal dominance, etc., which have been frequently reported in the literature (e.g., Hung & Anderson 1979, Rosenbrock 1969, Weller & Goodwin 1993, Desoer & Gündes 1986).

In all of the above references, limitations imposed by nonminimum phase zeros are discussed as an additional complication. Desoer & Gündes (1986) have made this more precise by showing that, in order to achieve diagonal decoupling, the multiplicity of nonminimum phase zeros may need to be increased. This, in turn, is associated with performance penalties such as increased undershoot and rise times.

Following Gómez & Goodwin (1995), we next study the cost of decoupling in the context of integral constraints. Note that, for a diagonally decoupled system, the elements S_{ik}, for $i \neq k$, are zero. It then follows from Corollary 4.3.3 that, for each k in $\mathcal{J}_\Psi$, the diagonal element S_{kk} satisfies the integral constraint

$$\int_{-\infty}^{\infty} \log|S_{kk}(j\omega)| \frac{\sigma_q}{\sigma_q^2 + (\omega_q - \omega)^2} \, d\omega \geq 0 \, . \tag{4.31}$$

Also, if the sensitivity function is diagonal, and the diagonal element S_{kk}, where k is in $\mathcal{J}_\Psi$, is designed to satisfy $|S_{kk}(j\omega)| \leq \alpha_{kk}$ for $\omega \in [-\omega_1, \omega_1]$, Corollary 4.3.4 gives

$$\|S_{kk}\|_\infty \geq \left(\frac{1}{\alpha_{kk}}\right)^{\frac{\Theta_q(\omega_1)}{\pi - \Theta_q(\omega_1)}} \, . \tag{4.32}$$

To investigate the cost of diagonal decoupling, we compare the constraint (4.28) with (4.31), and the lower bound (4.30) with (4.32).

We first study the case where the dimension of the null space corresponding to the zero at $s = q$ is one. In this case we note that, if the zero direction Ψ has more than one nonzero element, then it seems feasible to achieve a negative value for the RHS of (4.28) by making off-diagonal elements in the sensitivity function nonzero. Comparing this with the result in (4.31), where the RHS is zero, indicates that when Ψ has more than one nonzero element, it is possible to exploit nondiagonal sensitivity entries so as to *ameliorate the constraints on the diagonal element*. Thus, if Ψ has more than one nonzero element, it is possible to have a joint *spatial* (i.e., nondiagonal) and *frequency trade-off* in sensitivity allocation. Similar remarks apply to the bounds on $\|S_{kk}\|_\infty$ shown in (4.30) and (4.32). In the latter case, nondiagonal decoupling allows α_{ik}, for $i \neq k$, to be greater than zero and again, provided the vector Ψ has more than one nonzero element, then the RHS of (4.30) can be made less than the RHS of (4.32). Moreover, as already noted in §4.3.3, the LHS of (4.30) also contains positive terms related to the norms of the off-diagonal elements, which do not appear in (4.32). In this way, the cost induced by the ORHP zero is somehow *shared* among various elements of the column.

Interestingly, having more than one nonzero entry in the zero direction Ψ means that the particular nonminimum phase zero in question has its effects associated with a linear combination of more than one output. Conversely, if Ψ is canonical, i.e., it has only one nonzero element, then the zero affects only one output. In the latter case, one might expect that it is unhelpful to make the sensitivity nondiagonal and this is precisely the point made above.

The preceding arguments have focused on the case where the dimension of the null space is one. For cases where the dimension of the null space is greater than one, then the issue arises as to whether or not there exists a basis for the null space that is canonical, i.e., it uses only unit vectors with only one nonzero entry. In cases where such a canonical basis exists,[9] it would seem, by extension of the arguments used above for the one dimensional case, that there is no apparent benefit to be gained by exploiting spatial sensitivity trade-offs.

As a final remark, note that the inequality (4.32) is the same as for the scalar case (see §3.3.2 in Chapter 3), and can be used to describe the water-bed effect for nonminimum phase systems: reducing the sensitivity in one frequency range causes it to rise in other ranges. However, (4.30) shows that, if decoupling is removed as a constraint, then spatial as well as frequency trade-offs are possible. Thus, *the water-bed effect on the diagonal elements of the sensitivity function may be ameliorated* by allowing off-diagonal elements to be nonzero. Nevertheless, this requires directional conditions (noncanonical zero directions) so that the cost associated with the integral constraints can be shared by different outputs.

4.3.5 The Impact of Near Pole-Zero Cancelations

Cancelations of ORHP zeros and poles between plant and controller of the feedback loop of Figure 4.1 is highly undesirable because it leads to loss of internal stability (cf. §2.2.1 in Chapter 2). In the multivariable case, these cancelations involve not only frequency locations of zeros and poles, but also their directions. For example, if the plant and controller are expressed as in (4.2), then the closed loop will not be internally stable if $\tilde{D}_G$ and $\tilde{N}_K$ share an ORHP zero with the same input direction (Gómez & Goodwin 1995). It is easy to see that, if this is the case, then $\tilde{D}_G$ and $\tilde{N}_K$ are not right coprime. Loss of internal stability also occurs if $\tilde{N}_G$ and $\tilde{D}_K$ are not right coprime, if N_G and D_K are not left coprime, and if D_G and N_K are not left coprime (Vidyasagar 1985). All these four cases are multivariable extensions of unstable pole-zero cancelations. Note that Assumption 4.1 precludes any of these cancelations by not allowing even frequency coincidence of zeros and poles.

Approximate unstable cancelations are also problematic. Indeed, we have seen in Chapter 3 that, for scalar systems, severe sensitivity peaks

[9] One instance for the left null space of the plant to have a canonical basis is when the q-interactor (Weller & Goodwin 1993) associated with the zero q is diagonal (Gómez & Goodwin 1995).

arise if there exist open-loop zeros and poles at near frequencies. This can be seen clearly from the values of the Poisson integral constraints (3.29) and (3.30) in §3.3, where the Blaschke products on the RHSs tend to infinity as ORHP zeros approach ORHP poles. We will see next that similar conclusions hold in the multivariable case regarding approximate unstable cancelations — keeping in mind that these cancelations involve both frequency and directions. To analyze the effect of these unstable multivariable cancelations, we consider separately the case where the directions are the same but the poles and zeros are close in frequency, and the case where the locations of poles and zeros are the same but the directions are close.

Let $q \in \overline{\mathbb{C}^+}$ be a zero of G with output direction $\Psi \in \mathbb{C}^n$. It follows from Theorem 4.3.1 that, under Assumption 4.3, the sensitivity function satisfies the integral constraint (4.25). Consider first the case that the vector Ψ is also an output direction of a zero $q_1 \neq q$ of D_K, i.e., the controller has a pole at $s = q_1$ such that $\Psi^* D_K(q_1) = 0$. Inspection of the expression for S in (4.3) clearly shows that $s = q_1$ is a zero of the vector function $\Psi^* S$, i.e., each of its elements ρ_k defined in (4.24) has a zero at $s = q_1$. Thus, the Blaschke product of zeros of ρ_k, B_k, can be expressed as

$$B_k(s) = \left(\frac{s - q_1}{s + \overline{q}_1}\right)^{n_k} \tilde{B}_k(s) ,$$

where $n_k \geq 1$, and where $\tilde{B}_k(s)$ is the Blaschke product of the remaining zeros of ρ_k. Using the above expression in (4.25), and the fact that $|\tilde{B}_k^{-1}(s)| \geq 1$ at any point s in $\mathbb{C}^+$, we have that

$$\frac{1}{\pi}\int_{-\infty}^{\infty} \log\left|\sum_{i=1}^{n} \frac{\psi_i^*}{\psi_k^*} S_{ik}(j\omega)\right| \frac{\sigma_q}{\sigma_q^2 + (\omega_q - \omega)^2} d\omega \geq n_k \log\left|\frac{q + \overline{q}_1}{q - q_1}\right| , \tag{4.33}$$

for each k in $\mathcal{J}_\Psi$, i.e., such that $\psi_k \neq 0$. We can conclude from (4.33) that, similar to the scalar case, the lower bound on the RHS becomes arbitrarily large when q_1 approaches q. This, in turn, would lead to high peaks on the magnitudes of the elements of S (and of T since $S + T = I$) on the k-column. Similar arguments apply when an unstable pole of the plant has the same direction as a nonminimum phase zero of the controller. In this case, it is also possible to show that, as the location of the zero of the controller becomes close to that of the pole of the plant, large peaks in T will occur.

Another alternative for a near multivariable cancelation is when the controller has an unstable pole at the same frequency, q say, at which the plant has a nonminimum phase zero, but with a slightly different direction. Let $\Psi \in \mathbb{C}^n$ be the output direction of the zero of the plant, and let $\Psi + \epsilon$ be the direction of the pole of the controller. Then, cases (ii) and

(iii) of Lemma 4.1.1 hold, i.e.,

$$\Psi^* S(q) = \Psi^* , \qquad \text{and} \qquad \Psi^* T(q) = 0 ;$$
$$(\Psi^* + \epsilon^*) S(q) = 0 , \qquad \text{and} \qquad (\Psi^* + \epsilon^*) T(q) = \Psi^* + \epsilon^* .$$

The above interpolation conditions lead to

$$\epsilon^* S(q) = -\Psi^* , \qquad \text{and} \qquad \epsilon^* T(q) = \Psi^* + \epsilon^* ,$$

which, using any compatible matrix and vector norm, imply that

$$\|S(q)\| \geq |\Psi|/|\epsilon| ,$$
$$\|T(q)\| \geq |\Psi + \epsilon|/|\epsilon| .$$

Hence, if $|\epsilon| \ll |\Psi|$ then $\|S(q)\|$ and $\|T(q)\|$ must be large. Consider, for example, the Euclidean norm for vectors and the infinity norm for matrices, i.e., the maximum singular value of the matrix. Then the bounds above show that $\overline{\sigma}(S(q))$ and $\overline{\sigma}(T(q))$ will become very large if $|\epsilon|$ approaches zero. This, in turn, will harden the integral constraints on the whole frequency responses of $\overline{\sigma}(S)$ and $\overline{\sigma}(T)$, as can be seen from the Poisson integral inequalities obtained by Chen (1995), e.g.,

$$\frac{1}{\pi} \int_{-\infty}^{\infty} \log \overline{\sigma}(S(j\omega)) \frac{\sigma_q}{\sigma_q^2 + (\omega_q - \omega)^2} \, d\omega \geq \log \overline{\sigma}(S(q)) . \tag{4.34}$$

Summarizing, the analysis in this section has shown that unstable multivariable near cancelations lead to poor sensitivity properties. On the one hand, ORHP zeros and poles of the open-loop system having the same directions and near frequencies cause the elements of the sensitivity functions to exhibit large peaks, as demonstrated by (4.33). On the other hand, ORHP zeros and poles of the open-loop system at the same frequency and having directions close in norm produce peaks on the maximum singular value of the sensitivity functions, as seen from (4.34) and the above discussion. This suggests that multivariable approximate cancelations of ORHP zeros and poles should be avoided if possible.

4.3.6 *Examples*

In this subsection, we give examples that illustrate the issues discussed before for continuous-time multivariable systems.

Example 4.3.1. Consider the plant

$$G(s) = \begin{bmatrix} \dfrac{1-s}{(s+1)^2} & \dfrac{s+3}{(s+1)(s+2)} \\ \dfrac{1-s}{(s+1)(s+2)} & \dfrac{s+4}{(s+2)^2} \end{bmatrix} , \tag{4.35}$$

which has an ORHP zero at $s = 1$ with output direction equal to $\Psi^* = [5, -6]$. Since this direction is not canonical, we can argue from §4.3.4 that there will be a cost in sensitivity associated with achieving diagonal decoupling. We will investigate this cost.

Direct application of Corollary 4.3.2 leads to the following inequalities that must be satisfied by the entries of the sensitivity function, regardless of the design methodology used:

$$\frac{1}{\pi}\int_{-\infty}^{\infty} \log\left|S_{11}(j\omega) - \frac{6}{5}S_{21}(j\omega)\right| \frac{1}{1+\omega^2}\, d\omega \geq 0\,, \tag{4.36}$$

$$\frac{1}{\pi}\int_{-\infty}^{\infty} \log\left|\frac{5}{6}S_{12}(j\omega) - S_{22}(j\omega)\right| \frac{1}{1+\omega^2}\, d\omega \geq 0\,. \tag{4.37}$$

Assuming further that the sensitivity elements are required to satisfy (4.29), then Corollary 4.3.4 gives the following lower bounds on linear combinations of element norms:

$$\|S_{11}\|_\infty + \frac{6}{5}\|S_{21}\|_\infty \geq \left(\frac{1}{\alpha_{11} + 6/5\,\alpha_{21}}\right)^{\frac{\pi-\Theta_q(\omega_1)}{\Theta_q(\omega_1)}}\,, \tag{4.38}$$

$$\|S_{22}\|_\infty + \frac{5}{6}\|S_{12}\|_\infty \geq \left(\frac{1}{\alpha_{22} + 5/6\,\alpha_{12}}\right)^{\frac{\pi-\Theta_q(\omega_1)}{\Theta_q(\omega_1)}}\,. \tag{4.39}$$

Next, we add the restriction of diagonal decoupling. Then, using (4.31) and (4.32), we have, for $k = 1, 2$,

$$\int_{-\infty}^{\infty} \log|S_{kk}(j\omega)| \frac{1}{1+\omega^2}\, d\omega \geq 0\,, \tag{4.40}$$

$$\|S_{kk}\|_\infty \geq \left(\frac{1}{\alpha_{kk}}\right)^{\frac{\Theta_q(\omega_1)}{\pi-\Theta_q(\omega_1)}}\,. \tag{4.41}$$

Comparing (4.36) and (4.37) with (4.40), for $k = 1$ and $k = 2$, respectively, it is clear that the constraint on the diagonal elements is alleviated if the off-diagonal entries are nonzero. More interesting, however, is to compare the lower bounds on the peak sensitivity given by (4.38) and (4.39) with (4.41) for $k = 1$ and $k = 2$, respectively. Consider, for example, (4.38) and (4.41) for $k = 1$. These bounds are shown in Figure 4.2 as a function of the reduction level α_{11}, for $\omega_1 = 0.3$ and for different values of α_{21}. Observe that $\alpha_{21} = 0$ in (4.38) (curve 1 in Figure 4.2) corresponds to the lower bound in (4.41) for $k = 1$. Comparing curve 1 with the other curves in Figure 4.2, we can see that relaxing the requirement of diagonal decoupling and allowing the level of sensitivity reduction in the off-diagonal elements (S_{21} in this case) to be greater than zero, significantly reduces the constraint on the combination of peak sensitivity values.

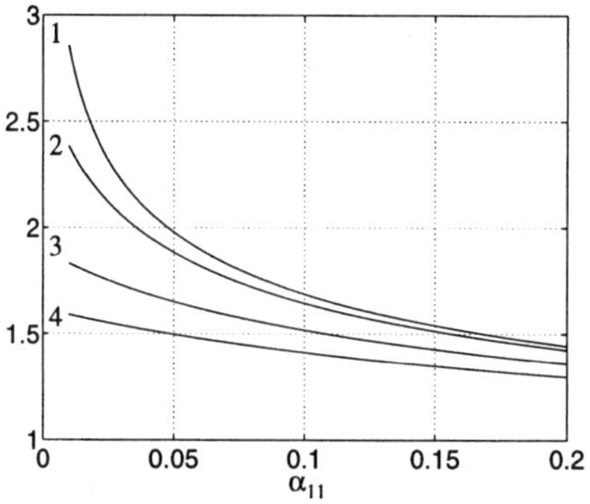

FIGURE 4.2. RHS of (4.38) as a function of α_{11}, for $\omega_1 = 0.3$ and for: 1- $\alpha_{21} = 0$; 2- $\alpha_{21} = 0.01$; 3- $\alpha_{21} = 0.05$; 4- $\alpha_{21} = 0.1$.

To further analyze the cost of decoupling, we next apply the particular design methodology proposed in Weller & Goodwin (1993). This design technique allows one to achieve different degrees of decoupling while keeping other design parameters essentially constant. Following this method with the parameter a_n — which very roughly defines the bandwidth — equal to 1, the resultant closed-loop sensitivity is

$$S(s) = \begin{bmatrix} \dfrac{s^3 + s^2(2+\lambda) + s(1+3\lambda)}{(s+1)^3} & 0 \\ \dfrac{(\lambda - 1)(3s^4 - 16s^3 - 21s^2)}{3(s+1)^5} & \dfrac{s^3 + 3s^2 + 4s}{(s+1)^3} \end{bmatrix}, \tag{4.42}$$

where $\lambda = 1$ gives diagonal decoupling and $\lambda < 1$ gives partial and static decoupling.[10]

Because of the structure of this particular design, which leads to $S_{12} = 0$, the integral constraints that must be satisfied are, for $\lambda < 1$: (4.40) for $k = 1, 2$, and for $\lambda = 0$: (4.36) and (4.40) for $k = 2$. Comparing (4.36) and (4.40) for $k = 1$ indicates that there should be an increase in the sensitivity peak of S_{11} if diagonalization is required. This is evident from Figure 4.3, where $|S_{11}(j\omega)|$ is plotted for different values of λ in the range $0 \leq \lambda \leq 1$.

Turning to the singular value approach, note that the integral constraint (4.34) holds due to the presence of the ORHP open-loop zero $q = 1$, i.e.,

$$\frac{1}{\pi}\int_{-\infty}^{\infty} \log \overline{\sigma}(S(j\omega)) \frac{1}{1+\omega^2}\, d\omega \geq \log \overline{\sigma}(S(1)) .$$

[10]The closed loop of Figure 4.1 is said to be: *statically decoupled* if it is internally stable and the complementary sensitivity function T is nonsingular and satisfies $T(0) = I$; *partially decoupled* if it is internally stable and the complementary sensitivity function is nonsingular and lower triangular.

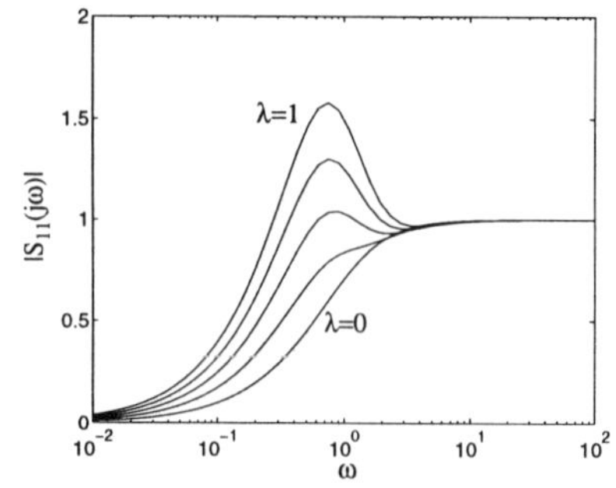

FIGURE 4.3. Effect of decoupling on $S_{11}(j\omega)$ for Example 4.3.1.

The above constraint, as it is stated, depends on the particular design methodology. However, it is easy to show that $\overline{\sigma}(S(1)) \geq 1$ for any design,[11] and hence the weighted integral of the logarithm of the maximum singular value of S is nonnegative. This implies, for example, that, for any design, the frequency response of $\overline{\sigma}(S)$ will necessarily achieve values greater than one.

We next consider again the (partial) decoupling design of Weller & Goodwin (1993), which led to the sensitivity function given in (4.42). Figure 4.4 shows the maximum (left) and minimum (right) singular values of S for different values of λ in the range $0 \leq \lambda \leq 1$.

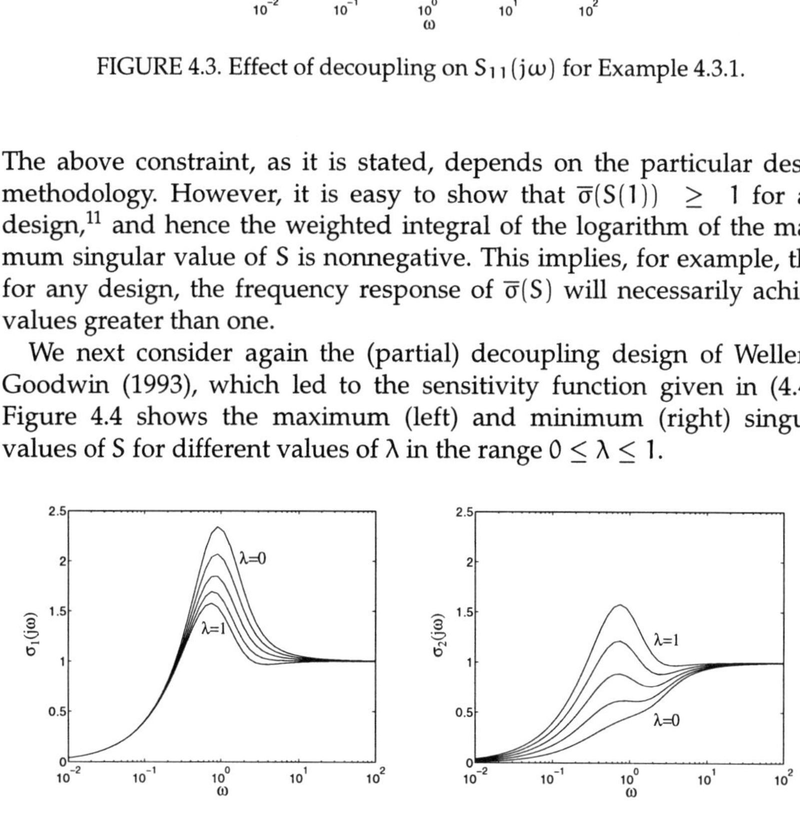

FIGURE 4.4. Maximum (left) and minimum (right) singular values of S in Example 4.3.1, for different degrees of decoupling.

In particular, the plot on the left confirms our prediction that $\overline{\sigma}(S(j\omega))$ would peak above one. An interesting observation is that total (diagonal) decoupling ($\lambda = 1$) is the best situation for the maximum singular value,

[11] This is because of norm properties and the fact that, if the open-loop system has an ORHP zero q, then there exists a vector Ψ such that $\Psi^* S(q) = \Psi^*$, see Theorem 13.1.1 in Chapter 13 for more details.

in the sense that $\lambda = 1$ gives the smallest peak value for its frequency response. This situation is desirable since, as seen in §2.2.4 of Chapter 2, having $\overline{\sigma}(S(j\omega))$ small is a requirement for robust performance against e.g., unstructured divisive perturbations of the plant. On the other hand, decreasing λ from 1 to 0, i.e., relaxing the degree of decoupling, increases the peak value of the maximum singular value. Comparing these conclusions with the previous ones about the hardening of the constraints on the element S_{11} as the degree of decoupling is increased, we can argue that, for this particular design methodology, the greater the degree of decoupling achieved, the worse the disturbance rejection properties of the diagonal elements become, while the performance is made more robust against unstructured perturbations of the plant. ∘

Example 4.3.2. Consider the plant

$$G(s) = \begin{bmatrix} \dfrac{(1-s)(s+2)}{(2s+1)(s+1)(s+4)} & \dfrac{(1-2)(s+2)}{(2s+1)(s+1)(s+3)} \\ \dfrac{-(s^4+7s^3+14s^2+6s-4)}{(2s+1)^2(s+1)(s+3)(s+4)} & \dfrac{-(s+2)(s^3+4s^2+3s+4)}{(2s+1)^2(s+1)(s+3)(s+5)} \end{bmatrix}, \tag{4.43}$$

which has an ORHP zero at $s = 1$ with output direction equal to $\Psi^* = [1, 0]$. This plant was considered in Weller & Goodwin (1993), where it was shown that the corresponding interactor matrix is nondiagonal and hence diagonal decoupling was achieved at the expense of introducing additional nonminimum phase zeros and increasing the relative degree of the closed-loop transfer functions.

From the point of view of sensitivity constraints, however, no increase in the peak value of $|S_{11}(j\omega)|$ is expected due to diagonal decoupling since the zero direction is canonical. Indeed, as discussed in §4.3.2, the element S_{11} satisfies the integral constraint

$$\int_{-\infty}^{\infty} \log|S_{11}(j\omega)| \frac{1}{1+\omega^2}\, d\omega \geq 0 ,$$

whether or not the system is decoupled. Thus, in terms of the sensitivity bounds presented in this chapter, there is no apparent cost associated with decoupling.

Using the methodology of Weller & Goodwin (1993), leads to the following sensitivity function:

$$S(s) = \begin{bmatrix} \dfrac{4s^4 + s^3(14-2\lambda) + 14s^2 + s(4+2\lambda)}{(s+1)^3} & 0 \\ \dfrac{(1-\lambda)s}{(2s+1)^2} & \dfrac{4s^3+8s^2+6s}{(2s+1)^2(s+1)} \end{bmatrix},$$

where, as before, $\lambda = 1$ gives diagonal decoupling and $\lambda < 1$ gives partial and static decoupling. Figure 4.5 shows the plot of $|S_{11}(j\omega)|$ versus frequency, for different values of λ in the range $0 \leq \lambda \leq 1$. It can be observed that the change in the peak of S_{11} for different values of λ is minimal and, in any case, is accompanied by a change in bandwidth. Comparing the results with those of Example 4.3.1 we conclude that here we essentially have only a frequency trade-off in sensitivity, whereas in Example 4.3.1 we also had a spatial trade-off; yet both have a nonminimum phase zero at the same location. The difference in these examples can be traced back to the directions associated with the zero.

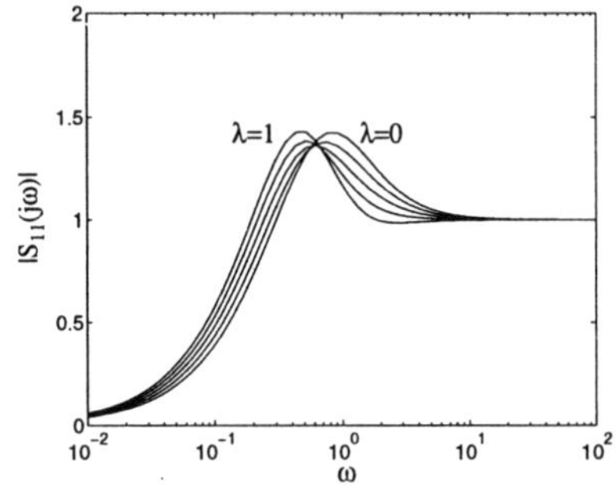

FIGURE 4.5. Effect of decoupling on $S_{11}(j\omega)$ for Example 4.3.2.

○

Example 4.3.3. Here we study the importance of directions when evaluating the impact of near cancelation of ORHP zeros and poles. Consider the plant $G = \tilde{D}_G^{-1}\tilde{N}_G = N_G D_G^{-1}$, where

$$\tilde{D}_G(s) = \begin{bmatrix} \frac{s-1}{s+1} & 0 \\ 1 & \frac{s+1}{s+2} \end{bmatrix} = D_G(s),$$

$$\tilde{N}_G(s) = \begin{bmatrix} \frac{2}{s+2} & 0 \\ 0 & \frac{2}{s+2} \end{bmatrix} = N_G(s).$$

The matrix $\tilde{D}_G$ has a zero at $s = 1$ with input direction $[2/3, -1]^*$, which corresponds to an unstable pole of the plant.

We will use the parametrization of all stabilizing controllers for the given plant. This parametrization possesses a free parameter, which will be used to study the effect of directions in near pole-zero cancelations. A straightforward calculation shows that the matrices

$$X(s) = Y(s) = \begin{bmatrix} \frac{s+2}{s+1} & 0 \\ -\frac{(s+2)^2}{(s+1)(s+3)} & \frac{s+2}{s+3} \end{bmatrix},$$

satisfy $XD + YN = I$. Following Vidyasagar (1985, Theorem 5.2.1), we have that all controllers that stabilize the closed loop have the form $K = (X + R\tilde{N}_G)^{-1}(Y - R\tilde{D}_G)$, where R is a proper and stable transfer matrix such that $X + R\tilde{N}_G$ is nonsingular and biproper. We next choose R to have the following special form

$$R(s) = \begin{bmatrix} \frac{r_1(s)}{s+1} & \frac{r_3(s)}{s+1} \\ \frac{r_2(s)}{s+1} & \frac{r_4(s)}{s+1} \end{bmatrix},$$

where the polynomials r_i, $i = 1, \ldots, 4$, have degree one and are such that the matrix R satisfies the following interpolation constraints

$$R(0) = \begin{bmatrix} -2 & 0 \\ \frac{4}{3} & -\frac{2}{3} \end{bmatrix}, \quad \text{and} \quad R(1) = \begin{bmatrix} 0 & \frac{27\psi_1}{2(6\psi_2 + 9\psi_1)} \\ 0 & \frac{9}{8} - \frac{81\psi_1}{4(6\psi_2 + 9\psi_1)} \end{bmatrix}.$$

It is possible to verify that this choice of R has the following properties:

- The controller stabilizes the closed loop.
- The closed loop achieves static decoupling.
- The controller has a nonminimum phase zero at $s = 1$ with input direction $\Psi = [\psi_1, \psi_2]^*$.

Thus, we can freely specify the direction of the zero of the controller, which is at the same frequency location as a pole of the plant.

First note that, as discussed in §4.3.5, internal stability will be lost if Ψ is aligned with $[2/3, -1]^*$. To continue, we will use Ψ to show that the sensitivity does not necessarily behave badly for coincident ORHP poles and zeros unless their directions are also (nearly) collinear. To do this, we consider several choices of Ψ, starting with $\Psi = [-1.94, 3.04]^*$, which is almost collinear with the direction of the pole, and then rotating Ψ clockwise up to $\Psi = [3, 2]^*$, which is orthogonal to the plant pole direction.

Figure 4.6 shows the peak values of $|T_{ij}(j\omega)|$, $i, j = 1, 2$ (in dB), as a function of the angle between zero and pole directions, which ranges from -0.99π (almost collinear) to $-\pi/2$ (orthogonal). It can be seen from this figure that the peaks increase significantly when the direction of the controller zero gets close to the corresponding direction of the unstable pole of the plant. Note, however, that the peak values are not large when the direction of the zero is not close to that of the pole. These results are in accord with the conclusions drawn in §4.3.5 and emphasize the role of directions in quantifying performance limits. ◦

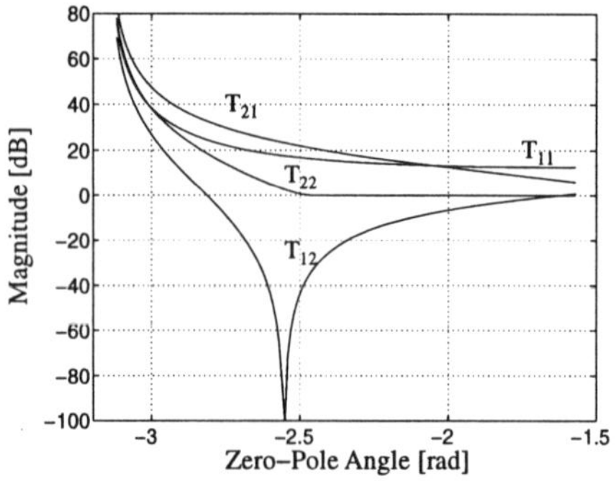

FIGURE 4.6. Peaks in complementary sensitivity versus angle between zero and pole directions.

4.4 Discrete Systems

In this section, we briefly address discrete-time systems. In particular, we give the discrete-time counterpart of the Poisson sensitivity integral for multivariable systems developed for continuous-time systems in §4.3.2.

4.4.1 *Poisson Integral for S*

Consider the unity feedback configuration of Figure 4.7, where the open-loop system, L, is an $n \times n$, proper transfer matrix of the complex variable z. As before, assume that L is formed by the cascade of plant, G, and controller, K, i.e., $L = GK$. Let the plant and controller have coprime factorizations[12] as in (4.2). Then, the sensitivity function in (3.43) can be expressed as in (4.3).

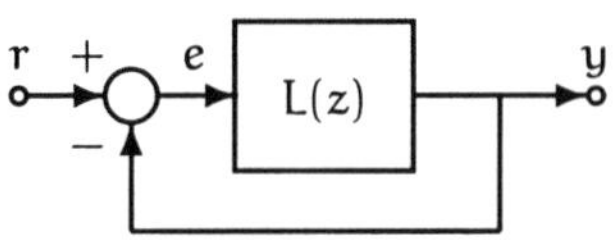

FIGURE 4.7. Discrete-time feedback control system.

To preclude unstable pole-zero cancelations in the open-loop system, we make the following assumption.

Assumption 4.4. The sets of frequency locations of zeros and poles of the open-loop system L outside the unit disk are disjoint. ◦

[12]Over the ring of proper transfer matrices having poles inside the unit disk (i.e., stable).

Under Assumption 4.4, if $q \in \overline{\mathbb{D}}^c$ is a zero of N_G (i.e., a nonminimum phase zero of G) with output direction $\Psi \in \mathbb{C}^n$, then S and T satisfy the following interpolation constraint

$$\Psi^* S(q) = \Psi^* , \qquad \text{and} \qquad \Psi^* T(q) = 0 . \tag{4.44}$$

Other interpolation constraints hold at the plant's unstable poles as well as at zeros and poles of the controller in $\overline{\mathbb{D}}^c$ (see Lemma 4.1.1).

Assume that S is proper and stable, and consider the notation of §4.3.1. In particular, ρ_k in (4.24) is now a proper, stable, scalar rational function of z. Let B_k be the Blaschke product of the zeros of ρ_k in $\overline{\mathbb{D}}^c$, defined as

$$B_k(z) = \prod_{i=1}^{n_k} \frac{\bar{z}_i}{|z_i|} \frac{z - z_i}{1 - \bar{z}_i z} .$$

Then, ρ_k can be factored as

$$\rho_k(z) = B_k(z)\, \tilde{\rho}_k(z)\, z^{-\delta_k} , \tag{4.45}$$

where δ_k is the relative degree of ρ_k, and $\tilde{\rho}_k$ is stable, minimum phase, and has relative degree zero. Note that, for the common situation of a strictly proper plant, the diagonal elements of S are biproper, and hence δ_k in (4.45) is zero in this case.

We then have the following result.

Theorem 4.4.1 (Poisson Integral for S). Let $q = r_q e^{j\theta_q}$, $r_q > 1$, be a zero of the plant G, and let $\Psi \in \mathbb{C}^n$, $\Psi \neq 0$, be its output direction. Assume that the sensitivity function, S, is proper and stable. Then, for each index k in $\mathcal{J}_\Psi$,

$$\frac{1}{2\pi}\int_{-\pi}^{\pi} \log\left|\sum_{i=1}^{n} \frac{\psi_i^*}{\psi_k^*} S_{ik}(e^{j\theta})\right| \frac{r_q^2 - 1}{1 - 2r_q\cos(\theta - \theta_q) + r_q^2}\, d\theta = \log|B_k^{-1}(q)| + \delta_k \log|q| , \tag{4.46}$$

where B_k and δ_k are as in (4.45).

Proof. Note that $\log|\tilde{\rho}_k|$, with $\tilde{\rho}_k$ defined in (4.45), satisfies the assumptions of Corollary A.6.4 in Appendix A. The result then follows by using this corollary on $\log|\tilde{\rho}_k|$ as in the proof of Theorem 3.4.2 in Chapter 3, and then using the interpolation constraint (4.44) in a similar fashion to the proof of Theorem 4.3.1. □

The following corollary establishes a constraint that is independent of the controller.

Corollary 4.4.2. Under the conditions of Theorem 4.4.1, the sensitivity function S satisfies, for each index k in $\mathcal{I}_\psi$,

$$\frac{1}{2\pi}\int_{-\pi}^{\pi} \log\left|\sum_{i=1}^{n} \frac{\psi_i^*}{\psi_k^*} S_{ik}(e^{j\theta})\right| \frac{r_q^2 - 1}{1 - 2r_q\cos(\theta - \theta_q) + r_q^2}\, d\theta \geq 0 \,. \qquad (4.47)$$

Proof. The proof follows from (4.46), on noting that $|B_k^{-1}(q)| \geq 1$ and $\delta_k \log|q| \geq 0$. □

The above results will be used in the following chapter to study limitations for periodic systems.

4.5 Summary

This chapter has discussed two different approaches to the investigation of sensitivity limitations in multivariable linear control. One of these approaches uses integral constraints on sensitivity vectors, whilst the other approach considers integral constraints on the logarithm of the singular values of the sensitivity functions. Each of these approaches to the problem emphasizes different aspects of multivariable feedback design, and, in combination, they provide a general view of multivariable design limitations imposed by open-loop ORHP zeros and poles. On the one hand, the analysis by means of integral constraints on sensitivity vectors has proven particularly useful to give insights into multivariable issues especially relating to the cost of decoupling, spatial and frequency domain trade-offs in sensitivity reduction, and the effect of directionality on the impact of near unstable pole-zero cancelations on performance. On the other hand, the integral relations on the logarithm of the singular values of the sensitivity function suggest a link between the directionality properties — in terms of singular vectors, see Chapter 2 — and sensitivity reduction. Indeed, all these integral relations exhibit a term involving the Laplacian of the logarithm of sensitivity, which, in turn, using a technique due to Freudenberg & Looze (1987, p. 184), can be expressed in a way that highlights the directionality properties of the sensitivity function. The understanding of the role of these Laplacians and their implications toward feedback design probably merits further investigation.

Notes and References

Bode Integral Formulae

§4.1 and §4.2 are largely based on Chen (1995). This work also obtained Poisson integral constraints on the singular values of the sensitivity function.

Poisson Integral Formulae

§4.3 is based on Gómez & Goodwin (1995). This latter work also presented design limitations for distributed systems, which allows one to consider transfer matrices with different time delays affecting each entry. The extension to this class of systems required additional technical results regarding zeros and the behavior at infinity of entire functions. The examples in §4.3.6 were also taken from Gómez & Goodwin (1995).

Other results on design limitations for multivariable systems were given by Boyd & Desoer (1985). This work obtained inequality versions of the Bode and Poisson integral formulae based on the recognition that the logarithm of the largest singular value of an analytic transfer function is a subharmonic function.

Related work has been reported by Freudenberg & Looze who convert the multivariable problem into a scalar one by pre and post multiplying the sensitivity function by vectors (Freudenberg & Looze 1985) or by the use of determinants (Freudenberg & Looze 1987). Similar ideas appear in the work of Sule & Athani (1991), who use directions associated with poles and zeros of the system, resulting in a directional study of trade-offs.

§4.4 follows Gómez & Goodwin (1995). Alternative approaches to the problem of integral constraints for MIMO discrete systems can be found in Chen & Nett (1995), and Hara & Sung (1989).

5

Extensions to Periodic Systems

Periodic dynamical systems frequently arise in applications. Examples include batch processes that are taken through a periodic operating cycle, and systems where periodic or multirate sampling strategies are employed (Feuer & Goodwin 1996). A periodic system is time-varying in nature; however, by using time or frequency domain raising techniques, it is possible to reduce the analysis to that of a special LTI multivariable system.

Here we focus on discrete-time linear periodic feedback systems. For these systems, the integral constraints for MIMO discrete-time systems of §4.4 in Chapter 4, can be used to quantify sensitivity limitations. However, the implications of the latter results in the present context are different and reflect the intrinsic structure of periodic systems.

5.1 Periodic Discrete-Time Systems

Consider a state space description for an n-input n-output discrete-time periodic system given by

$$\begin{aligned} x_{k+1} &= A_k x_k + B_k u_k \,, \\ y_k &= C_k x_k + D_k u_k \,, \end{aligned} \tag{5.1}$$

where $\{A_k\}$, $\{B_k\}$, $\{C_k\}$ and $\{D_k\}$ are periodic sequences of period N. With each of these periodic sequences, $\{A_k\}$ for example, we associate its *Fourier series expansion,*[1] given by

$$\bar{A}_n = \sum_{k=0}^{N-1} A_k e^{-j\theta_N kn} , \qquad n = 0, 1, \ldots, N-1 , \tag{5.2}$$

where $\theta_N = 2\pi/N$. For future use, we arrange the Fourier coefficients $\bar{A}_n$ in (5.2) to form the following matrix

$$\bar{A} \triangleq \frac{1}{N} \begin{bmatrix} \bar{A}_0 & \bar{A}_1 & \ldots & \bar{A}_{N-1} \\ \bar{A}_{N-1} & \bar{A}_0 & \ldots & \bar{A}_{N-2} \\ \vdots & \vdots & \ddots & \vdots \\ \bar{A}_1 & \bar{A}_2 & \ldots & \bar{A}_0 \end{bmatrix} . \tag{5.3}$$

5.1.1 Modulation Representation

We will study the system (5.1) in the frequency domain. To do this, we make use of the *double-sided $\mathcal{Z}$-transform*[2], which, for a sequence $\{x_k\}$, is defined by

$$X(z) = \sum_{k=-\infty}^{\infty} x_k z^{-k} . \tag{5.4}$$

Let $U(z)$ and $Y(z)$ be the $\mathcal{Z}$-transforms of the input and output sequences in (5.1), $\{u_k\}$ and $\{y_k\}$, respectively. We introduce the *modulation representation* of U and Y, denoted by $\bar{U}$ and $\bar{Y}$, and given by

$$\bar{U}(z) = \begin{bmatrix} U(z) \\ U(z e^{-j\theta_N}) \\ \vdots \\ U(z e^{-j(N-1)\theta_N}) \end{bmatrix} , \qquad \bar{Y}(z) = \begin{bmatrix} Y(z) \\ Y(z e^{-j\theta_N}) \\ \vdots \\ Y(z e^{-j(N-1)\theta_N}) \end{bmatrix} . \tag{5.5}$$

I.e., the modulation representation of a signal is obtained by "stacking" into a vector the signal and its $N-1$ versions modulated by the roots of unity of order N (except the root equal to 1). The modulation representation is sometimes called *frequency domain raising* of a signal, as opposed to the more common time domain raising.

Let $\bar{A}$, $\bar{B}$, $\bar{C}$, $\bar{D}$, defined as in (5.3), be the matrices of Fourier series coefficients of the sequences $\{A_k\}$, $\{B_k\}$, $\{C_k\}$ and $\{D_k\}$ in (5.1). Applying the

[1]Note that the Fourier series expansion of a periodic sequence is another periodic sequence.

[2]Throughout the rest of this chapter, the double-sided $\mathcal{Z}$-transform will be called just the $\mathcal{Z}$-transform.

$\mathcal{Z}$-transform to (5.1), and using the matrices $\bar{A}$, $\bar{B}$, $\bar{C}$, $\bar{D}$, it is easy to show that $\bar{U}$ and $\bar{Y}$ defined in (5.5) are related by

$$\bar{Y}(z) = \bar{H}(z)\bar{U}(z) , \tag{5.6}$$

where

$$\bar{H}(z) = \bar{C}\bar{W}^{-1}(zI - \bar{A}\bar{W}^{-1})^{-1}\bar{B} + \bar{D} , \tag{5.7}$$

and where

$$\bar{W} \triangleq \begin{bmatrix} I & 0 & \dots & 0 \\ 0 & e^{-j\theta_N}I & 0 & \dots \\ \vdots & \vdots & \ddots & \vdots \\ 0 & \dots & \dots & e^{-j(N-1)\theta_N} \end{bmatrix} . \tag{5.8}$$

We will refer to the matrix $\bar{H}$ in (5.7) as the *modulated transfer matrix* of system (5.1). Note that the use of this matrix allows one to treat a periodic system as a particular MIMO LTI system.

An alternative to the frequency domain input-output representation given in (5.6) is found by using a time domain raising technique. This latter technique takes the system (5.1) into an equivalent time-invariant system by stacking a set of N samples of the appropriate signal into a vector. This is a form of series to parallel conversion. The resultant vector can be (single-sided) $\mathcal{Z}$-transformed in the usual way to obtain an input-output transfer matrix, H^R say. It is shown in e.g., Feuer & Goodwin (1996) that $\bar{H}$ in (5.7) is related to this "time domain raised" transfer matrix H^R by a linear transformation of the form

$$\bar{H}(z) = W_s^{-1}\Lambda^{-1}(z)H^R(z^N)\Lambda(z)W_s , \tag{5.9}$$

where Λ and W_s are given by

$$\Lambda(z) = \begin{bmatrix} I & 0 & \dots & 0 \\ 0 & zI & 0 & \vdots \\ \vdots & \vdots & \ddots & \vdots \\ 0 & \dots & \dots & z^{N-1}I \end{bmatrix} ,$$

$$W_s = \begin{bmatrix} I & I & \dots & \dots & I \\ I & W & W^2 & \dots & W^{N-1} \\ \vdots & \vdots & \vdots & \ddots & \vdots \\ I & W^{N-1} & W^{2(N-1)} & \dots & W^{(N-1)^2} \end{bmatrix} ,$$

with $W = e^{-j\theta_N}I$ and $\theta_N = 2\pi/N$.

In view of (5.9) we may conclude that both $\bar{H}$ and H^R are equally valid representations of periodic systems and, hence, contain all the relevant frequency domain information. Yet, the time domain raised transfer function has a major disadvantage from a frequency domain point of view, because it describes the output sequence at each N samples, rather than at each sample. However, we are usually interested in the response at each sample, which requires us to (somehow) combine the information coded in the successive rows of the time domain raised signals. This is precisely the interleaving operation captured in (5.9). Note that H^R will, in general, be difficult to interpret whereas the response of the modulated transfer matrix $\bar{H}$ in (5.7) can be directly interpreted as the normal frequency response at unity sampling period.[3]

Our analysis in the following sections will depend on the "poles" and "zeros" of functions that are modulated transfer matrices such as $\bar{H}$ in (5.7). Note that these are not standard transfer functions due to the frequency shifting. Not withstanding this, we define poles, zeros and their associated directions in the usual way as would be done if $\bar{H}$ were a transfer function. For example, we say that p is a nonminimum phase zero of $\bar{H}$ if $|p| > 1$ (i.e., $p \in \overline{\mathbb{D}}^c$) and $\operatorname{rank}\bar{H}(p) < \operatorname{rank}\bar{H}(z)$ for almost all z.

We end this introductory section with a property of modulated transfer matrices that will be useful in subsequent analysis.

Proposition 5.1.1. The modulated transfer matrix $\bar{H}$ in (5.7) has the following property

$$Q\bar{H}(z e^{-j\theta_N}) = \bar{H}(z)Q ,$$

where Q is given by

$$Q = \begin{bmatrix} 0 & \cdots & \cdots & \cdots & I \\ I & 0 & \cdots & \cdots & 0 \\ 0 & I & \ddots & \ddots & \vdots \\ \vdots & \ddots & \ddots & \ddots & \vdots \\ 0 & \cdots & \cdots & I & 0 \end{bmatrix} .$$

Proof. From (5.9), the matrix $\bar{H}(z e^{-j\theta_N})$ can be expressed as

$$\bar{H}(z e^{-j\theta_N}) = W_s^{-1}\Lambda^{-1}(z e^{-j\theta_N})H^R(z^N e^{-jN\theta_N})\Lambda(z e^{-j\theta_N})W_s .$$

But $N\theta_N = 2\pi$ and thus $H^R(z^N e^{-jN\theta_N}) = H^R(z)$. It is also straightforward to verify that $\Lambda(z e^{-j\theta_N})W_s = \Lambda(z)W_sQ$. The result then follows. □

Note that Proposition 5.1.1 shows that the matrix $Q\bar{H}(z e^{-j\theta_N})$ is obtained by shifting the columns of $\bar{H}(z)$ to the left.

---

[3]To further illustrate this claim, say that the N-periodic system happens to be time-invariant. In this case $\bar{H}$ in (5.7) turns out to be a diagonal matrix formed from the usual frequency response of the time-invariant system, whereas the time domain raised transfer function will be related to the usual frequency response in a very indirect fashion (see Goodwin & Gómez (1995) for more details).

5.2 Sensitivity Functions

Consider the feedback configuration of Figure 5.1, where the plant and the controller are n-input n-output discrete-time N-periodic systems.

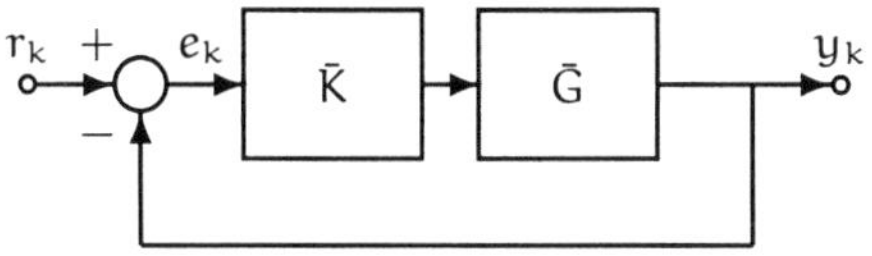

FIGURE 5.1. Feedback control system.

We represent the plant and controller by their corresponding modulated transfer matrices, $\bar{G}$ and $\bar{K}$, respectively. The signals $\{r_k\}$, $\{e_k\}$ and $\{y_k\}$ are sequences representing the reference input, error, and output, respectively, which have modulated representation $\bar{R}$, $\bar{E}$, and $\bar{Y}$.

We write $\bar{G}$ and $\bar{K}$ using coprime factorizations, over the ring of proper and stable transfer functions, as[4]

$$\begin{aligned} \bar{G} &= \tilde{D}_G^{-1}\tilde{N}_G = N_G D_G^{-1} , \\ \bar{K} &= \tilde{D}_K^{-1}\tilde{N}_K = N_K D_K^{-1} . \end{aligned} \tag{5.10}$$

By an argument similar to the one used to establish (5.9), it can be shown that N_G, for example, is related to a coprime factor N_G^R corresponding to the time domain raised transfer function G^R by

$$N_G(z) = W_s^{-1}\Lambda^{-1}(z)N_G^R(z^N)\Lambda(z)W_s , \tag{5.11}$$

and similar relations hold for the remaining factors of the coprime factorizations of $\bar{G}$ and $\bar{K}$ in (5.10).

We next consider the mappings, in the modulated signal space, between $\bar{R}$ and $\bar{E}$, and between $\bar{R}$ and $\bar{Y}$. We will refer to these matrices as the *modulated sensitivity*, $\bar{S}$, and the *modulated complementary sensitivity*, $\bar{T}$. In terms of the coprime factors in (5.10), $\bar{S}$ and $\bar{T}$ are given by

$$\begin{aligned} \bar{S} &= D_K(\tilde{D}_G D_K + \tilde{N}_G N_K)^{-1}\tilde{D}_G , \\ \bar{T} &= N_G(\tilde{N}_K N_G + \tilde{D}_K D_G)^{-1}\tilde{N}_K . \end{aligned} \tag{5.12}$$

Note that these are the same expressions used in Chapter 4, but now $\bar{S}$ and $\bar{T}$ are $Nn \times Nn$ transfer matrices that map modulated representations of signals.

In the sequel, we make the following assumption.

---

[4]To simplify notation, we eliminate the use of "bars" on the coprime factors of modulated transfer matrices.

Assumption 5.1. The sets of frequency locations of zeros and poles of the open-loop system $\bar{G}\bar{K}$ outside the unit disk are disjoint. ◦

Under Assumption 5.1, if $q \in \overline{\mathbb{D}}^c$ is a zero of N_G (i.e., a nonminimum phase zero of $\bar{G}$) with output direction $\Psi \in \mathbb{C}^{Nn}$, then $\bar{S}$ and $\bar{T}$ satisfy the following interpolation constraint

$$\Psi^* \bar{S}(q) = \Psi^* , \qquad \text{and} \qquad \Psi^* \bar{T}(q) = 0 . \tag{5.13}$$

Other interpolation constraints hold at plant unstable poles as well as at zeros and poles of the controller in $\overline{\mathbb{D}}^c$ (see Lemma 4.1.1 in Chapter 4).

It is interesting at this stage to reflect on the nature of the zeros of N_G. From (5.11), we have that $N_G(q) = W_s^{-1}\Lambda^{-1}(q)N_G^R(q^N)\Lambda(q)W_s$ and, therefore, the nonminimum phase zeros of N_G are the Nth-roots of the nonminimum phase zeros of N_G^R. Thus, if $q^N = r_q^N e^{jN\theta_q}$ is a nonminimum phase zero of N_G^R, then the set

$$\mathcal{Z} \triangleq \{r_q e^{j\theta_q}, r_q e^{j(\theta_q - \theta_N)}, \dots, r_q e^{j[\theta_q - (N-1)\theta_N]}\} , \tag{5.14}$$

is a set of nonminimum phase zeros of N_G. We will see in the following section, however, that every element in $\mathcal{Z}$ gives the same information in terms of integral constraints on the modulated sensitivity function.

5.3 Integral Constraints

We will use the ideas of §4.4 in Chapter 4, applied here to the modulated sensitivity function. For convenience, we first recall some notation. Consider the modulated sensitivity function, $\bar{S}$, in (5.12). We denote by $\bar{S}_{ik}$, the element in the i-row and k-column of $\bar{S}$. If Ψ is a vector in $\mathbb{C}^{Nn}$, we denote its elements by ψ_i, $i = 1, \dots, Nn$, i.e., $\Psi = [\psi_1^*, \psi_2^*, \dots, \psi_{Nn}^*]^*$. Also, we use the index set $\mathcal{I}_\Psi \triangleq \{i \in \mathbb{N} : \psi_i \neq 0\}$ as the set of indices of the nonzero elements of Ψ.

Given a set $\mathcal{Z}_k = \{z_i, i = 1, \dots, n_k\}$ of complex numbers in $\overline{\mathbb{D}}^c$, we define its Blaschke product as the function

$$B_k(z) = \prod_{i=1}^{n_k} \frac{\bar{z}_i}{|z_i|} \frac{z - z_i}{1 - \bar{z}_i z} . \tag{5.15}$$

If $\mathcal{Z}_k$ is empty, we define $B_k(z) = 1$, $\forall z$.

Next, let $\Psi \in \mathbb{C}^{Nn}$, $\Psi \neq 0$, be the output direction of a nonminimum phase zero of the plant $\bar{G}$, which thus satisfies (5.13). Assume that $\bar{S}$ is proper and stable, and consider the vector function $\Psi^*\bar{S} : \mathbb{C} \to \mathbb{C}^{Nn}$, whose Nn elements are proper, stable, scalar rational functions. Pick one of these elements, say

$$\rho_k(z) \triangleq \sum_{i=1}^{Nn} \psi_i^* \bar{S}_{ik}(z) , \tag{5.16}$$

where k is in $\mathcal{J}_\Psi$, and let B_k be the Blaschke product of the zeros of ρ_k in $\overline{\mathbb{D}}^c$, defined as in (5.15). Then, ρ_k can be factored as

$$\rho_k(z) = B_k(z)\, \tilde{\rho}_k(z)\, z^{-\delta_k} , \tag{5.17}$$

where δ_k is the relative degree of ρ_k, and $\tilde{\rho}_k$ is stable, minimum phase, and has relative degree zero.

We then have the following result.

Theorem 5.3.1 (Poisson Integral for $\bar{S}$). Let $q = r_q e^{j\theta_q}$, $r_q > 1$, be a zero of the plant $\bar{G}$, and let $\Psi \in \mathbb{C}^{Nn}$, $\Psi \neq 0$, be its output direction. Assume that the modulated sensitivity function, $\bar{S}$, is proper and stable. Then, for each index k in $\mathcal{J}_\Psi$,

$$\frac{1}{2\pi}\int_{-\pi}^{\pi} \log\left|\sum_{i=1}^{Nn} \frac{\psi_i^*}{\psi_k^*}\bar{S}_{ik}(e^{j\theta})\right| \frac{r_q^2 - 1}{1 - 2r_q\cos(\theta - \theta_q) + r_q^2}\, d\theta = \log|B_k^{-1}(q)| + \delta_k \log|q| , \tag{5.18}$$

where B_k and δ_k are as in (5.17).

Proof. Follows by direct application of Theorem 4.4.1 in Chapter 4. □

Note that the above result makes use of B_k and δ_k, which are in general not known unless a design has already been carried out. Regarding the relative degree δ_k, it can be said that it is zero for strictly proper plants, but in the general case, all that can be said based on knowledge of the plant is that it will be nonnegative. Since we are interested in fundamental constraints that apply regardless of the controller used, we give the following corollary that establishes a constraint that is independent of the controller.

Corollary 5.3.2. Under the conditions of Theorem 4.4.1, the sensitivity function $\bar{S}$ satisfies, for each index k in $\mathcal{J}_\Psi$,

$$\frac{1}{2\pi}\int_{-\pi}^{\pi} \log\left|\sum_{i=1}^{Nn} \frac{\psi_i^*}{\psi_k^*}\bar{S}_{ik}(e^{j\theta})\right| \frac{r_q^2 - 1}{1 - 2r_q\cos(\theta - \theta_q) + r_q^2}\, d\theta \geq 0 . \tag{5.19}$$

Proof. The proof follows from (5.18), on noting that $|B_k^{-1}(q)| \geq 1$ and $\delta_k \log|q| \geq 0$. □

As discussed at the end of §5.2, each nonminimum phase zero $r_q e^{j\theta_q}$ of N_G defines the set $\mathcal{Z}$ in (5.14) of nonminimum phase zeros of N_G. In principle, we should use all of them to compute the integral constraints associated with the system. However, we will show next that every element in $\mathcal{Z}$ gives the same set of constraints.

Suppose that the zero $q = r_q e^{j\theta_q}$, with output direction Ψ, is used to compute the integral constraints. Then there is an integral of the form (5.18) corresponding to each ρ_k in (5.16), such that k in $\mathcal{I}_\Psi$. Let

$$\tilde{q} = r_q e^{j(\theta_q - \theta_N)}$$

be used to compute another set of constraints. Because of Proposition 5.1.1,

$$QN_G(\tilde{q}) = N_G(q)Q ,$$

and we see that

$$\tilde{\Psi} = (\Psi^* Q)^* \tag{5.20}$$

is an output direction associated with $\tilde{q}$.

Proposition 5.1.1 also shows that $Q\bar{S}(e^{j\theta}) = \bar{S}(e^{j(\theta+\theta_N)})Q$, and thus, using (5.20),

$$\begin{aligned} \tilde{\Psi}^* \bar{S}(e^{j\theta}) &= \Psi^* Q\bar{S}(e^{j\theta}) \\ &= \Psi^* \bar{S}(e^{j(\theta+\theta_N)})Q . \end{aligned} \tag{5.21}$$

Consider any element $\rho_{\tilde{k}}$, $\tilde{k}$ in $\mathcal{I}_{\tilde{\Psi}}$, of $\tilde{\Psi}^*\bar{S}$, i.e., $\rho_{\tilde{k}}(z) \triangleq \sum_{i=1}^{Nn} \tilde{\psi}_i^* \bar{S}_{i\tilde{k}}(z)$. It then follows using (5.21) that

$$\rho_{\tilde{k}}(e^{j\theta}) = \rho_k(e^{j(\theta+\theta_N)}) ,$$

where $k = \tilde{k} - n$ if $n < \tilde{k} \le Nn$, and $k = (N-1)n + \tilde{k}$ if $1 \le \tilde{k} \le n$.

It is now a matter of a simple computation to show that the integral constraint of the form (5.18) corresponding to $\rho_{\tilde{k}}$ and the zero $\tilde{q}$ is the same as that corresponding to ρ_k and the zero q. Furthermore, the relationship between k and $\tilde{k}$ is clearly bijective and, hence, every one of the integrals obtained from the zero $\tilde{q}$ can be obtained from the zero q. The converse is also true. The same also happens between the integrals associated with the zeros $r_q e^{j(\theta_q - \theta_N)}$ and $r_q e^{j(\theta_q - 2\theta_N)}$, and so on. In conclusion, only one of the zeros is needed to compute the integral constraints.

An alternative to compute the integral constraints on $\bar{S}$ for only one of the zeros in the set $\mathcal{Z}$ is to compute the integral constraints on the first n columns of $\bar{S}$ but for all the zeros in $\mathcal{Z}$. This also follows from Proposition 5.1.1. The importance of this fact is that, unlike time-invariant systems, fundamental limitations of periodic feedback control systems expressed in terms of integral constraints involve different terms, each one of them associated with how one single frequency of the output is affected by multiple frequencies of the input, or how multiple frequencies of the output are affected by a single frequency of the input.

Finally, we point out that similar results can be obtained for the modulated complementary sensitivity function.

5.4 Design Interpretations

In this section we use the integral constraints developed above to explore two aspects of the performance of linear periodic feedback systems. In particular, we apply the integrals to the problem of control of periodic systems where it is required to have a time-invariant map for the closed loop, and to the problem of periodic control of a LTI system.

5.4.1 Time-Invariant Map as a Design Objective

We have remarked earlier (see the footnote on page 122) that a time-invariant system has a diagonal modulated transfer matrix. Hence, there is a clear connection between having a time-invariant system as a design objective and achieving diagonal decoupling for a MIMO time-invariant system. The latter problem was studied in §4.3.4 of Chapter 4, where it was shown that there is usually a sensitivity cost associated with diagonal decoupling. We will see next that the same is true for the problem of having a time-invariant closed loop for periodic systems.

Consider the periodic feedback system of Figure 5.1, which, for simplicity, is assumed to be a SISO system. Assume that, inter-alia, the control objective is to achieve a time-invariant map from the reference to the output. This implies that, in the frequency domain raised system, $\bar{T}(e^{j\theta})$ and $\bar{S}(e^{j\theta})$ are required to be block diagonal. We then have the following result.

Theorem 5.4.1. Let $q = r_q e^{j\theta_q}$, $r_q > 1$, be a zero of the plant $\bar{G}$, and let $\Psi \in \mathbb{C}^{Nn}$, $\Psi \neq 0$, be its output direction. Assume that the modulated sensitivity function, $\bar{S}$, is proper and stable, and such that $\bar{S}(e^{j\theta})$ is diagonal. Then, for each index k in $\mathcal{J}_\Psi$,

$$\frac{1}{2\pi}\int_{-\pi}^{\pi} \log|\bar{S}_{kk}(e^{j\theta})| \frac{r_q^2 - 1}{1 - 2r_q\cos(\theta - \theta_q) + r_q^2}\, d\theta \geq 0\,, \tag{5.22}$$

or equivalently

$$\frac{1}{2\pi}\int_{-\pi}^{\pi} \log|\bar{S}_{11}(e^{j\theta})| \frac{r_q^2 - 1}{1 - 2r_q\cos(\theta - \theta_q + (k-1)\theta_N) + r_q^2}\, d\theta \geq 0\,, \tag{5.23}$$

for each index k in $\mathcal{J}_\Psi$.

Proof. The constraint (5.22) is immediate from (5.19), on noting that, if $\bar{S}(e^{j\theta})$ is diagonal, then $\bar{S}_{ik}(e^{j\theta}) = 0$ for $i \neq k$.

For the constraint (5.23), note that $\bar{S}_{kk}(e^{j\theta}) = \bar{S}_{11}(e^{j(\theta-(k-1)\theta_N)})$, which follows from Proposition 5.1.1. We then obtain, using (5.22), that

$$\frac{1}{2\pi}\int_{-\pi}^{\pi} \log|\bar{S}_{11}(e^{j(\theta-(k-1)\theta_N)})|\frac{r_q^2-1}{1-2r_q\cos(\theta-\theta_q)+r_q^2}\,d\theta \geq 0\,.$$

Making the change of variables $\phi = \theta-(k-1)\theta_N$, and invoking periodicity of the integrand, we finally arrive at the constraint (5.23). □

Note that $\bar{S}_{11}(e^{j\theta})$ is the frequency response of the error to the reference signal, i.e., $E(e^{j\theta}) = \bar{S}_{11}(e^{j\theta})R(e^{j\theta})$, and there are no terms involving shifted functions of $R(e^{j\theta})$ because it is assumed that the closed loop has a time-invariant input-output map.

The inequality (5.23) implies that $\bar{S}_{11}$ satisfies as many integral constraints as there are nonzero elements of the zero direction Ψ (i.e., for each k in $\mathcal{J}_\Psi$). Moreover, in each of this integrals, the weighting function

$$W_q(\theta+(k-1)\theta_N) \triangleq \frac{r_q^2-1}{1-2r_q\cos(\theta-\theta_q+(k-1)\theta_N)+r_q^2} \tag{5.24}$$

is shifted in angle for different values of k, and thus, their corresponding maximum values are located at different angles.

The impact of this constraint is more evident if we assume also that $|\bar{S}_{11}(e^{j\theta})|$ is intended to be reduced in some set $\Theta_1 \subset (-\pi,\pi)$, i.e.,

$$|\bar{S}_{11}(e^{j\theta})| \leq \alpha < 1\,, \qquad \forall\theta \in \Theta_1 \triangleq [-\theta_1,\theta_1]\,, \quad \theta_1 < \pi\,. \tag{5.25}$$

Let the weighted length of the interval Θ_1 be defined, for each k in $\mathcal{J}_\Psi$, as (cf. (3.50) in Chapter 3)

$$\Theta_q^k(\theta_1) \triangleq \int_{-\theta_1}^{\theta_1} W_q(\theta+(k-1)\theta_N)\,d\theta\,.$$

Then, from (5.23), we obtain

$$[2\pi-\Theta_q^k(\theta_1)]\log\|\bar{S}_{11}\|_\infty + \Theta_q^k(\theta_1)\log\alpha \geq 0\,, \quad \forall k \in \mathcal{J}_\Psi\,,$$

which is equivalent to

$$\|\bar{S}_{11}\|_\infty \geq \left(\frac{1}{\alpha}\right)^{\frac{\Theta_q^k(\theta_1)}{2\pi-\Theta_q^k(\theta_1)}}\,, \quad \forall k \in \mathcal{J}_\Psi\,. \tag{5.26}$$

Note that the value of $\Theta_q^k(\theta_1)$ (for some $k \in \mathcal{J}_\Psi$) will, in general, be large if the frequency range Θ_1 where sensitivity reduction is required includes the maximum value of the corresponding weighting function W_q in (5.24). Therefore, in view of (5.26), peaks in $|\bar{S}_{11}(e^{j\theta})|$ are more likely to be large when sensitivity reduction is required over ranges that include the maximum value of some of the different weighting functions.

We then note that this situation is, in general, worse for periodic systems than it is for time-invariant systems. Indeed, in the latter case, we need consider only one weighting function for each nonminimum phase zero rather than as many as there are nonzero elements in the zero direction. We also note that the situation becomes potentially worse as the number of elements in $\mathcal{I}_\Psi$ increases or as the dimension of the null space associated with the zero grows (see §4.3.4 in Chapter 4).

We conclude from the previous analysis that, for periodic systems, the problem of sensitivity reduction with the additional requirement that the closed-loop map from reference to output be time-invariant is very likely to result, in general, in large peaks for $|\bar{S}(e^{j\theta})|$.

As an illustration of the above results we study a simple situation where it will turn out that there is no extra sensitivity cost associated with having a time-invariant closed-loop map.

Example 5.4.1. Consider a SISO plant described by

$$\begin{aligned} x_{k+1} &= Ax_k + B_k u_k \,, \\ y_k &= Cx_k + D_k u_k \,, \end{aligned} \tag{5.27}$$

where A and C are constant matrices, and $\{B_k\}$ and $\{D_k\}$ are N-periodic sequences. Assume further that the very special condition holds that there exists a periodic scalar gain $K_k \neq 0$, $\forall k$, such that

$$B_k K_k = B \,, \qquad \text{and} \qquad D_k K_k = D \,,$$

where B and D are constant. It is then clear that one can achieve time-invariance by simply redefining the input. We will use our previous analysis to show that, in this particular case, there is no penalty in achieving time-invariance for the closed-loop system.

Let $\bar{G}$ be as in (5.7), i.e., $\bar{G}$ is the modulated transfer matrix associated with the plant (5.27). Similarly, let $\bar{K}$ be the modulated transfer matrix associated with the periodic gain K_k. Using the relationship (5.9) between time and frequency domain raising, it is easy to see that $\bar{K}$ is given by[5]

$$\bar{K} = W_s^{-1} \begin{bmatrix} K_0 & 0 & \dots & 0 \\ 0 & K_1 & \dots & \vdots \\ \vdots & \ddots & \ddots & \vdots \\ 0 & \dots & \dots & K_{N-1} \end{bmatrix} W_s \,,$$

[5] The time domain raised transfer matrix corresponding to the scalar gain K_k is a diagonal matrix having the elements $K_0, K_1, \dots, K_{N-1}$ in its diagonal (see Feuer & Goodwin (1996)).

and we see that $\bar{K}$ is nonsingular since $K_k \neq 0$ for all k.

Let $N_G D_G^{-1}$ be a right coprime factorization of $\bar{G}$. We will show that the output zero directions of N_G are canonical, i.e., for each zero q of N_G, there is a basis for the left null space of $N_G(q)$ constructed of vectors having only one nonzero element.

We first note that there are no unstable pole-zero cancelations in $\bar{G}\bar{K}$, and, clearly $N_G(\bar{K}^{-1}D_G)^{-1}$ is a right coprime factor of $\bar{G}\bar{K}$. Moreover, since K_k was chosen so that $B_k K_k$ and $D_k K_k$ are constant, it is clear that $\bar{G}\bar{K}$ is diagonal. Because of this, it is possible to build a right coprime fraction $\bar{N}\bar{D}^{-1} = \bar{G}\bar{K}$ where $\bar{N}$ and $\bar{D}$ are diagonal. We then conclude that[6]

$$N_G = \bar{N}R\,, \tag{5.28}$$

where R is biproper and bistable.

Since $\bar{N}$ is diagonal, it is straightforward to build a canonical basis for the left null space of $\bar{N}(q)$ and, because of (5.28), this basis will also be a canonical basis for the left null space of $N_G(q)$.

Let Ψ be a vector in the left null space of $N_G(q)$; it follows that Ψ is canonical, i.e., $\mathcal{J}_\Psi$ has only one nonzero element. Hence, there is only one weighting function of the form (5.24), which implies that there is only one integral constraint (5.23) associated with the zero with direction Ψ. This situation is no worse than that corresponding to a time-invariant system. ∘

5.4.2 *Periodic Control of Time-invariant Plant*

A question that has been the subject of an ongoing discussion is whether or not it is advantageous to use linear time-varying controllers (e.g., periodic controllers) for time-invariant plants. From some points of view, periodic control of time-invariant plants seems to yield improved performance over time-invariant control, whilst from other points of view there appear to be inherent disadvantages.[7] We give here a particular view of this problem based on the integral constraints derived in §5.3.

Let G be the (usual) transfer function of the time-invariant plant, and let $\bar{G}$ be the modulated transfer matrix corresponding to G, based on a period of N, i.e., assuming that an N-periodic controller will be used. Because of the time-invariant characteristic of the plant, $\bar{G}$ will have the following

---

[6]This is because any two coprime factorizations are related by units of the ring of proper and stable transfer matrices, i.e., biproper and bistable matrices (Vidyasagar 1985, Theorem 4.1.43, p.75).

[7]See section on notes and references at the end of the chapter.

structure

$$\bar{G}(z) = \begin{bmatrix} G(z) & 0 & \cdots & 0 \\ 0 & G(ze^{-j\theta_N}) & \cdots & \vdots \\ \vdots & \ddots & \ddots & \vdots \\ 0 & \cdots & \cdots & G(ze^{-j(N-1)\theta_N}) \end{bmatrix}.$$

Then, if $\tilde{d}_g^{-1}\tilde{n}_g$ and $n_g d_g^{-1}$ are left and right coprime factorizations[8] of the plant G, respectively, we can construct coprime factorizations of $\bar{G} = \tilde{D}_G^{-1}\tilde{N}_G = N_G D_G^{-1}$, where the different factors contain shifted versions of $\tilde{d}_g$, $\tilde{n}_g$, n_g, and d_g. For example, N_G and $\tilde{D}_G$ can be constructed as

$$N_G(z) = \begin{bmatrix} n_g(z) & 0 & \cdots & 0 \\ 0 & n_g(ze^{-j\theta_N}) & \cdots & \vdots \\ \vdots & \ddots & \ddots & \vdots \\ 0 & \cdots & \cdots & n_g(ze^{-j(N-1)\theta_N}) \end{bmatrix}, \tag{5.29}$$

and

$$\tilde{D}_G(z) = \begin{bmatrix} \tilde{d}_g(z) & 0 & \cdots & 0 \\ 0 & \tilde{d}_g(ze^{-j\theta_N}) & \cdots & \vdots \\ \vdots & \ddots & \ddots & \vdots \\ 0 & \cdots & \cdots & \tilde{d}_g(ze^{-j(N-1)\theta_N}) \end{bmatrix}. \tag{5.30}$$

Let $q = r_q e^{j\theta_q}$ be a nonminimum phase zero of n_g, with output direction $\Psi \in \mathbb{C}^{Nn}$. Then every element in $\mathcal{Z}$ defined in the set (5.14) is a nonminimum phase zero of N_G in (5.29). However, we know from §5.3 that only one of the zeros in this set is needed to obtain the associated integral constraints, because the rest lead to the same integrals. We will use the zero q and its corresponding direction Ψ. We observe that, a particular feature of the modulated representation of the plant is that the direction, $\bar{\Psi}$ say, of q as a zero of N_G in (5.29) has the form

$$\bar{\Psi}^* = [\Psi^*, 0, \ldots, 0]\,. \tag{5.31}$$

We next study the structure of the modulated sensitivity function. First note that, under the assumption of internal stability of the closed loop, the modulated form of the controller has a left coprime factorization (lcf) satisfying $\tilde{D}_G D_K + \tilde{N}_G N_K = I$ (Vidyasagar 1985). Hence, from (5.12), the modulated sensitivity is simply $\bar{S} = D_K \tilde{D}_G$. Thus, splitting D_K into blocks of

[8]We use here lower case letters to avoid confusion with the factorizations of the modulated matrices.

appropriate dimensions, and using the expression for $\tilde{D}_G$ in (5.30), we obtain that $\bar{S}(z) =$

$$\begin{bmatrix} D_K^{(11)}(z)\tilde{d}_g(z) & D_K^{(12)}(z)\tilde{d}_g(ze^{-j\theta_N}) & \dots & D_K^{(1N)}(z)\tilde{d}_g(ze^{-j(N-1)\theta_N}) \\ D_K^{(21)}(z)\tilde{d}_g(z) & D_K^{(22)}(z)\tilde{d}_g(ze^{-j\theta_N}) & \dots & \vdots \\ \vdots & \ddots & \ddots & \vdots \\ D_K^{(N1)}(z)\tilde{d}_g(z) & \dots & \dots & D_K^{(NN)}(z)\tilde{d}_g(ze^{-j(N-1)\theta_N}) \end{bmatrix}.$$

According to Theorem 5.3.1, we have only to consider those columns of $\bar{S}$ corresponding to nonzero entries of $\bar{\Psi}$. Then it is clear from the structure of $\bar{\Psi}$ given in (5.31) that the problem is equivalent to the multivariable integral constraints for $\bar{S}_{11} = D_K^{(11)}\tilde{d}_g$. This, in turn, generates the same integral constraints as if one uses a time-invariant controller having $D_K^{(11)}$ as the denominator of a lcf. Thus, the use of periodic control is no less constrained than the use of time-invariant control.

In conclusion, we see that periodic controllers offer no apparent advantage over time-invariant ones from the point of view of fundamental integral constraints that must be satisfied by the sensitivity function. We remark that similar conclusions hold, mutatis-mutandis, for the complementary sensitivity function.

5.5 Summary

This chapter has studied Poisson integral constraints for discrete periodic feedback systems. The integrals have been developed on a modulated representation of the sensitivity function. This modulated representation associates a transfer function with periodic systems, which can then be dealt with as one would a MIMO time-invariant system. One has to pay particular attention, however, to the structure of these modulated matrices and the nature of its zeros and poles.

Using the resultant integral constraints, it is possible to analyze design limitations inherent to linear, periodically time-varying systems. In particular, we have shown that there is generally an additional cost associated with having a time-invariant target closed loop for a periodic open-loop plant. It was also shown that nonminimum phase zeros and/or unstable poles of a discrete LTI plant continue to impose design limitations even if a periodic time-varying controller is used.

Notes and References

This chapter is mainly based on Goodwin & Gómez (1995).

Frequency Domain Raising

The origins of this tool can be related to early work of Zadeh (1950), who gave a frequency domain description of general time-varying systems. The modulation representation has been extensively used in the signal processing literature, see e.g., Shenoy, Burnside & Parks (1994), Vetterli (1987) and Vetterli (1989).

The modulated transfer matrix, when evaluated along the unit circle, is sometimes referred to as the *alias component matrix* (Smith & Barnwell 1987, Ramstad 1984, Vetterli 1989).

Time Domain Raising

For a description of time domain raising and its utility see e.g., Khargonekar, Poolla & Tannenbaum (1985), Meyer (1990), Ravi, Kharghonekar, Minto & Nett (1990) and Feuer & Goodwin (1996). This latter reference establishes the relation between time and frequency domain raising.

Transform Techniques

For a more detailed exposition of Fourier techniques see e.g., Feuer & Goodwin (1996). The double-sided $\mathcal{Z}$-transform and its properties is studied, for example, in Franklin, Powell & Workman (1990).

Periodic Control of LTI Systems

Khargonekar et al. (1985) argue that, for an important class of robustness problems, discrete periodic compensators are superior to time-invariant ones. This superiority is explained in terms of improved gain and phase margins. Similar ideas are presented in Lee, Meerkov & Runolfsson (1987) for continuous-time systems, and in Francis & Georgiou (1988) for sampled-data systems. On the other hand, Khargonekar et al. (1985) showed that time-varying controllers offer no advantage over time-invariant ones in the problem of weighted sensitivity minimization. Furthermore, in Shamma & Dahleh (1991) it is argued that time-varying compensation does not improve optimal rejection of persistent bounded disturbances, and also it does not help in the bounded-input bounded-output robust stabilization of time-invariant plants with unstructured uncertainty.

An analysis of the use of periodic controllers, based on frequency domain arguments, has been given by Goodwin & Feuer (1992) and Feuer (1993). These works showed that the use of periodic control faces two problems: inherent presence of high frequency components, and sensitivity to high frequency system uncertainty.

6
Extensions to Sampled-Data Systems

This chapter deals with fundamental limitations for sampled-data (SD) feedback systems. By the term SD, we refer to a system with both continuous-time and discrete-time signals — as is the case of digital control of an analogue plant — but which is studied in continuous-time. This contrasts with the approach taken in §3.4 in Chapter 3, where we were concerned only with the *sampled* behavior of the system. In this chapter, the full intersample behavior will be taken into account.

Unfortunately, the discrete results in Chapter 3 are insufficient to describe fundamental limitations in SD systems, since good sampled behavior is clearly necessary but not sufficient for good overall behavior.

Because a SD system is intrinsically time-varying due to the sampling process, one cannot use transfer functions to describe its input-output properties. However, it is possible to calculate the Laplace transform of the response of a SD system to a particular input, and hence one may evaluate the steady-state response of a stable SD system to a sinusoidal input of given frequency. For analogue systems, the response to such an input is a sinusoid of the same frequency as the input, but with amplitude and phase modified according to the transfer function of the system evaluated at the input frequency (see §2.1.3 in Chapter 2). The response of a stable SD system to an input sinusoidal signal, on the other hand, consists of a sum of infinitely many sinusoids spaced at integer multiples of the sampling frequency away from the frequency of the input. We will refer to the component having the same frequency as the input as the *fundamental*, and the other components as the *harmonics*. In fact, the fundamental corresponds to the first harmonic, which will be predominant in most applications. This is because higher frequency components will normally be suppressed in a well-designed SD system.

The fundamental, and each of the harmonics, is governed by what we term the *frequency response function,* which has many properties similar to those of a transfer function. In particular, these frequency response functions have sufficient structure to allow complex analysis to be applied to derive a set of formulae analogous to the Bode and Poisson integrals. As in the LTI case, these integrals describe trade-offs between system properties in different frequency ranges. Our analysis in this chapter concentrates on the fundamental components of the SD response, but similar results may be derived for the harmonics (Braslavsky 1995).

6.1 Preliminaries.

6.1.1 Signals and System

We consider the SISO SD feedback system shown in Figure 6.1, where G and F are the transfer functions of the plant and anti-aliasing filter, K_d is the digital controller, and $\mathcal{S}_\tau$ and H represent the sampler and hold device respectively.

The plant and controller are assumed to be proper, and the filter strictly proper and stable,[1] and they are all free of unstable hidden modes. The plant may include a pure time delay, denoted by $\tau_G \geq 0$. The relative degree of the controller is denoted by $\mathrm{RD}\, K_d$.

The signals in Figure 6.1 are as follows:

r : reference input,	$\{u_k\}$: discrete control sequence,
y : system output,	u : analogue control input,
d : output disturbance	v : analogue output of the filter,
w : measurement noise,	$\{v_k\}$: sampled output of the filter.

Continuous-time signals are assumed scalar functions from $[0, \infty)$ to $\mathbb{R}$, and discrete sequences are defined for $k = 0, 1, 2, \ldots$, taking values on $\mathbb{R}$.

A class of signals that will be quite useful for our purposes is connected with the class of functions of *bounded variation.*

Definition 6.1.1 (Function of Bounded Variation). A function h, defined on the closed real interval $[a, b]$, is of *bounded variation* if the *total variation*

[1] The assumption that the filter is strictly proper is standard and guarantees that the sampling operation is well defined (cf. Sivashankar & Khargonekar 1993, Dullerud & Glover 1993). The assumption of stability is only made for simplicity of exposition and may be removed.

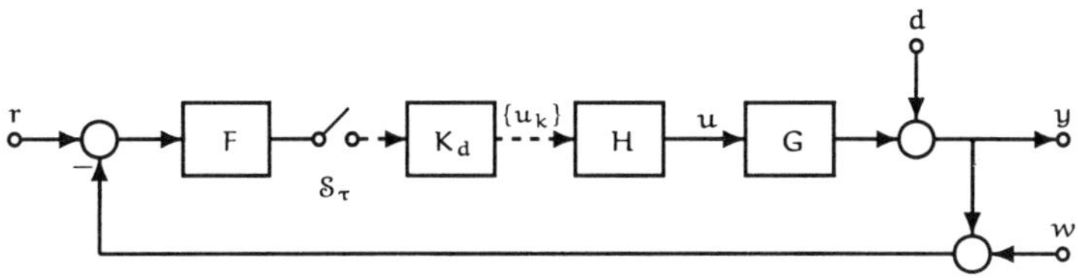

FIGURE 6.1. SD control system.

of h on [a, b],

$$V_h(a,b) \triangleq \sup \sum_{k=1}^{n} |h(t_k) - h(t_{k-1})| \tag{6.1}$$

is finite. The supremum here is taken over every $n \in \mathbb{N}$ and every partition of the interval $[a, b]$ into subintervals $\{t_k, t_{k-1}\}$, where $k = 1, 2, \dots n$, and $a = t_0 < t_1 < \cdots < t_n = b$.

A function h defined on $[0, \infty)$ is of *uniform bounded variation* (UBV) (Braslavsky, Meinsma, Middleton & Freudenberg 1995) if, given $\delta > 0$, the total variation $V_h(t, t+\delta)$ on intervals $[t, t+\delta]$ of length δ is uniformly bounded, that is, if

$$\sup_{t\in[0,\infty)} V_h(t, t+\delta) < \infty \,. \tag{6.2}$$

∘

A function of BV is not necessarily continuous, but it is differentiable almost everywhere. Moreover, the limits

$$h(t^{\pm}) \triangleq \lim_{\epsilon \downarrow 0} h(t \pm \epsilon), \quad \epsilon > 0 \tag{6.3}$$

are well defined at every t, which means that h can have discontinuities of at most the "finite-jump" type. This is particularly appropriated for signals involved in sampling operations, as we will see.

We will assume throughout that the input signals r, d, and w, are such that, when multiplied by some exponentially decaying term $e^{-\sigma t}$, are functions of UBV. It is straightforward to verify that steps, ramps, sinusoids and exponentials are all signals satisfying this condition. However, signals like $\sin(e^{t^2})$, and signals that contain impulses are excluded.

6.1.2 *Sampler, Hold and Discretized System*

The implementation of a controller for a continuous-time system by means of a digital device, such as a computer, implies the process of sampling and reconstruction of analogue signals. By the sampling process, an analogue signal is converted into a sequence of numbers that can then be digitally manipulated. The hold device performs the inverse operation, translating the output of the digital controller into a continuous-time signal. We will assume throughout that nonlinearities associated with the process of discretization, such as finite memory word-length, quantization, etc., have no significant effect on the sampled-data system.

We assume also that the sampling is regular, i.e., if τ is the *sampling period*, sampling is performed at instants $t = k\tau$, with $k = 0, \pm 1, \pm 2, \ldots$. Associated with τ, we define the *sampling frequency* $\omega_s = 2\pi/\tau$. By $\omega_N = \omega_s/2$ we denote the Nyquist frequency, and by Ω_N the *Nyquist range* of frequencies $[-\omega_N, \omega_N]$.

We consider an idealized model of the sampler. If ν is a continuous-time signal, we define the sampling operation with period τ by

$$\mathcal{S}_\tau \nu = \{\nu_k\}_{k=0}^{\infty},$$

where the sequence $\{\nu_k\}_{k=0}^{\infty}$ represents the sampled version of ν, with $\nu_k = \nu(k\tau^+)$, and $k \in \mathbb{N}_0$. The z-transform operator is denoted by $\mathcal{Z}$, i.e.,

$$\mathcal{Z}\{u_k\} \triangleq \sum_{k=0}^{\infty} u_k z^{-k},$$

and the Laplace transform operator is denoted by $\mathcal{L}$, i.e., $\mathcal{L}u = U$, where

$$U(s) = \int_0^{\infty} e^{-st} u(t) dt .$$

The hold device, H, is a generalized SD hold function (GSHF) *a la* Kabamba (1987), defined by

$$u(t) = h(t - k\tau)u_k, \quad k\tau \le t < (k+1)\tau, \quad k \in \mathbb{N}_0. \tag{6.4}$$

The function h in the above definition, which characterizes the hold, is assumed to be of bounded variation, and with support on the interval $[0, \tau]$. In particular, the choice $h(t) = 1$ for $t \in [0, T]$ yields the zero order hold (ZOH). By considering GSHFs we extend the scope of our analysis to a wide class of SD control schemes that do not necessarily rely on ZOH reconstruction.

Associated with the hold we define its *frequency response function* by $H = \mathcal{L}h$. Since h is supported on a finite interval, it follows that H is an entire function, i.e., it has no singularities at any finite s in $\mathbb{C}$. For example, in the case of the ZOH we get the familiar response $H(s) = (1 - e^{-s\tau})/s$. Frequency responses for other types of hold functions are derived in Middleton & Freudenberg (1995).

The frequency response of the hold is useful in computing the Laplace transform of the output of the hold device.

Lemma 6.1.1. Consider a hold device as defined in (6.4) and its frequency response function H. Let U_d denote the z-transform of the input sequence $\{u_k\}$, and U the Laplace transform of the hold output to this sequence. Then

$$U(s) = H(s)\, U_d(e^{s\tau})\,. \tag{6.5}$$

Proof. Using (6.4), we can express the output of the hold device to $\{u_k\}$ as

$$u(t) = \sum_{k=0}^{\infty} h(t-k\tau)\, u_k \left(\mathbf{1}(t-(k+1)) - \mathbf{1}(t-k\tau)\right), \tag{6.6}$$

where $\mathbf{1}$ denotes the unit step function $\mathbf{1}(t) = 1$ if $t \in [0,\tau)$, and 0 otherwise. Hence,

$$\begin{aligned} U(s) &= \int_0^{\infty} e^{-s\tau} u(t)\, dt \\ &= \sum_{k=0}^{\infty} \int_{k\tau}^{(k+1)\tau} e^{-s\tau}\, h(t-k\tau)\, u_k\, dt \\ &= \sum_{k=0}^{\infty} e^{-sk\tau} u_k \int_0^{\tau} e^{-st}\, h(t)\, dt\,. \end{aligned}$$

The result then follows from the definition of the frequency response function of the hold, H, and the definition of the z-transform. □

We denote by $(FGH)_d$ the *discretized plant*, defined as

$$(FGH)_d \triangleq \mathcal{Z}\mathcal{S}_\tau\mathcal{L}^{-1}FGH,$$

where $\mathcal{L}^{-1}FGH$ denotes the inverse Laplace transform of FGH. The assumptions on G, H and F stated above are sufficient to guarantee (e.g., Freudenberg, Middleton & Braslavsky 1995, Braslavsky, Meinsma, Middleton & Freudenberg 1995) that the discretized plant $(FGH)_d$ satisfies the well-known formula[2]

$$(FGH)_d(e^{j\omega\tau}) = \frac{1}{\tau} \sum_{k=-\infty}^{\infty} H_k(j\omega)G_k(j\omega)F_k(j\omega), \tag{6.7}$$

where the notation $Y_k(\cdot)$ represents $Y(\cdot + jk\omega_s)$, with k an integer, and $\omega_s = 2\pi/\tau$. This notation will be frequently used in the sequel.

In relation to $(FGH)_d$, we also recall the definitions of the discrete sensitivity and complementary sensitivity functions, here given by

$$S_d = \frac{1}{1+K_d(FGH)_d} \quad \text{and} \quad T_d = \frac{K_d(FGH)_d}{1+K_d(FGH)_d}\,.$$

[2]Sometimes called the *impulse modulation formula* (e.g., Araki, Ito & Hagiwara 1993), this identity is closely related to the *Poisson summation formula* (e.g., Rudin 1987).

An important feature of SD systems is evident from (6.7), namely, the response of the discretized plant at a frequency $\omega \in \Omega_N$ depends upon the response of the analogue plant, filter, and hold function at an infinite number of frequencies. Indeed, it is well known that the steady-state response of a stable SD system to a sinusoidal input consists of a fundamental component and infinitely many aliases shifted by multiples of the sampling frequency. As we will see in §6.2, analogous expressions can also be obtained for the response to more general inputs.

6.1.3 *Closed-loop Stability*

As with the case of a ZOH, closed-loop stability is guaranteed by the assumptions that sampling is nonpathological and that the discretized system is closed-loop stable. The next lemma is a generalization of the well-known result of Kalman, Ho & Narendra (1963) to the case of GSHFs (Middleton & Freudenberg 1995).[3]

Lemma 6.1.2 (Nonpathological Sampling). Suppose that G and F are as defined in §6.1.1 and assume further that

(i) if p_i and p_k are CRHP poles of G, then

$$p_i \neq p_k + jn\omega_s, \quad n = \pm 1, \pm 2, \cdots \tag{6.8}$$

(ii) if p_i is a CRHP pole of G, then $H(s)$ has no zeros at $s = p_i$.

Then the discretized plant (6.7) is free of unstable hidden modes. ∘

Since the response of a GSHF may have zeros in $\overline{\mathbb{C}^+}$, Lemma 6.1.2 says, in particular, that it may be necessary to ensure that none of these zeros coincides with an unstable pole of the analogue plant. [4] Under the nonpathological sampling hypothesis, it is straightforward to extend the exponential and $\mathcal{L}_2$ input-output stability results of Francis & Georgiou (1988) and Chen & Francis (1991) to the case of GSHF.

Lemma 6.1.3. Suppose that G, F, K_d, and H are as defined in §6.1.1 and §6.1.2, that the nonpathological sampling conditions (i) - (ii) are satisfied, that the product $K_d(FGH)_d$ has no pole-zero cancelations in $\mathbb{D}^c$, and that all poles of S_d lie in $\mathbb{D}$. Then the feedback system in Figure 6.1 is exponentially stable and $\mathcal{L}_2$ input-output stable.

Proof. The proof may be obtained by simple modification of the proofs of Theorem 4 in Francis & Georgiou (1988), and Theorem 6 in Chen & Francis (1991). □

---

[3]See Middleton & Xie (1995) for the multivariable case.

[4]Note that this *is* necessary in the SISO case.

Lemma 6.1.3 establishes the conditions for the nominal stability of the SD system of Figure 6.1, which will be assumed throughout the rest of the chapter.

6.2 Sensitivity Functions

6.2.1 *Frequency Response*

The steady-state response of a stable SD feedback system to a complex sinusoidal input consists of a fundamental component at the frequency of the input as well as additional harmonics located at integer multiples of the sampling frequency away from the fundamental. This well-known fact is discussed in textbooks (cf. Åström & Wittenmark 1990, Franklin et al. 1990), and has been emphasized in several recent research papers (e.g., Araki et al. 1993, Goodwin & Salgado 1994).

We next present expressions for the output response y in Figure 6.1 to disturbances and noise. Analogous expressions may be stated for the response to the reference input, and for the response of the control u to these signals.

Lemma 6.2.1. Denote the responses of y to each of d and w by y^d and y^w respectively. Then the Laplace transforms of these signals are given by

$$\begin{aligned} Y^d(s) = &\left[I - \frac{1}{\tau} G(s)\,H(s)\,S_d(e^{s\tau})\,K_d(e^{s\tau})\,F(s)\right] D(s) \\ &- \sum_{\substack{k=-\infty \\ k\neq 0}}^{\infty} \left[\frac{1}{\tau} G(s)\,H(s)\,S_d(e^{s\tau})\,K_d(e^{s\tau})\,F_k(s)\right] D_k(s), \end{aligned} \tag{6.9}$$

and

$$\begin{aligned} Y^w(s) = &-\left[\frac{1}{\tau} G(s)\,H(s)\,S_d(e^{s\tau})\,K_d(e^{s\tau})\,F(s)\right] W(s) \\ &- \sum_{\substack{k=-\infty \\ k\neq 0}}^{\infty} \left[\frac{1}{\tau} G(s)\,H(s)\,S_d(e^{s\tau})\,K_d(e^{s\tau})\,F_k(s)\right] W_k(s). \end{aligned} \tag{6.10}$$

Proof. These formulae may be derived using standard techniques from SD control theory (e.g., Franklin et al. 1990, Åström & Wittenmark 1990). We present only a derivation of (6.10). Assume that r and d are zero. Block diagram algebra in Figure 6.1 and (6.5) yield

$$Y^w(s) = G(s)\,H(s)\,U_d(e^{s\tau}) \tag{6.11}$$

and

$$U_d(z) = -S_d(z)\,K_d(z)\,V_d(z). \tag{6.12}$$

The sampled output of the antialiasing filter can be written as

$$\begin{aligned} V_d(z) &= \mathcal{Z}\mathcal{S}_\tau\mathcal{L}^{-1}V(s) \\ &= \mathcal{Z}\mathcal{S}_\tau\mathcal{L}^{-1}F(s)W(s). \end{aligned}$$

The assumptions on w, and the fact that F is strictly proper guarantee the validity of the relation (Braslavsky, Meinsma, Middleton & Freudenberg 1995)

$$V_d(e^{s\tau}) = \frac{1}{\tau}\sum_{k=-\infty}^{\infty} F_k(s)W_k(s). \tag{6.13}$$

Substituting (6.12) - (6.13) into (6.11) and rearranging yields the desired result. □

Under the assumption of closed-loop stability, the preceding formulae may be used to derive the steady-state response of the system to a periodic input. As noted above, the response will be equal to the sum of infinitely many harmonics of the input frequency. The magnitude of each component is governed by a function analogous to the usual sensitivity or complementary sensitivity function for LTI systems.

Definition 6.2.1 (SD Sensitivity Functions). We define the *fundamental sensitivity* and *complementary sensitivity functions* by

$$S^0(s) \triangleq I - \frac{1}{\tau}G(s)\,H(s)\,S_d(e^{s\tau})\,K_d(e^{s\tau})\,F(s) \tag{6.14}$$

and

$$T^0(s) \triangleq \frac{1}{\tau}G(s)\,H(s)\,S_d(e^{s\tau})\,K_d(e^{s\tau})\,F(s) \tag{6.15}$$

respectively. For $k \neq 0$ define the *k-th harmonic response function* by

$$T^k(s) \triangleq \frac{1}{\tau}G_k(s)\,H_k(s)\,S_d(e^{s\tau})\,K_d(e^{s\tau})\,F(s). \tag{6.16}$$

∘

These SD response functions are not rational functions, since their definition involves functions of the variable $e^{s\tau}$, like $H(s)$, $K_d(e^{s\tau})$, and $S_d(e^{s\tau})$. In addition, note that they are *not* transfer functions in the usual sense, because they do not equal the ratio of the transforms of output to input signals. However, these functions do govern the steady-state frequency response of the SD system.

Lemma 6.2.2 (Steady-State Frequency Response). Let the hypotheses of Lemma 6.1.3 be satisfied and assume that $d(t) = e^{j\omega t}$, $t \geq 0$, and $w(t) = e^{j\omega t}$, $t \geq 0$. Then as $t \to \infty$, we have that

$$y^d(t) \to y^d_{ss}(t) \quad \text{and} \quad y^w(t) \to y^w_{ss}(t),$$

where

$$y_{ss}^{d}(t) = S^0(j\omega)e^{j\omega t} - \sum_{\substack{k=-\infty \\ k\neq 0}}^{\infty} T^k(j\omega)e^{j(\omega+k\omega_s)t}, \tag{6.17}$$

and

$$y_{ss}^{w}(t) = -T^0(j\omega)e^{j\omega t} - \sum_{\substack{k=-\infty \\ k\neq 0}}^{\infty} T^k(j\omega)e^{j(\omega+k\omega_s)t}. \tag{6.18}$$

Proof. See §B.2.1 in Appendix B. □

Notice from (6.17) and (6.18) that the fundamental components of the disturbance and noise responses can potentially be reduced over some frequency ranges by shaping S^0 and T^0 adequately. These facts correspond to the LTI case, as seen in Chapter 2.

Yet, a feature that is distinctive of SD systems is also evident from (6.17)-(6.18), and this is the presence of harmonics at frequencies other than that of the input. The existence of these harmonics is due to the use of SD feedback, and is reflection of the fact that SD feedback has no counterpart in analogue systems.

6.2.2 *Sensitivity and Robustness*

The fundamental sensitivity and complementary sensitivity functions, together with the harmonic response functions, may be also used to describe differential sensitivity and robustness properties of a SD feedback system.

As seen in §2.2.4 of Chapter 2, the sensitivity function of a LTI feedback system governs the relative change in the reference response of the system with respect to small changes in the plant. Derivations similar to those of Lemma 6.2.2 show that the steady-state response of the system in Figure 6.1 to a reference input $r(t) = e^{j\omega t}$, $t \geq 0$, is given by

$$y_{ss}^{r}(t) = T^0(j\omega)e^{j\omega t} - \sum_{\substack{k=-\infty \\ k\neq 0}}^{\infty} T^k(j\omega)e^{j(\omega+k\omega_s)t}. \tag{6.19}$$

Since $T^0(j\omega)$ depends upon $S_d(e^{j\omega\tau})$, it follows from (6.7) that the fundamental component of the reference response at a particular frequency is sensitive to variations in the plant response at infinitely many frequencies.

Lemma 6.2.3 (Differential Sensitivity). At each frequency ω, the relative sensitivity of the steady-state reference response (6.19) to variations in $G(j(\omega + l\omega_s))$ is given by

(i) For $l = 0$,

$$\frac{G(j\omega)}{T^0(j\omega)} \frac{\partial T^0(j\omega)}{\partial G(j\omega)} = S^0(j\omega). \tag{6.20}$$

(ii) For all $l \neq 0$,

$$\frac{G_l(j\omega)}{T^0(j\omega)} \frac{\partial T^0(j\omega)}{\partial G_l(j\omega)} = -T^0(j(\omega + l\omega_s)). \tag{6.21}$$

Proof. The proof is a straightforward calculation, keeping in mind the dependence of $S_d(e^{j\omega\tau})$ upon $G_l(j\omega)$. □

These results may best be interpreted by considering frequencies in the Nyquist range. Fix $\omega \in \Omega_N$. Then (6.20) states that the sensitivity of the fundamental component of the reference response to small variations in the plant *at that frequency* is governed by the fundamental sensitivity function. On the other hand, (6.21) states that the sensitivity of the fundamental component to *higher frequency* plant variations is governed by the fundamental complementary sensitivity function evaluated at the higher frequency.

Sensitivity functions S^0 and T^0 also have implications on the stability robustness properties of the SD system. Suppose that the plant is subject to a multiplicative perturbation, i.e., the perturbed plant, denoted $\tilde{G}$, can be written as

$$\tilde{G} = G(1 + W_1\Delta) , \tag{6.22}$$

where Δ is a stable and proper perturbation, and W_1 is a stable minimum phase weighting function used to represent frequency dependence of the modeling error, and such that GW_1 is proper.[5]

Lemma 6.2.4 (Robust Stability against Multiplicative Perturbations). Consider the system of Figure 6.1 and let the assumptions of Lemma 6.1.3 hold. A necessary condition for the system to remain stable when the plant is perturbed as in (6.22) for all Δ such that $\|\Delta\|_\infty < 1$ is that

$$\|T^0(j\omega)W_1(j\omega)\|_\infty \leq 1. \tag{6.23}$$

Proof. See Appendix B. □

Unsurprisingly, a similar result holds for S^0 when we consider a *divisive perturbation model*, where the perturbed plant is expressed by

$$\tilde{G} = (1 + W_2\Delta)^{-1}G , \tag{6.24}$$

[5]See also §2.2.4 in Chapter 2, where we gave an equivalent description of multiplicative uncertainty.

where Δ is as for the multiplicative perturbation model, and W_2 is stable, minimum phase, and such that GW_2 is proper.

Lemma 6.2.5 (Robust Stability against Divisive Perturbations). Under the assumptions of Lemma 6.1.3, a necessary condition for the system of Figure 6.1 to remain stable when the plant is perturbed as in (6.24) for all Δ such that $\|\Delta\|_\infty < 1$ is that

$$\|S^0(j\omega)W_2(j\omega)\|_\infty \leq 1. \tag{6.25}$$

Proof. See Appendix B. □

It follows from Lemmas 6.2.4 and 6.2.5 that if $|T^0(j\omega)|$ or $|S^0(j\omega)|$ are very large at any frequency, then the SD system will exhibit poor robustness to unstructured uncertainty in the analogue plant at that frequency.

6.3 Interpolation Constraints

In this section we present a set of interpolation constraints that must be satisfied by the SD sensitivity functions defined in (6.14)-(6.16). SD sensitivity responses have fixed values on $\mathbb{C}^+$ that are determined by the open-loop zeros and poles of the plant, hold response, and digital controller. As we will see later, a significant difference between the SD case and the continuous-time only or discrete-time only cases is that the poles and zeros of the controller yield different constraints than do those of the plant.

We introduce the following notation for the ORHP zeros of S^0 and T^0, respectively given by

$$\mathcal{Z}_S \triangleq \{s \in \mathbb{C}^+ : S^0(s) = 0\}, \tag{6.26}$$
$$\mathcal{Z}_T \triangleq \{s \in \mathbb{C}^+ : T^0(s) = 0\}. \tag{6.27}$$

Two definitions concerning the mapping between the z-plane and the s-plane will be handy in the sequel.

Definition 6.3.1 (Unfolded Images and Periodic Reflections). Given a number $z_0 \in \mathbb{C}$, we say that $s_{k0} = \frac{1}{\tau}\log z_0 + jk\omega_s$, with $k = 0, \pm 1, \pm 2, \cdots$, is an *unfolded image* of z_0.

Given a number $s_0 \in \mathbb{C}$, we say that $s_{k0} = s_0 + jk\omega_s$, with $k = \pm 1, \pm 2, \cdots$, is a *periodic reflection* of s_0. ∘

The following theorem describes the interpolation relations for the fundamental sensitivity and complementary sensitivity functions.

Theorem 6.3.1 (Interpolation Constraints for S^0 and T^0). Consider the system of Figure 6.1 and assume that the hypotheses of Lemma 6.1.3 are satisfied. Then,

(i) S^0 and T^0 have no poles in $\overline{\mathbb{C}^+}$.

(ii) $\mathcal{Z}_S \supset \{p \in \mathbb{C}^+ : p \text{ is a pole of } G\}$.

(iii) $\mathcal{Z}_T = \{q \in \mathbb{C}^+\}$, where q is either

  (a) an ORHP zero of G, i.e., $G(q) = 0$;
  (b) an ORHP zero of H, i.e., $H(q) = 0$;
  (c) an unfolded image of a nonminimum phase zero of K_d, i.e., $K_d(e^{q\tau}) = 0$;
  (d) a periodic reflection of an ORHP pole p of G, i.e., $q = p + jk\omega_s$ for some integer $k \neq 0$.

(iv) If $b \in \mathbb{D}^c$ is a pole of K_d, and b_k is an unfolded image of b, i.e., $b_k = \frac{1}{\tau}\log b + jm\omega_s$, $m = 0, \pm 1, \pm 2, \cdots$, then

$$S^0(b_m) = 1 - \frac{G(b_m)\,H(b_m)\,F(b_m)}{\tau(FGH)_d(b)}, \tag{6.28}$$

$$T^0(b_m) = \frac{G(b_m)\,H(b_m)\,F(b_m)}{\tau(FGH)_d(b)}. \tag{6.29}$$

Proof. Introduce factorizations

$$G(s)\,F(s) = e^{-s\tau}\frac{N(s)}{D(s)},$$

where N and D are coprime rational functions with no poles in $\overline{\mathbb{C}^+}$, and

$$(FGH)_d(z) = \frac{N_d(z)}{D_d(z)}, \tag{6.30}$$

where N_d and D_d are coprime rational functions with no poles in $\overline{\mathbb{D}}^c$. By the Youla parametrization, all controllers K_d that stabilize (6.30) have the form[6]

$$K_d = \frac{Y_d + D_d Q_d}{X_d - N_d Q_d}, \tag{6.31}$$

where Q_d, X_d, and Y_d are stable, and X_d and Y_d satisfy the Bezout identity

$$D_d X_d + N_d Y_d = 1. \tag{6.32}$$

It follows that $S_d = D_d(X_d - N_d Q_d)$ and

$$K_d S_d = D_d(Y_d + D_d Q_d). \tag{6.33}$$

[6]We suppress dependence on the transform variable when convenient; the meaning will be clear from the context.

Using (6.33) in (6.14) and (6.15) yields

$$S^0(s) = 1 - \frac{1}{\tau} e^{-s\tau} \frac{N(s)H(s)}{D(s)} D_d(e^{s\tau}) \, [Y_d(e^{s\tau}) + D_d(e^{s\tau}) Q_d(e^{s\tau})] \quad (6.34)$$

and

$$T^0(s) = \frac{1}{\tau} e^{-s\tau} \frac{N(s)H(s)}{D(s)} D_d(e^{s\tau}) \, [Y_d(e^{s\tau}) + D_d(e^{s\tau}) Q_d(e^{s\tau})] \; . \quad (6.35)$$

Then:

(i): T^0 is stable because each factor in the numerator of (6.35) is stable, and because the assumption of nonpathological sampling guarantees that any unstable pole of $1/D$ must be canceled by a corresponding zero of $D_d(e^{s\tau})$.

(ii): It follows from (6.32) that $Y_d(e^{p\tau}) = 1/N_d(e^{p\tau})$. Using this fact, and evaluating (6.34) in the limit as $s \to p$ yields

$$S^0(s) \longrightarrow 1 - \lim_{s \to p} \frac{F(s)G(s)H(s)}{\tau (FGH)_d(e^{s\tau})} \, .$$

Replace $(FGH)_d(e^{s\tau})$ using (6.7):

$$S^0(s) \longrightarrow 1 - \lim_{s \to p} \frac{H(s)G(s)F(s)}{\sum_{k=-\infty}^{\infty} F_k(s) \, G_k(s) \, H_k(s)} \, . \quad (6.36)$$

By the assumptions that F is stable and that sampling is nonpathological, G and F have no poles at $p + jk\omega_s$, $k \neq 0$. Using this fact, and the fact that H has no finite poles, yields that each term in the denominator of (6.36) remains finite as $s \to p$ except the term $k = 0$. Then (ii) follows.

(iii): Observe first that (6.35) implies that T^0 has ORHP zeros only at the ORHP zeros of N, H, $D_d(e^{s\tau})$, or $[Y_d(e^{s\tau}) + D_d(e^{s\tau}) Q_d(e^{s\tau})]$.

From the assumptions in §6.1.1, G, F, and GF are each free of unstable hidden modes. Hence N and D can have no common ORHP zeros and (iii)a follows. By the assumption of nonpathological sampling, neither can H and D. Hence (iii)b follows. Note next that the zeros of $D_d(e^{s\tau})$ lie at $p + jk\omega_s$, $k = 0, \pm 1, \pm 2, \cdots$, where p is any ORHP pole of G and hence a zero of D. It follows from this fact that the ratio $D_d(e^{s\tau})/D(s)$ can have zeros only for $k = \pm 1, \pm 2, \cdots$. By the assumption of nonpathological sampling, no other cancelations occur, and all these zeros are indeed zeros of T^0. This proves (iii)d. By (6.31), the ORHP zeros of $[Y_d(e^{s\tau}) + D_d(e^{s\tau}) Q_d(e^{s\tau})]$ are identical to the ORHP zeros of $K_d(e^{s\tau})$. By the hypotheses of Lemma 6.1.3, none of these zeros can coincide with those of $D_d(e^{s\tau})$, and thus with those of D. This proves (iii)c.

(iv): It follows from (6.33) that $K_d(b)S_d(b) = 1/(FGH)_d(b)$. Substitution of this identity into (6.14)-(6.15) yields (6.28)-(6.29).

□

Theorem 6.3.1 establishes ORHP values of S^0 and T^0 that are fixed by open-loop ORHP poles and zeros of the plant, which hold irrespective of the controller.

There are a number of differences between the interpolation constraints for the SD and the continuous-time cases; we describe these in the following remarks.

Remark 6.3.1 ***(Unstable Plant Poles).*** Each ORHP plant pole yields constraints directly analogous to the continuous-time case; i.e., at each ORHP pole p, $S^0(p) = 0$ and $T^0(p) = 1$. Furthermore, each of these poles yields the additional constraints (iii)d, i.e., $S^0(p + jk\omega_s) = 1$ and $T^0(p + jk\omega_s) = 0$, $k \neq 0$, which arise from the periodically spaced zeros of $S_d(e^{s\tau})$ and the fact that nonpathological sampling precludes all but one of these zeros from being canceled by a pole of G. ○

Remark 6.3.2 ***(Nonminimum Plant Zeros).*** Each ORHP zero of the plant, q, yields constraints directly analogous to the continuous-time case; i.e., $S^0(q) = 1$ and $T^0(q) = 0$. Note in particular that these constraints are present *independent* of the choice of the hold function. The zeros of the discretized plant lying in $\mathbb{D}^c$, on the other hand, do not impose any inherent constraints on S^0. Indeed, suppose that $\nu \in \mathbb{D}^c$ is a zero of $(FGH)_d$. Then for each $\nu_k \triangleq \frac{1}{\tau}\log(\nu) + jk\omega_s$, $k = 0, \pm 1, \pm 2, \cdots$, it follows that

$$S^0(\nu_k) = 1 - \frac{1}{\tau}G(\nu_k)H(\nu_k)K_d(\nu)F(\nu_k), \tag{6.37}$$

and thus the size of $S^0(\nu_k)$ is *not* independent of the choice of controller.

○

Remark 6.3.3 ***(Unstable Controller Poles).*** In the analogue case, unstable plant and controller poles yield identical constraints on the sensitivity and complementary sensitivity functions; namely, when evaluated at such a pole, the sensitivity must equal zero and the complementary sensitivity must equal one. Comparing (ii) and (iv) in Theorem 6.3.1, we see that in a SD system unstable plant and controller poles will generally yield different constraints on sensitivity and complementary sensitivity. In particular, unstable controller poles will yield corresponding zeros of S^0 only in special cases. ○

Remark 6.3.4 ***(Zeros of*** K_d***).*** Each zero of the controller lying in $\mathbb{D}^c$ imposes infinitely many interpolation constraints upon the continuous-time system because there are infinitely many points in the s-plane that map to the location of the zero in the z-plane. These constraints are due to the fact that a pole at *any* of these points will lead to an unstable discrete pole-zero cancelation. ○

Remark 6.3.5 ***(Zeros of Hold Response).*** By (iii)b, zeros of H lying in the ORHP impose constraints on the sensitivity function identical to those imposed by ORHP zeros of the plant. A ZOH has zeros only on the $j\omega$-axis, but GSHF response functions may have zeros in the *open* right half plane (Braslavsky, Middleton & Freudenberg 1995). ∘

Remark 6.3.6 ***(Zeros of*** S^0***).*** The list of ORHP zeros for T^0 is exhaustive; however, that for S^0 is not. It is interesting to contrast this situation with the analogue case. For analogue systems, the ORHP zeros of the sensitivity function consist precisely of the union of the ORHP poles of the plant and controller. On the other hand, by Theorem 6.3.1 (ii) and (iv), unstable plant poles yield zeros of S^0 while unstable controller poles generally do not. Furthermore, as the following example shows, S^0 may have ORHP zeros even if both plant and controller are stable.

Example 6.3.1. Consider the plant $G(s) = 1/(s+1)$. Discretize with a ZOH, sample period $\tau = 1$, and no anti-aliasing filter (i.e., $F(s) = 1$). This yields $(FGH)_d(z) = .6321/(z - .3670)$. A stabilizing discrete controller for this plant is

$$K_d(z) = \frac{(4.8158)(z^2 + .1z + 0.3988)}{(z^2 - 1.02657z + 0.9025)}.$$

Both plant and controller are stable; yet it may be verified that S^0 has zeros at $s = 0.2 \pm j$ (see Figure 6.2).

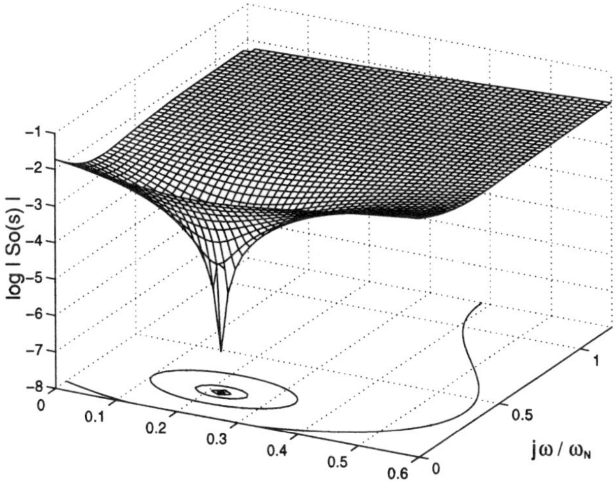

FIGURE 6.2. Fundamental sensitivity for Example 6.3.1

∘

6.4 Poisson Integral formulae

The interpolation constraints derived in the preceding section fix the values of the SD response functions at points of the ORHP. In this section, we translate these constraints into Poisson integral relations that must be satisfied by the SD sensitivity functions. These relations are equivalent to those presented in Chapter 3.

We introduce some Blaschke products first. For convenience, we denote the ORHP zeros of S^0 and T^0 respectively as

$$\mathcal{Z}_S = \{p_i : i = 1, 2, \ldots, n_p\}, \quad \text{and} \quad \mathcal{Z}_T = \{q_i : i = 1, 2, \ldots, n_q\},$$

where n_S and n_T may be infinite. Correspondingly, let their Blaschke products be

$$B_S(s) \triangleq \prod_{i=1}^{n_p} \frac{s - p_i}{s + \bar{p}_i}, \quad \text{and} \quad B_T(s) \triangleq \prod_{i=1}^{n_q} \frac{s - q_i}{s + \bar{q}_i}. \tag{6.38}$$

We also denote by[7]

$B_{G\curlywedge}$: the Blaschke product of the ORHP poles of G,

$B_{G\curlyvee}$: the Blaschke product of the ORHP zeros of G,

B_H: the Blaschke product of the ORHP zeros of H,

B_{K_d}: the Blaschke product of the unfolded images of the ORHP zeros of K_d, and

$B_{G\pitchfork}$: the Blaschke product of the periodic reflections of the ORHP poles of G.

6.4.1 *Poisson Integral for S^0*

The following theorem presents a Poisson integral *inequality* that must be satisfied by $\log|S^0(j\omega)|$.

Theorem 6.4.1 (Poisson integral for S^0). Assume that the hypotheses of Lemma 6.1.3 are satisfied. Let $q = \sigma_q + j\omega_q$, with $\sigma_q > 0$, be an ORHP zero of T^0, as described in Theorem 6.3.1. Then,

$$\int_{-\infty}^{\infty} \log|S^0(j\omega)| \frac{\sigma_q}{\sigma_q^2 + (\omega - \omega_q)^2} \, d\omega \geq \pi \log|B_{G\curlywedge}^{-1}(q)|. \tag{6.39}$$

---

[7]Since there are three Blaschke products emanating from G, we use in the notation some kind of graphical indication, i.e., $\curlyvee$ to indicate a zero, $\curlywedge$ to indicate a pole, and $\pitchfork$ to suggest periodic reflection.

Proof. Factorize S^0 as $S^0 = \tilde{S}B_S$, where B_S is given in (6.38), and where $\tilde{S}$ has no poles or zeros in the ORHP. Then applying Corollary A.6.3 in Appendix A to $\tilde{S}$ implies that (6.39) holds with *equality* if $B_{G\lambda}(q)$ is replaced by $B_S(q)$. Since the set of ORHP zeros of S^0 due to the ORHP poles of G is generally a *proper* subset of all such zeros (cf. Remark 6.3.6) inequality (6.39) follows. □

Theorem 6.4.1 has several design implications, which we describe in a series of remarks.

Remark 6.4.1 **(*Nonminimum Plant Zeros*).** As in the continuous time case, if the plant is nonminimum phase, then requiring that $|S^0(j\omega)| < 1$ over a frequency range Ω implies that, necessarily, $|S^0(j\omega)| > 1$ at other frequencies. The severity of this trade-off depends upon the relative location of the ORHP zero and the frequency range Ω. We now discuss this in more detail.

We recall the definition of the weighted length of an interval by the Poisson kernel for the half plane, introduced in Chapter 3, (3.33). Let $s_0 = \sigma_0 + j\omega_0$ be a point lying in $\mathbb{C}^+$, and consider the frequency interval $\Omega_1 = [-\omega_1, \omega_1)$. Then, we had that

$$\Theta_{s_0}(\omega_1) = \int_{-\omega_1}^{\omega_1} \frac{\sigma_0}{\sigma_0^2 + (\omega - \omega_0)^2}\, d\omega.$$

We have also seen in §3.3.2 that $\Theta_{s_0}(\omega_1)$ equals the negative of the phase lag contributed by a Blaschke product of s_0 at the upper end point of the interval Ω_1. With this notation, the following result is an immediate consequence of (6.39).

Corollary 6.4.2 (Water-Bed Effect). Suppose that q is an ORHP zero of the plant, and suppose that

$$|S^0(j\omega)| \le \alpha, \quad \text{for all } \omega \text{ in } \Omega_1.$$

Then

$$\sup_{\omega > \omega_1} |S^0(j\omega)| \ge (1/\alpha)^{\frac{\Theta_q(\omega_1)}{\pi - \Theta_q(\omega_1)}} \left|B_{G\lambda}^{-1}(q)\right|^{\frac{\pi}{\pi - \Theta_q(\omega_1)}} \tag{6.40}$$

∘

The bound (6.40) shows that if disturbance attenuation is required throughout a frequency interval in which the ORHP zero contributes significant phase lag, then disturbances will be greatly amplified at some higher frequency. The factor due to the Blaschke product in (6.40) shows that plants with approximate ORHP pole-zero cancelations yield particularly sensitive feedback systems. ∘

Remark 6.4.2 **(*Nonminimum Hold Zeros*).** An ORHP zero of the hold response imposes precisely the same trade-off as does a zero of the plant in the same location. This trade-off is exacerbated if the ORHP hold zero is near an unstable plant pole. Poor sensitivity in this case is to be expected, as an exact pole-zero cancelation yields an unstable hidden mode in the discretized plant[8]. ∘

Remark 6.4.3 ***(Unstable Plant Poles).*** Using analogue control, the sensitivity function of a system with an unstable, but minimum phase, plant can be made arbitrarily small over an arbitrarily wide frequency range (Zames & Bensoussan 1983) while keeping sensitivity bounded outside this range. This is no longer true for digital controllers and the fundamental sensitivity function. The following result is an immediate consequence of (6.39). ∘

Corollary 6.4.3.

(i) Assume that the plant has a real ORHP pole, $p = \sigma_p$. Then

$$\|S^0\|_\infty \geq \sqrt{1 + \left(\frac{\sigma_p}{\omega_N}\right)^2}\,. \tag{6.41}$$

(ii) Assume that the plant has an ORHP complex conjugate pole pair, $p = \sigma_p + j\omega_p, \bar{p} = \sigma_p - j\omega_p$. Then for $k = \pm 1, \pm 2 \ldots$

$$\|S^0\|_\infty \geq \sqrt{1 + \left(\frac{\sigma_p}{k\omega_N}\right)^2}\sqrt{1 + \left(\frac{\sigma_p}{\omega_p - k\omega_N}\right)^2}\,. \tag{6.42}$$

∘

Proof. We show only (i); (ii) is similar. Evaluate (6.39) with q equal to one of the periodic reflections of p, i.e., $q = \sigma_p + jk\omega_s$, with $k = \pm 1, \pm 2, \ldots$ Then

$$\begin{aligned} \pi \log \|S^0\|_\infty &\geq \int_{-\infty}^{\infty} \log |S^0(j\omega)| \frac{\sigma_p}{\sigma_p^2 + (\omega - k\omega_s)^2}\, d\omega \\ &\geq \pi \log \left| \frac{2\sigma_p + jk\omega_s}{-jk\omega_s} \right|. \end{aligned} \tag{6.43}$$

From (6.43) follows

$$\begin{aligned} \|S^0\|_\infty &\geq \sqrt{1 + [\sigma_p/(k\omega_N)]^2} \\ &\geq \sqrt{1 + (\sigma_p/\omega_N)^2}. \end{aligned}$$

□

[8]See the conditions for nonpathological sampling in Lemma 6.1.2.

In either case of Corollary 6.4.3, the fundamental sensitivity function necessarily has a peak strictly greater than one.

For a real pole, achieving good sensitivity requires that the sampling rate be sufficiently fast with respect to the time constant of the pole; e.g., achieving $\|S^0\|_\infty < 2$ requires that $\omega_N > \sigma_p/\sqrt{3}$. This condition is also necessary for a complex pole pair. Furthermore, sensitivity will be poor if $\omega_p \approx k\omega_N$ for some $k \neq 0$. The reason for poor sensitivity in this case is clear; if $\omega_p = k\omega_N$, then the complex pole pair violates the nonpathological sampling condition (6.8), and the discretized plant will have an unstable hidden mode.

More generally, we have

Corollary 6.4.4. Assume that the plant has unstable poles p_i and p_k with $p_i \neq \bar{p}_k$. Then

$$\|S^0\|_\infty \geq \max_{k \neq 0} \left| \frac{\bar{p}_i + p_k + jk\omega_s}{p_i - p_k - jk\omega_s} \right| \tag{6.44}$$

and

$$\|S^0\|_\infty \geq \max_{k \neq 0} \left| \frac{p_i + p_k + jk\omega_s}{\bar{p}_i - p_k - jk\omega_s} \right| \tag{6.45}$$

○

It follows that if sampling is "almost pathological", in that $p_i - p_k \approx jk\omega_s$, or $\bar{p}_i - p_k \approx jk\omega_s$, then sensitivity will be large.

*Remark 6.4.4 **(Approximate Discrete Pole Zero Cancelations)**.* Let the discrete controller have an ORHP zero a. Then (6.39) holds with q equal to one of the unfolded images of a in the s-plane. If the plant has an unstable pole near one of these points, then the right hand side of (6.39) will be large, and S^0 will have a large peak. Poor sensitivity is plausible, because this situation corresponds to an approximate pole-zero cancelation between an ORHP zero of the controller and a pole of the discretized plant. ○

6.4.2 Poisson Integral for T^0

We now derive a result for T^0 dual to that for S^0 obtained in the previous section. An important difference is that we can characterize *all* ORHP zeros of T^0, and thus obtain an integral *equality*.

First, we note an additional property of the hold response function.

Lemma 6.4.5. The hold response function may be factored as

$$H(s) = \tilde{H}(s)e^{-s\tau_H}B_H(s), \tag{6.46}$$

where $\tau_H \geq 0$, B_H is as defined on page 150, and $\log|\tilde{H}|$ satisfies the Poisson integral relation.

Proof. Follows directly from the definition of H, which implies that H is a function in $\mathcal{H}_2$, and thus the factorization of Hoffman (1962, pp. 132-133) applies. □

Theorem 6.4.6 (Poisson integral for T^0). Suppose that the conditions of Lemma 6.1.3 are satisfied. Let $p = \sigma_p + j\omega_p$ be an ORHP pole of G. Then

$$\begin{aligned}\int_{-\infty}^{\infty} \log|T^0(j\omega)| \frac{\sigma_p}{\sigma_p^2 + (\omega - \omega_p)^2} d\omega = \pi\sigma_p(\tau_G + \tau_H + \mathrm{RD}\, K_d\, \tau) \\ + \pi \log|B_{G_Y}^{-1}(p)| \\ + \pi \log|B_H^{-1}(p)| \\ + \pi \log|B_{G\tilde{h}}^{-1}(p)| \\ + \pi \log|B_{K_d}^{-1}(p)|\,,\end{aligned} \tag{6.47}$$

where RD K_d denotes the relative degree of K_d.

Proof. Note that T^0 has an inner-outer factorization

$$T^0(s) = \tilde{T}(s)\, e^{-s\tau_P}\, e^{-s\tau_H}\, e^{-s\tau\, \mathrm{RD}\, K_d}\, B_{G_Y}(s)\, B_H(s)\, B_{G\tilde{h}}(s)\, B_{K_d}(s)$$

where $\tilde{T}$ satisfies Corollary A.6.3 in Appendix A. Since $\log|\tilde{T}(j\omega)| = \log|T^0(j\omega)|$, the result follows. □

We comment on the design implications of Theorem 6.4.6 in a series of remarks.

Remark 6.4.5. The first term on the right hand side of (6.47) shows that $|T^0(j\omega)|$ will display a large peak if there is a long time delay in the plant, digital controller, or hold function. ∘

Remark 6.4.6. The third and fourth terms on the right hand side of (6.47) show that $|T^0(j\omega)|$ will display a large peak if there is an approximate unstable pole-zero cancelation in the plant, or between the plant and the hold function. By the nonpathological sampling condition (ii) in Lemma 6.1.2, the latter peak corresponds to an approximate unstable pole-zero cancelation in the *discretized* plant. ∘

The following result is analogous to Corollary 6.4.4 for T^0.

Corollary 6.4.7.

(i) Assume that $p = \sigma_p$, a real pole. Then

$$\|T^0\|_\infty \geq \frac{\sinh\left(\dfrac{\pi\sigma_p}{\omega_N}\right)}{\left(\dfrac{\pi\sigma_p}{\omega_N}\right)}\,. \tag{6.48}$$

(ii) Assume that $p = \sigma_p + j\omega_p$, a complex pole. Then

$$\|T^0\|_\infty \geq \frac{\sinh\left(\frac{\pi\sigma_p}{\omega_N}\right)}{\left(\frac{\pi\sigma_p}{\omega_N}\right)} \left|\frac{\sinh\left(\frac{\pi p}{\omega_N}\right)}{\left(\frac{\pi p}{\omega_N}\right)}\right| \left|\frac{\left(\frac{\pi\omega_p}{\omega_N}\right)}{\sin\left(\frac{\pi\omega_p}{\omega_N}\right)}\right| . \tag{6.49}$$

Proof. Consider the Blaschke product corresponding to the periodic reflections of p, which can be also written as

$$B_p(s) = \prod_{k=1}^{\infty} \frac{1 - \left(\frac{s-p}{jk\omega_s}\right)^2}{1 - \left(\frac{s+\bar{p}}{jk\omega_s}\right)^2}$$

Using the identities (Levinson & Redheffer 1970, p. 387),

$$\frac{\sin \pi\alpha}{\pi\alpha} = \prod_{k=1}^{\infty}\left(1 - \frac{\alpha^2}{k^2}\right)$$

and $\sin j\alpha = j \sinh \alpha$ yields

$$B_p(s) = \frac{\sinh \pi\left(\frac{s-p}{\omega_s}\right)}{\pi\left(\frac{s-p}{\omega_s}\right)} \frac{\pi\left(\frac{\bar{p}+s}{\omega_s}\right)}{\sinh \pi\left(\frac{\bar{p}+s}{\omega_s}\right)} \tag{6.50}$$

Note that the first factor on the right hand side of (6.50) converges to one as $s \to p$. It follows that

$$B_p(p) = \frac{\left(\frac{\pi\sigma_p}{\omega_N}\right)}{\sinh\left(\frac{\pi\sigma_p}{\omega_N}\right)} \tag{6.51}$$

Inverting yields (6.48). Furthermore

$$B_{\bar{p}}(p) = \frac{\sin\left(\frac{\pi\omega_p}{\omega_N}\right)}{\left(\frac{\pi\omega_p}{\omega_N}\right)} \frac{\left(\frac{\pi p}{\omega_N}\right)}{\sinh\left(\frac{\pi p}{\omega_N}\right)} \tag{6.52}$$

Together (6.51)-(6.52) yield (6.49). □

Figure 6.3(a) gives plots of the bound (6.49) versus $\sigma_p = \operatorname{Re} p$ for various values of $\omega_p = \operatorname{Im} p$, and Figure 6.3(b) gives plots of the bound (6.49) versus ω_p for various values of σ_p. The pole location has been normalized by the Nyquist frequency. Note in Figure 6.3(b) that for a complex pole $\|T^0\|_\infty$ will become arbitrarily large as $\omega_p \to k\omega_N, k = \pm 1, \pm 2 \ldots$, because sampling becomes pathological at such frequencies. It follows from these plots that to achieve good robustness the Nyquist frequency should be chosen several times larger than the radius of any unstable pole.

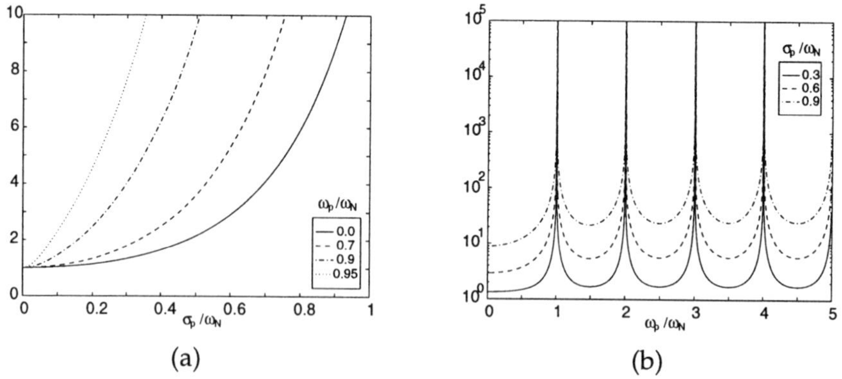

FIGURE 6.3. Lower bounds on $\|T^0\|_\infty$.

6.5 Example: Robustness of Discrete Zero Shifting

We illustrate the use of the tools presented in this chapter by analyzing an example of design originally presented in Er & Anderson (1994). The plant is given by

$$P(s) = \frac{s-5}{(s+1)(s+3)} .$$

It is required that the closed-loop bandwidth, ω_b, satisfy the lower bound $\omega_b \geq 15.3$ rad/s. Notice that the nonminimum phase zero of the analogue plant lies well within the desired closed-loop bandwidth. Hence, by the results of Chapter 3, if this bandwidth is achieved with an analogue controller, then the resulting closed-loop system will have very poor sensitivity and robustness properties.

Briefly, the design suggested in Er & Anderson (1994) consists of first using a GSHF to relocate the discrete zeros so that the discretized plant is minimum phase, and then applying a standard discrete-time LQG/LTR

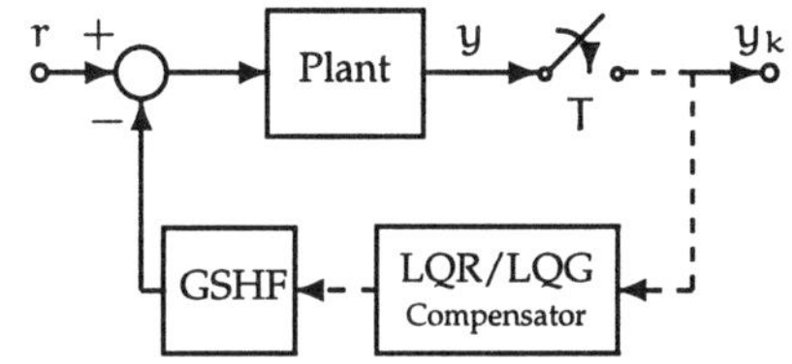

FIGURE 6.4. Structure for the GSHF-based LTR of Er & Anderson (1994).

procedure (e.g., Zhang & Freudenberg 1993). It follows that perfect LTR *at the sampling times* is feasible independent of the zero distribution of the analogue plant. The proposed scheme is shown in Figure 6.4. The sampling period is chosen as $\tau = 0.04$s, and an appropriate GSHF is given by

$$h(t) = \begin{cases} -1957 & \text{for } 0 \leq t < 0.02\ , \\ 1707 & \text{for } 0.02 \leq t < 0.04\ . \end{cases} \tag{6.53}$$

This yields a *minimum phase* discretized plant. In Figure 6.5 we plot $|S_d(e^{j\omega\tau})|$ and $|T_d(e^{j\omega\tau})|$ obtained with the compensator suggested in Er & Anderson (with their parameter $q = 3$). The closed-loop specification is achieved and both $|S_d(e^{j\omega\tau})|$ and $|T_d(e^{j\omega\tau})|$ are well behaved.

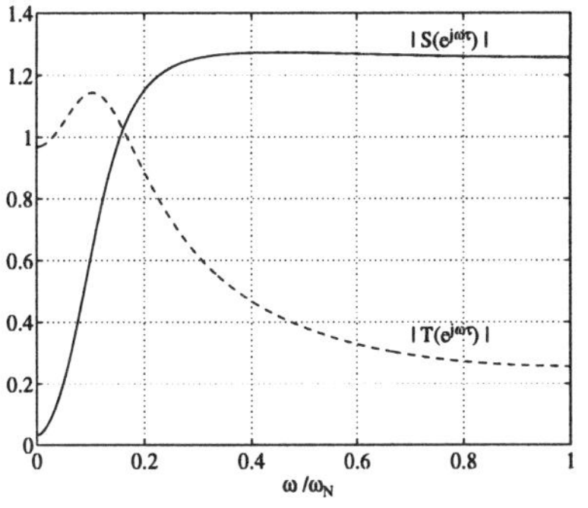

FIGURE 6.5. Discrete sensitivity and complementary sensitivity functions.

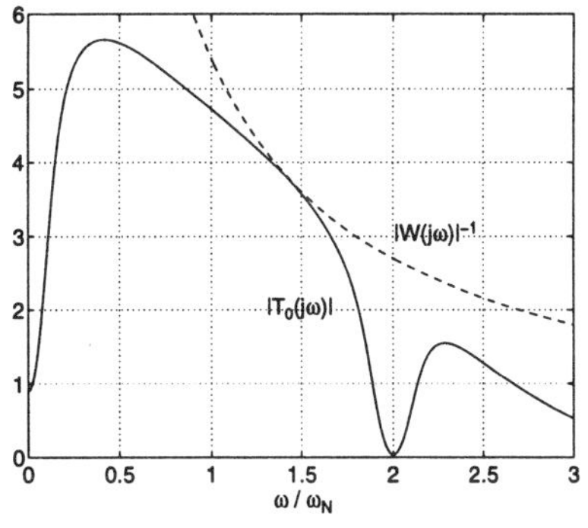

FIGURE 6.6. Fundamental complementary sensitivity and weight W_1.

However, as the results in this chapter predict, the closed-loop system will still be prone to sensitivity and robustness problems. Figure 6.6 shows a plot of $|T^0(j\omega)|$, which exhibits large values within and outside the Nyquist interval. Following Lemma 6.2.4, we should expect poor sensitivity to perturbations in the analogue plant for which W_1 in (6.22) satisfies

$$|T^0(j\omega)| > \frac{1}{|W_1(j\omega)|}\ ,$$

at some ω. Indeed, suppose that the plant has an unmodeled time delay τ_G. This may be represented by a multiplicative perturbation model as in (6.22), with

$$\Delta(s) = \frac{1 - e^{-s\tau_G}}{s\tau_G(1+\delta)}, \quad \text{and} \quad W_1(s) = -\tau_G s(1+\delta),$$

where $0 < \delta \ll 1$. In Figure 6.6 we have also plotted $1/W_1(j\omega)|$ for $\tau_G = 0.0024$s, and it can be seen that there is a frequency for which $|T^0(j\omega)| \approx 1/|W_1(j\omega)|$. In fact, as can be checked by simulation, the system becomes unstable even for this small time delay. Note that this extreme sensitivity to small errors in the analogue plant model is not apparent from the Bode plots of the discrete sensitivities.

6.6 Summary

This chapter has developed performance limitations for SISO SD systems. The approach relies on the analysis of the SD sensitivity functions S^0 and T^0, which serve to quantify the sensitivity and robustness properties of the system taking intersample behavior into account.

It follows from the results in this chapter that SD systems do not escape the difficulty imposed upon analogue feedback design by open-loop nonminimum phase zeros, unstable poles, and time delays. This difficulty, then, persists when the controller is implemented digitally and, furthermore, independently of the type of hold function used.

In addition, a number of limitations are unique to digital controller implementations. First, there are design limitations due to potential nonminimum phase zeros of the hold function. Second, and perhaps most interesting, are the design limitations due to unstable plant poles. If the sample rate is "almost pathological" and/or is slow with respect to the time constant of the pole, then sensitivity, robustness, and response to exogenous inputs will all be poor.

Notes and References

The contents of this chapter are largely based on Freudenberg et al. (1995). In this paper, integral constraints on S^0 and T^0 of the Bode-type have also been derived. Extensions of these results to multirate SD systems have been reported in Xie & Middleton (1996), while applications to the analysis of robustness properties of zero shifting control schemes have been presented in Freudenberg, Middleton & Braslavsky (1994), from which the example in §6.5 was taken; see also Braslavsky (1995) for a detailed discussion. More on performance limitations for SD systems may be found in Araki (1993), Feuer & Goodwin (1994) and Zhang & Zhang (1994).

Fundamental sensitivity functions were first introduced in Goodwin & Salgado (1994) to study SD systems considering full intersample behavior. A frequency domain framework embodying all harmonics has been presented in Araki et al. (1993). Recent references on the frequency response of SD systems include Yamamoto & Araki (1994), Hagiwara, Ito & Araki (1995), Rosenvasser (1995), and Yamamoto & Khargonekar (1996).

A rigorous and self-contained derivation of the key sampling formula that yields (6.7) may be found in Braslavsky, Meinsma, Middleton & Freudenberg (1995).

For further reading on SD systems see the recent books by Chen & Francis (1995) and Feuer & Goodwin (1996).

Part III

Limitations in Linear Filtering

7
General Concepts

This chapter sets up the general framework for the discussion in Part III regarding sensitivity limitations in filter design. In particular, two main concepts are introduced here: *filtering sensitivity functions* and *bounded error estimators*. These concepts are the starting point of a theory of design limitations for a broad class of linear filtering problems, as we will see in the following chapters.

7.1 General Filtering Problem

This section presents the filtering problem that we will be primarily concerned with the next few chapters.

Consider the general filtering configuration of Figure 7.1, where the signals are as follows,

v : process input, in $\mathbb{R}^m$,	y : measured output, in $\mathbb{R}^\ell$,
w : measurement noise, in $\mathbb{R}^\ell$,	$\hat{z}$: estimate, in $\mathbb{R}^n$,
z : signal to be estimated, in $\mathbb{R}^n$,	$\tilde{z}$: estimation error, in $\mathbb{R}^n$.

The plant, G, is a LTI system given by the following state-space representation (cf. the notation in (2.4)):

$$G \overset{s}{=} \left[\begin{array}{c|cc} A & B & 0 \\ \hline C_1 & D_1 & 0 \\ C_2 & D_2 & I \end{array}\right], \tag{7.1}$$

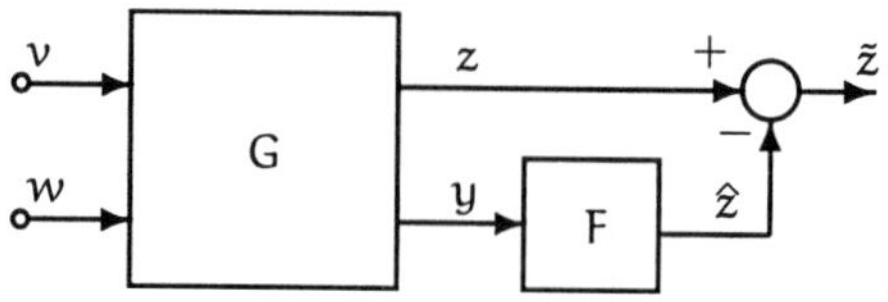

FIGURE 7.1. General filtering configuration.

where the pair (A, C_2) is assumed to be detectable. For simplicity, we will also assume that the pair (A, B) is stabilizable, and indicate the main differences with the nonstabilizable case when relevant.[1] The process input v and the measurement noise w are nonmeasured external inputs to the plant.

The filter is represented by a proper transfer function F. The general problem of filtering then focuses on the design of a suitable F such that the target signal z can be estimated from the noisy measurements y. In optimal filtering, for example, F is selected to minimize a certain performance criteria, typically involving the "smallness" of the estimation error in some sense based on particular assumptions about v and w.

Sometimes the process has, in addition to v, a control or measured input u. In this case the filter will generally use both measured input and output in the construction of the estimate. An important class of filters that use both u and y in the estimate is the class of unbiased filters, which will be briefly considered in §7.3.1. The following example illustrates the construction of a filter when a measured input is available.

Example 7.1.1. Suppose that the plant G in Figure 7.1 has an additional measured input u, i.e., G is given by the state-space description

$$\begin{aligned} \dot{x} &= Ax + B_u u + Bv, \qquad x(0) = x_0, \\ z &= C_1 x + D_1 v, \\ y &= C_2 x + D_2 v + w. \end{aligned} \tag{7.2}$$

One way to estimate the signal z using both u and y is to build a full-state observer for the state x. Then, the filter F obtained from this observer will be given by a realization of the following general form:

$$\begin{aligned} \dot{\xi} &= \hat{A}\xi + K_u u + K_y y, \qquad \xi(0) = \xi_0, \\ \hat{x} &= \hat{C}\xi, \\ \hat{z} &= C_1 \hat{x}, \end{aligned} \tag{7.3}$$

where ξ is the state of the filter and $\hat{x}$ is an estimate of the plant's state x. Different choices of the matrices in (7.3) will lead to filters having different

[1] See Seron (1995) for the complete analysis of this latter case.

characteristics, such as boundedness of the estimation error, unbiasedness, etc. More on this in §7.3. ∘

For convenience, we partition the plant as shown in Figure 7.2, with the obvious definitions for the transfer functions G_z and G_y appearing there. Note that, by the assumption of detectability from y and stabilizability from v, the function G_y contains all the unstable modes of G, some or all of which may not appear in G_z.

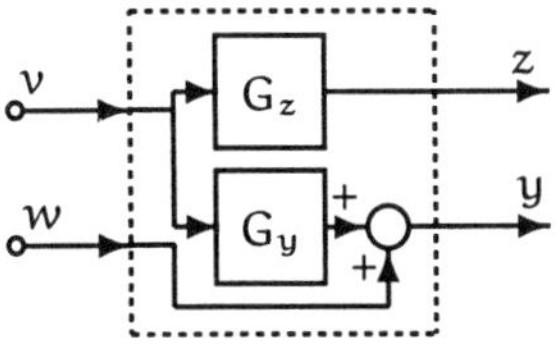

FIGURE 7.2. Structure of the plant.

The following standing assumptions about the transfer functions F and G_z are needed in the sequel.

Assumption 7.1.

(i) F is stable.

(ii) G_z is right invertible.

∘

Condition (i) in Assumption 7.1 is clear, and the least requirement that we can ask for the filter; namely, that it should be a stable operation on the available signal, y. Condition (ii) is necessary for our definition of filtering sensitivity functions, in the section coming. In relation to this condition, recall that $G_z(s)$ is right invertible if it has full row rank at almost all $s \in \mathbb{C}$, which requires that there be at least as many process inputs as signals to be estimated, i.e., $m \geq n$.

For future use, we introduce a notation for generic input-output operators in the setting of Figure 7.1. If a and b are two generic signals (i.e., either of v, w, z, etc.), we denote by the symbol H_{ba} the direct – i.e., generally physical and open-loop – mapping from signal a to signal b. For example, H_{zv} denotes the map from process input v to signal z, i.e., the operator given by the function G_z.

7.2 Sensitivity Functions

The principal focus in the filtering literature has been on various optimization procedures. Until recently, by far the greatest emphasis was on quadratic performance criteria. These criteria measure performance in terms of variance of the estimation error (Anderson &

Moore 1979, Kailath 1974). However, in later years, attention has focused on other optimization criteria. For example, the recently introduced robust design methods employ the infinity norm (Yaesh & Shaked 1989, Nagpal & Kharghonekar 1991, Bernstein & Haddad 1989, de Souza, Shaked & Fu 1995). Whilst optimization methods play an invaluable role in determining the best solution under a given set of conditions, they fail to pin-point why a desired performance level is not achievable. They also do not allow one to evaluate the benefits of changing the measurement system, for example, by relocating a sensor or by adding additional sensors.

One way to address these issues is to identify complementary operators — such as S and T in the control case — that represent relevant properties of the filtering system, and then use integral constraints to quantify the inherent sensitivity trade-offs associated with filtering problems. This is the approach taken in Goodwin, Mayne & Shim (1995) and Seron & Goodwin (1995), and is the one that we will develop in this chapter.

Consider the operators $H_{\tilde{z}v}$, $H_{\hat{z}v}$, and H_{zv} mapping process input to estimation error, estimate, and estimated signal, respectively, in the diagram of Figure 7.1. Then, we have the following definition.

Definition 7.2.1 (Filtering Sensitivity Functions). Let the conditions in Assumption 7.1 hold. Then, the *filtering sensitivity* and *complementary sensitivity* functions, denoted by P and M, respectively, are defined as

$$P(s) \triangleq H_{\tilde{z}v}(s)H_{zv}^{-1}(s), \qquad \text{and} \qquad M(s) \triangleq H_{\hat{z}v}(s)H_{zv}^{-1}(s). \tag{7.4}$$

○

The filtering sensitivity P represents the relative effect of the process input on the estimation error, while the filtering complementary sensitivity M represents the relative effect of the process input on the estimate. In terms of the transfer functions defined in Figure 7.2, we can express P and M as follows:

$$P = (G_z - FG_y)G_z^{-1}, \qquad \text{and} \qquad M = FG_yG_z^{-1}. \tag{7.5}$$

The following complementarity constraint holds for the filtering sensitivity and complementary sensitivity functions.

Theorem 7.2.1. The filtering sensitivity and complementary sensitivity defined in (7.4) satisfy the relation

$$P(s) + M(s) = I, \tag{7.6}$$

at any complex frequency s that is not a pole of P and M.

Proof. It follows from Figure 7.1 that

$$H_{\tilde{z}v} + FH_{yv} = H_{zv}. \tag{7.7}$$

Then (7.6) follows immediately using the definitions (7.4) and noting that $H_{\hat{z}v} = FH_{yv}$. □

Equation (7.6) is reminiscent of the complementary result of feedback control, i.e., (3.7), that the sum of sensitivity, S, and complementary sensitivity, T, is the identity operator. However, P and M are not the direct 'filtering' versions of S and T; this will be discussed in more detail in §7.2.2. The following subsection studies the use of the filtering sensitivities in measuring important properties related to the filtering problem.

7.2.1 *Interpretation of the Sensitivities*

Consider, for simplicity, that P and M in (7.4) are scalar, and assume further that the ratio of system transfer functions H_{yv}/H_{zv} is proper.

From the mere definition of P in (7.4), it is evident that $|P(j\omega)|$ measures the magnitude of the frequency response of the estimation error relative to the magnitude of the frequency response of the signal to be estimated. This suggests a criterion to specify an appropriate shape for $|P(j\omega)|$ in design. Indeed, suppose for example that the frequency distribution of the process input v is known. Then, achieving $|P(j\omega)|$ small at those frequencies where v has a significant magnitude will be desirable, since it will reduce the relative effect of v on the estimation error.

On the other hand, the filtering complementary sensitivity M represents the ratio between the transfer function from the process input to the estimated signal and the transfer function from the process input to the actual signal. Thus, it is desirable to achieve $|M(j\omega)|$ close to one over the frequency range where the spectrum of v is concentrated. This will result in an impact (in magnitude) of the process input on the estimate similar to the impact that it has on the signal to be estimated.

Typically, the spectrum of v is concentrated at low frequencies, so appropriated shapes for P and M in this case will be as shown in Figure 7.3.

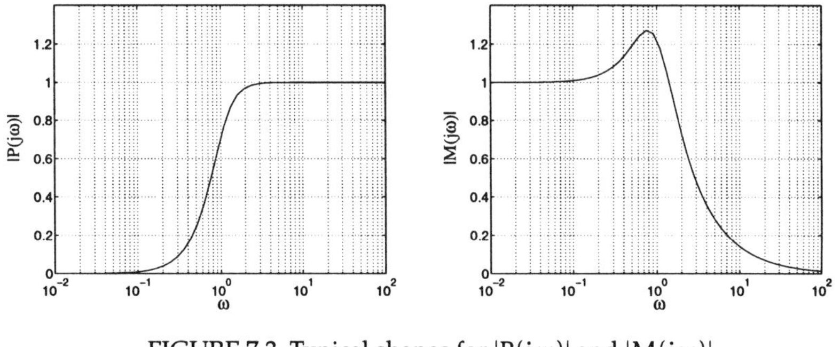

FIGURE 7.3. Typical shapes for $|P(j\omega)|$ and $|M(j\omega)|$.

In addition, note from (7.5) that M it is the product of a proper transfer function and the filter transfer function F. Clearly, the effect of the output disturbance, w, on the estimation error (and on the estimate itself) will be reduced if $|F(j\omega)|$ is small over the range of frequencies where w is strong. This desired reduction of $|F|$ can be readily translated into the shape of $|M(j\omega)|$, after normalizing through the factor $|H_{yv}(j\omega)/H_{zv}(j\omega)|$. Note that, since F is a strictly proper transfer function, its output disturbance rejection properties are typically aimed at high-frequency disturbances, such as measurement noise.

The foregoing discussion analyzed the sensitivities as relative gains, and gave some insights into desirable shapes for their frequency responses. However, dynamic interpretations of the filtering sensitivities can also be obtained for particular cases.

Indeed, suppose that we express the filter as driven by the *innovations process* (Kailath 1968), denoted ι, and defined as

$$\iota \triangleq y - \hat{z}, \tag{7.8}$$

where $\hat{z}$ is, in this case, an estimate of the noise-free output, i.e., $z = y - w$ (notice that here $H_{zv} = H_{yv}$). Then it is possible to express the filtering system as a feedback loop, as shown in Figure 7.4.

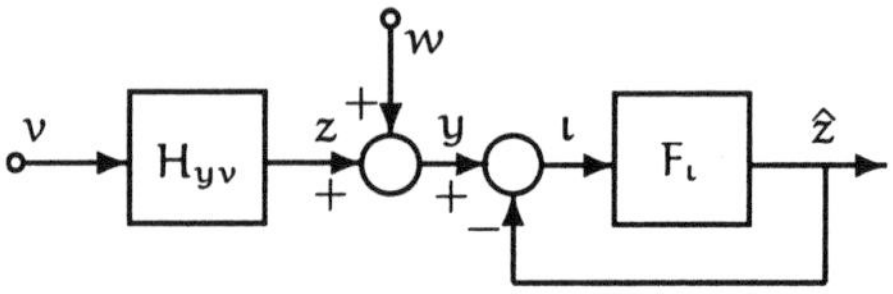

FIGURE 7.4. Filtering Feedback Loop.

Loosely speaking, then, the sensitivities will represent dynamic responses to system disturbances in those filtering problems where the estimate, $\hat{z}$, is fed-back into this feedback loop.

The following examples analyze two particular cases where the filtering sensitivities have dynamic meaning.

Example 7.2.1 ('Output-Filtering' Sensitivities). Assume that the variable to be estimated is the disturbance-free output, i.e., z in Figure 7.1 is given by

$$z = y - w.$$

In this case, we have that $G_y = G_z$ in Figure 7.2, and hence the sensitivities (7.5) become

$$\begin{aligned} P^y &= 1 - F, \\ M^y &= F. \end{aligned} \tag{7.9}$$

We call these functions the *output-filtering* sensitivities. It is instructive to express these sensitivities in terms of the filter function F_ι, shown in Figure 7.4, which maps innovations into estimate. We have,

$$P^y = (1 + F_\iota)^{-1}, \qquad M^y = F_\iota(1 + F_\iota)^{-1}, \tag{7.10}$$

which are similar to the control sensitivities S and T. It is then evident that P^y is the dynamic transfer between the measurement noise, w, and the innovations, ι, whilst M^y is the dynamic transfer between w and the estimated variable, $\hat{z}$. ○

Example 7.2.2. Alternatively assume that the transfer function H_{yv} can be expressed as

$$H_{yv} = H_{yz}H_{zv},$$

such that $z = H_{zv}v$. Assume further that H_{yz} is proper. Then, it is natural to factor the filter function F_ι in Figure 7.4 as

$$F_\iota = F_y F_z,$$

for appropriate transfer functions F_y and F_z, and take $\hat{z} = F_z \iota$ as an estimate for z. The corresponding filtering loop is shown in Figure 7.5.

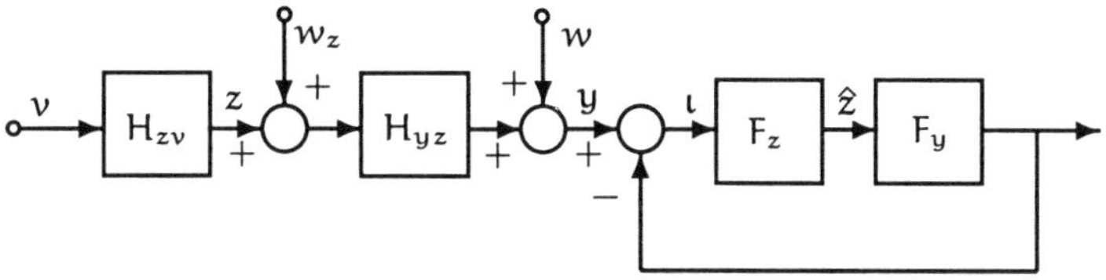

FIGURE 7.5. Particular Filtering Loop.

It is easy to see that, in this case, the filtering sensitivities are given by

$$P = [1 + F_z(F_y - H_{yz})]P^y, \qquad M = F_z P^y H_{yz}, \tag{7.11}$$

where P^y is the output-filtering sensitivity given by (7.10). The sensitivities in (7.11) are the responses of ι and $\hat{z}$ to the internal disturbance w_z, respectively. Note that they collapse into P^y and M^y if $F_y = H_{yz}$. The latter condition would be a natural choice for F_y. ○

7.2.2 *Filtering and Control Complementarity*

The filtering sensitivities P and M are not the direct counterpart of the control sensitivities S and T. This is because the estimation problem depicted in Figure 7.1, when set in a control framework, is more general than the one-degree of freedom control loop where S and T are defined. Indeed, the "filtering" versions of S and T would correspond to the output-filtering sensitivities, i.e., the sensitivities arising in the problem of estimating the system noise-free output ($z = y - w$), as discussed in Example 7.2.1.

We will next derive a complementarity constraint for linear feedback control that parallels the filtering constraint given in (7.6). Consider the control loop shown in Figure 7.6, where G is the plant transfer function given by

$$G \triangleq \begin{bmatrix} G_{zv} & G_{zu} \\ G_{yv} & G_{yu} \end{bmatrix}, \tag{7.12}$$

and K is a controller that stabilizes the feedback loop.

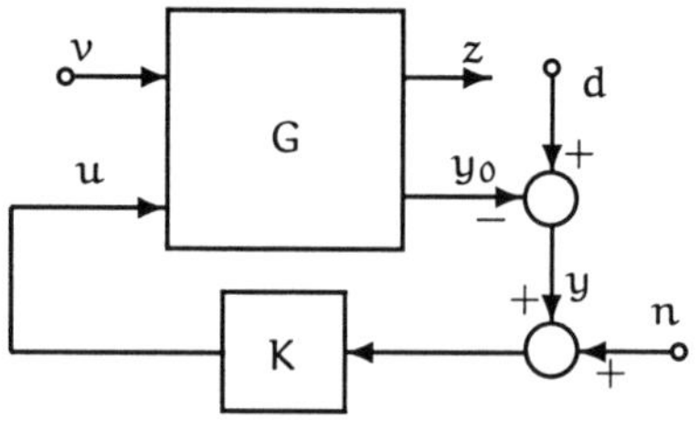

FIGURE 7.6. Control Loop.

This configuration is standard, e.g., in the $\mathcal{H}_\infty$ control literature.[2] The plant G has two sets of inputs: the external inputs, v, and the control inputs, u. Also, it has a set outputs, y_0, that are available to the controller K after being corrupted with output disturbances, d, and sensor noise, n, and a set of signals of interest, z, which are generally the signals to be controlled or regulated.

The "classical" sensitivity and complementary sensitivity, S and T, are defined for a simpler unity feedback control loop (see Figure 3.2 in Chapter 3), where they are directly connected with performance, disturbance rejection and robustness properties (cf. §2.2.2 in Chapter 2). However, as we see next, they can also be defined for the feedback loop in Figure 7.6.

Introduce first the notation $\mathbf{H}_{ba}$ to indicate the total mapping – possibly nonphysical – from signal a to signal b, i.e., after combining (adding, composing, etc.) all the ways in which a is a function of b and solving any feedback loops that may exist.[3] Then, we have that

$$\begin{aligned} S &= \mathbf{H}_{yd} = (I + G_{yu}K)^{-1}, \\ T &= -\mathbf{H}_{yn} = G_{yu}K(I + G_{yu}K)^{-1}. \end{aligned} \tag{7.13}$$

[2]In the standard configuration, in fact, d and n are absorbed into v. Here we choose to draw them explicitly to emphasize the way in which they affect the output y.

[3]Cf. the notation H_{ba} introduced at the end of §7.1.

As seen in Chapter 3, S and T satisfy the complementarity constraint

$$S + T = I.$$

This relation involves mappings from external inputs that are directly injected into the loop (i.e., without intermediate dynamics), to system signals that are fed-back in the same loop (see Figure 7.6). A more general result, comparable to the filtering relation given in (7.7), can be developed for the configuration of Figure 7.6. This structural constraint involves mappings from external inputs that are injected dynamically in the loop, to some internal variables, e.g., the system (combination of) states, z, which are not directly fed-back. This result is stated below.

Theorem 7.2.2. Consider the general control loop of Figure 7.6, where the plant G is given by (7.12). The total mappings, $\mathbf{H}_{zv}$ and $\mathbf{H}_{zd}$, from the external inputs v and d to the internal variables z, satisfy the following structural relation:

$$\mathbf{H}_{zv} + (-\mathbf{H}_{zd})G_{yv} = G_{zv}. \tag{7.14}$$

Proof. From Figure 7.6, and the definitions in (7.12) and (7.13), we have, in the Laplace transform domain

$$Y_0 = SG_{yv}V + TD,$$

and

$$\begin{aligned} Z &= G_{zv}V + G_{zu}K(D - Y_0) \\ &= (G_{zv} - G_{zu}KSG_{yv})V + G_{zu}KSD. \end{aligned} \tag{7.15}$$

Then

$$\begin{aligned} \mathbf{H}_{zv} &= G_{zv} - G_{zu}KSG_{yv}, \\ \mathbf{H}_{zd} &= G_{zu}KS, \end{aligned}$$

from which (7.14) follows. □

Notice that, when considering mappings to z, the sensor noise, n, in Figure 7.6 can be set to zero since its influence on z is the same as that of d.

A complementarity constraint that parallels (7.6) can be readily obtained by multiplying through both sides of (7.14) by G_{zv}^{-1} and making similar definitions to those in (7.4).

The relation given in (7.14) is clearly analogous to that of (7.7). This indicates that the filtering problem considered in §7.2 fits into the structure of the control configuration of Figure 7.6. In the search for generality, however, we have gone too far: we now need to select the particular choices of G and K that will give the exact control counterpart of (7.7).

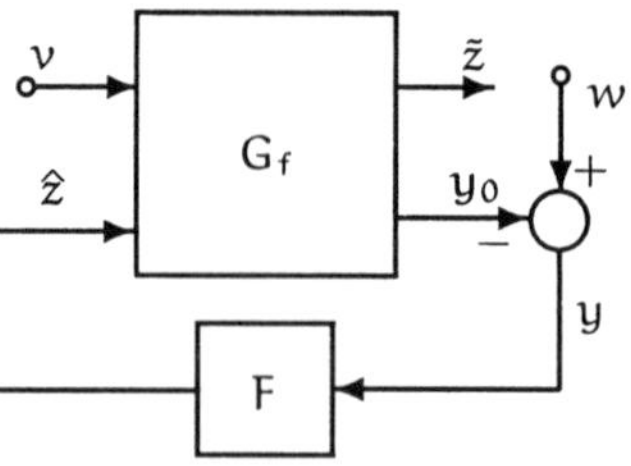

FIGURE 7.7. Equivalent Filtering Loop.

Making the choices of inputs and outputs as shown in Figure 7.7, we conclude that the controller K corresponds to the filter F, whilst the plant G corresponds to the matrix

$$G_f \triangleq \begin{bmatrix} G_{zv} & -I \\ G_{yv} & 0 \end{bmatrix}.$$

Observe that setting the entry (2,2) of G_f equal to zero amounts to opening the loop ($S = I$, $T = 0$ in (7.13)), but this is the way in which we have set up the filtering problem in §7.2. In the context of bounded error filtering, however, we will soon see that F satisfies conditions that ensure boundedness of $\tilde{z}$. There is then, ideally, no need for feedback if these conditions hold.

Although a feedback loop can be made explicit by considering the filter driven by the innovations, as we have seen in §7.2.1, note that the mere presence of feedback does not make the observer a closed loop in general (i.e., when the signal to be estimated is not just the disturbance-free output, i.e., $z \neq y - w$). Indeed, it was shown by Bhattacharyya (1976) that an observer is a closed-loop system if and only if it can be expressed as a system *driven by the estimation error* $\tilde{z} = z - \hat{z}$. Moreover, Bhattacharyya (1976) proved that an observer must be a closed-loop system if it is to provide observer action despite arbitrarily small perturbations of the system state space matrices.

7.3 Bounded Error Estimators

It is natural to require that a state estimator, in addition to being stable, should produce a bounded estimation error when all the signals entering the system are bounded. Thus, from the general class of stable estimators, we will make a mild restriction to *bounded error estimators* (BEEs) according to the following definition.

Definition 7.3.1 (BEE). We say that a stable filter is a BEE if, for all finite initial states of the plant and the filter, the estimation error $\tilde{z}$ is bounded whenever the system inputs v and w are bounded. ○

Let x_0 be the initial state of the plant, and let V and W be the Laplace transforms of the process input and measurement noise, respectively. It then follows from Figures 7.1 and 7.2, and (7.1) that — modulo exponentially decaying terms due to the filter initial state — the Laplace transform of the estimation error, denoted by $\tilde{Z}$, is given by

$$\tilde{Z} = \tilde{G}_0 x_0 + (G_z - FG_y)V - FW, \tag{7.16}$$

where

$$\tilde{G}_0(s) \triangleq [C_1 - F(s)C_2](sI - A)^{-1} \tag{7.17}$$

leads the estimation error component due to the plant initial state.

Thus, since the filter F is stable and the plant G — i.e., the pair (A, B) — is stabilizable, a necessary and sufficient condition for F to be BEE is that the transfer function $G_z - FG_y$ be stable. Note that this means that

- unstable modes shared by G_z and G_y must be canceled[4] by a nonminimum phase zero of the difference $G_z - FG_y$, while
- unstable poles of G_y that do not appear in G_z must be canceled by nonminimum phase zeros of F.

If (A, B) is not stabilizable then a necessary and sufficient condition for F to be BEE is that the transfer function $\tilde{G}_0$ in (7.17) be stable. In Chapters 10 and 11, we will consider predictors and smoothers based on BEEs that perform full-state filtering. For future reference then, we note that, if F is intended to estimate the full state, F is a BEE if and only if the transfer function

$$\tilde{G}_0(s) = [I - F(s)C_2](sI - A)^{-1} \tag{7.18}$$

is stable.

The BEE assumption is the minimum that can be required from a filter and, therefore, it underlies numerous state estimation problems in the literature. The class of BEEs includes, for example, observers of the form (7.3) that assign pre-specified stable dynamics to the estimation error. Indeed, assume that the external inputs to the plant in (7.2) are zero and write the corresponding dynamics of the state estimation error, $\tilde{x} \triangleq x - \hat{C}\xi$, as

$$\dot{\tilde{x}} = (A - \hat{C}K_y C_2)x - \hat{C}\hat{A}\xi .$$

Assume next that we desire that $\tilde{x}$ evolve as

$$\dot{\tilde{x}} = \tilde{A}\tilde{x}, \tag{7.19}$$

[4]In frequency and direction in the multivariable case.

where $\tilde{A}$ is a stability matrix of the same dimensions as A. This will occur if and only if the matrices $\hat{A}$, K_y and $\hat{C}$ satisfy

$$A - \hat{C}K_yC_2 = \tilde{A} \qquad \text{and} \qquad \hat{C}\hat{A} = \tilde{A}\hat{C}. \tag{7.20}$$

It is clear that a filter satisfying (7.19) and (7.20) is a BEE.

BEEs do not, in general, assign the error dynamics, since the estimation error component due to the initial state of the observer may decay with different dynamics than the component due to the plant initial state. To further see the difference, note that the estimation error corresponding to a general BEE for full-state estimation satisfies the following equations

$$\begin{aligned} \begin{bmatrix} \dot{x} \\ \dot{\xi} \end{bmatrix} &= \begin{bmatrix} A & 0 \\ K_yC_2 & \hat{A} \end{bmatrix} \begin{bmatrix} x \\ \xi \end{bmatrix}, \\ \tilde{x} &= \begin{bmatrix} I & -\hat{C} \end{bmatrix} \begin{bmatrix} x \\ \xi \end{bmatrix}, \end{aligned} \tag{7.21}$$

while the estimation error corresponding to an observer with pre-specified error dynamics $\tilde{A}$ satisfies (7.19). We can then argue that the concept of bounded error estimation corresponds to making the unstable modes of the plant unobservable from the estimation error, whilst the subclass that achieves pre-specified error dynamics actually assigns the eigenvalues of the estimation error. The following examples illustrate these points.

Example 7.3.1. We want to estimate the state of the scalar plant $\dot{x} = x, \ y = x$. A filter F_1, of the form (7.3), given by

$$\begin{aligned} \dot{\xi} &= \begin{bmatrix} 0 & 1 \\ -1 & -2 \end{bmatrix} \xi + \begin{bmatrix} -2 \\ 2 \end{bmatrix} y, \\ \hat{x} &= \begin{bmatrix} 0 & 1 \end{bmatrix} \xi, \end{aligned}$$

is a BEE for this plant, since $\tilde{G}_0$ in (7.16) becomes, using the above data,

$$\begin{aligned} \tilde{G}_0(s) &= \frac{(s^2 - 1)}{(s^2 + 2s + 1)} \frac{1}{(s - 1)} \\ &= \frac{1}{s + 1}, \end{aligned}$$

which is stable. However, this filter does not assign error dynamics, since (7.20) is not satisfied, i.e.,

$$\hat{C}\hat{A} - A\hat{C} = [-1 \quad -3] \neq -\hat{C}K_yC_2\hat{C} = [0 \quad 2].$$

On the other hand, the filter F_2 given by

$$\begin{aligned} \dot{\xi} &= \begin{bmatrix} 0 & 1 \\ -1 & -2 \end{bmatrix} \xi + \begin{bmatrix} 0 \\ 2 \end{bmatrix} y, \\ \hat{x} &= \begin{bmatrix} 1 & 1 \end{bmatrix} \xi. \end{aligned}$$

is also a BEE and, moreover, achieves an estimation error with dynamics $\dot{\tilde{x}} = -\tilde{x}$, and thus satisfies (7.20) with $\tilde{A} = -1$. Figure 7.8 shows the time response of both filters to an initial state error.

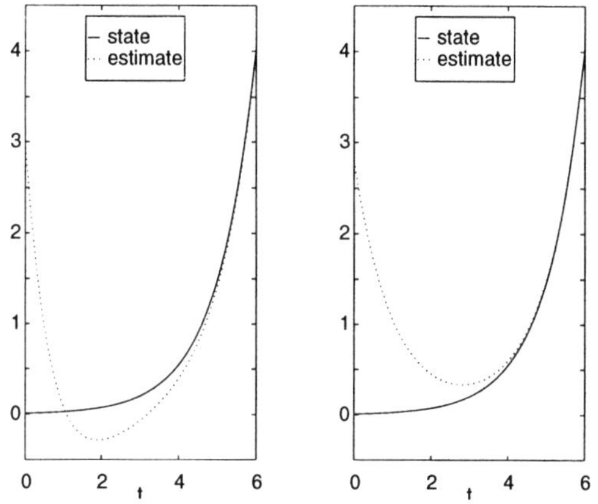

FIGURE 7.8. Estimates of $\dot{x} = x$: general BEE (left) and observer that assigns the error dynamics (right).

$\circ$

Example 7.3.2. Consider the special case of an error driven stable observer for the system (7.2), having the same order as the system. Such an observer may be of the form (7.3) where K_y is such that $\hat{A} = A - K_y C_2$ is a stability matrix and $\hat{C} = I$, i.e.,

$$\dot{\hat{x}} = (A - K_y C_2)\hat{x} + K_u u + K_y y. \tag{7.22}$$

The observer (7.22) is easily seen to be a BEE for full-state estimate since

$$\begin{aligned} \tilde{G}_0 &= [I - (sI - A + K_y C_2)^{-1} K_y C_2](sI - A)^{-1} \\ &= [I + (sI - A)^{-1} K_y C_2]^{-1} (sI - A)^{-1} \\ &= (sI - A + K_y C_2)^{-1}, \end{aligned} \tag{7.23}$$

which is stable. Moreover, the error dynamics are given by $\tilde{A} = \hat{A}$. $\circ$

General BEEs do not specify the way in which the filter is to treat the measured input, if available. Thus, they may be biased, i.e., the estimation error has a component driven by the measured input. On the other hand, unbiased filters are designed to process the measured input in the same way as the plant does, as seen in the following subsection.

7.3.1 *Unbiased Estimators*

In this section we discuss a special subclass of BEEs, namely, that of *unbiased estimators*. The distinguishing feature of this type of BEE is that they achieve a total decoupling of the estimation error from the measured input. The formal definition of this subclass is given below (Middleton & Goodwin 1990).

Definition 7.3.2 (Unbiased Estimator). We say that a stable filter for the plant (7.2) is unbiased if, whenever $v \equiv 0$ and $w \equiv 0$, for all measured inputs u, and for all finite initial states of the plant and the filter, the estimation error $\tilde{z}$ decays asymptotically to zero. ∘

We then have the following result.

Lemma 7.3.1. A stable state estimator for the system (7.2) is unbiased if and only if

(i) it is a BEE and

(ii) the transfer function from the measured input u to the estimation error, $H_{\tilde{x}u}$, is identically zero.

Proof. Immediate from Definition 7.3.2 on writing the expression for the estimation error. □

Example 7.3.3. Consider again Example 7.3.2. Note that the special observer (7.22) is an unbiased estimator if the particular choice $K_u = B_u$ is made. This can be seen on noting that

$$H_{\tilde{x}u}(s) = (sI - A + K_y C_2)^{-1} B_u - (sI - \hat{A})^{-1} K_u \equiv 0$$

using (7.23). This particular observer is generally known as an *identity observer*. ∘

As the above example suggests, unbiased estimators of the form (7.3) for (7.2) are observers that assign the error dynamics which, in addition to (7.20), satisfy the constraint

$$\hat{C} K_u = B_u. \tag{7.24}$$

A nice feature of unbiased filters is that *separation* applies, that is, if used to control the plant, then the resulting closed loop is automatically stable provided the state estimates are fed-back with a gain such that, if the same gain were to be used on the true states, then stability would result. This is why the unbiasedness requirement is frequently presupposed in the definitions of observers appearing in the control literature. However, we will show in subsequent chapters that there are inherent trade-offs in filter design that affect all BEEs, whether unbiased or not.

7.4 Summary

This chapter has introduced the notation and definitions necessary for the development of the remaining chapters in this part. The main concepts here are the filtering sensitivity functions and the class of BEEs. The filtering sensitivity functions play a role similar to their control counterparts; we have discussed in this chapter some important interpretations in the context of linear filtering design, and their connections to a general control paradigm.

The filtering sensitivities are the basis for the theory of design limitations developed in the next chapters. This theory of limitations holds for a wide class of linear filtering problems, namely, those based on BEEs, which are estimators that produce a bounded estimation error under any possible input driving the system.

Notes and References

The definition of filtering sensitivity functions is taken from Seron & Goodwin (1995). Goodwin et al. (1995) introduced similar functions for the case of unbiased estimators. BEEs were also introduced in Seron & Goodwin (1995); see Seron (1995) for a more detailed discussion.

In view of its generality, the class of linear BEEs encompasses various specific state estimators that have been proposed in contemporary literature. For example, robust $\mathcal{H}_\infty$ optimization designs, in general lead to BEEs that are not necessarily unbiased (see e.g., de Souza et al. 1995). The bias results from the fact that, in the presence of plant parameter uncertainty, the component of the estimation error due to the known input signal u cannot, in general, be completely canceled. In fact, in de Souza et al. (1995) the authors have shown, through comparison with Kalman filter and nominal $\mathcal{H}_\infty$ filter design, that the biased component of the estimation error may be minimized by allowing the filter to be biased from the outset.

For further reading on filters and observers see e.g., the textbooks Anderson & Moore (1979) and O'Reilly (1983).

8
SISO Filtering

In this chapter we examine the fundamental design trade-offs that apply to linear scalar filtering problems based on bounded error estimators. We will see that, due to the condition of bounded error estimation, the filtering sensitivity functions introduced in the last chapter are necessarily constrained at points in the complex plane determined by ORHP poles and zeros of the plant. These interpolation constraints, in turn, translate into Poisson and Bode-type integral relations, which show essential limitations in the achievable performance, and induce clear trade-offs in filter design.

8.1 Interpolation Constraints

We have seen in Chapter 3 that, for the case of feedback control and under minimality assumption, S has zeros at the unstable open-loop plant poles and T has zeros at the nonminimum phase open-loop plant zeros. In this section we consider the filtering sensitivities resulting from *any scalar BEE*, and show that right half plane poles and zeros of the plant transfer matrices G_y and G_z impose constraints on particular complex values of P and M given in (7.5) of Chapter 7.

First, it is interesting to observe that P and M in (7.5) are not necessarily analytic in the ORHP since G_z may have ORHP zeros (that are not canceled in the division). This fact differs from the common assumption in control theory that the sensitivity and complementary sensitivity are analytic in the ORHP. Note, however, that the more general control sensitivities that can be defined from (7.14) in Chapter 7 are potentially nonanalytic in the ORHP if H_{zv} is nonminimum phase.

For the problem of filtering with BEEs, we have the following result on the analyticity of P and M in the ORHP.

Lemma 8.1.1. Consider the filtering sensitivities given in (7.5) and assume that F is a BEE. Then P and M are analytic in the ORHP if and only if either of the following conditions hold.

(i) G_z is minimum phase.

(ii) Every ORHP zero of G_z is also a zero of the product FG_y.

Proof. Immediate from (7.5). □

The following lemma formalizes the interpolation constraints that P and M must satisfy at the ORHP poles and zeros of G_y and G_z.

Lemma 8.1.2 (Interpolation Constraints). Assume that F in (7.5) is a BEE. Then P and M must satisfy the following conditions.

(i) If $p \in \mathbb{C}^+$ is a pole of G_z, then

$$P(p) = 0, \qquad \text{and} \qquad M(p) = 1.$$

(ii) If $q \in \mathbb{C}^+$ is a zero of G_y that is not also zero of G_z, then

$$P(q) = 1, \qquad \text{and} \qquad M(q) = 0.$$

(iii) If $q \in \mathbb{C}^+$ is a zero of G_z that is not also zero of FG_y, then $P(s)$ and $M(s)$ have a pole at $s = q$.

Proof. Case (i) readily follows from (7.5) by noting that the term $G_z - FG_y$ is stable if F is a BEE. Case (ii) is immediate from (7.5) since F is stable. Finally, case (iii) follows from Lemma 8.1.1. □

We introduce for convenience the following notation for the ORHP zeros of P and M:[1]

$$\begin{aligned} \mathcal{Z}_P &\triangleq \{s \in \mathbb{C}^+ : P(s) = 0\}, \\ \mathcal{Z}_M &\triangleq \{s \in \mathbb{C}^+ : M(s) = 0\}. \end{aligned} \tag{8.1}$$

Then, Lemma 8.1.2 establishes that $\mathcal{Z}_P$ includes the set of ORHP poles of G_z and $\mathcal{Z}_M$ contains ORHP zeros of G_y. As in the control case, we have then translated the plant characteristics of instability and "nonminimum

[1]Recall that, when defining a set of zeros of a transfer function, the zeros are repeated according to their multiplicities.

phaseness" into properties that the functions P and M must satisfy in the ORHP. Note that $\mathcal{Z}_P$ also contains those ORHP zeros of $(G_z - FG_y)$ that are not also zeros of G_z and $\mathcal{Z}_M$ also contains those ORHP zeros of F that are not also zeros of the (possibly nonproper) transfer function G_z/G_y.

Notice that since the plant is detectable from y, the transfer function G_y has all the unstable poles of G_z. In consequence, G_y/G_z — and hence M — has no zeros at any unstable pole of G.

The interpolation constraints given in Lemma 8.1.2 will be translated into constraints on the frequency response of the sensitivities in the following section.

8.2 Integral Constraints

In this section we use the interpolation constraints of Lemma 8.1.2 to derive Poisson and Bode-type integral relations satisfied by P and M. These relations constrain the values of the filtering sensitivities on the imaginary axis. The Poisson integral constraints on P and M are derived using the Poisson integral formulae for the recovery of a function, analytic in the ORHP, from its values on the imaginary axis (see §A.6.1 in Appendix A).

Let the sets (8.1) be given by

$$\mathcal{Z}_P = \{p_i : i = 1, \dots, n_p\} ,$$
$$\mathcal{Z}_M = \{q_i : i = 1, \dots, n_q\} ,$$

and let the corresponding Blaschke products be[2]

$$B_P(s) = \prod_{i=1}^{n_p} \frac{s - p_i}{s + \overline{p}_i} , \qquad \text{and} \qquad B_M(s) = \prod_{i=1}^{n_q} \frac{s - q_i}{s + \overline{q}_i} . \tag{8.2}$$

We also introduce the following sets of zeros.

$$\begin{aligned} \mathcal{Z}_z &\triangleq \{s \in \mathbb{C}^+ : G_z(s) = 0 \quad \text{and} \quad F(s)G_y(s) \neq 0\} , \\ \mathcal{Z}_y &\triangleq \{s \in \mathbb{C}^+ : G_y(s) = 0 \quad \text{and} \quad G_z(s) \neq 0\} , \\ \mathcal{Z}_{1/z} &\triangleq \{s \in \mathbb{C}^+ : G_z^{-1}(s) = 0\} . \end{aligned} \tag{8.3}$$

Note that the set of zeros of P includes $\mathcal{Z}_{1/z}$, i.e.,

$$\mathcal{Z}_P \supset \mathcal{Z}_{1/z} , \tag{8.4}$$

the set of zeros of M includes $\mathcal{Z}_y$, i.e.,

$$\mathcal{Z}_M \supset \mathcal{Z}_y , \tag{8.5}$$

---

[2]Recall that, if the set $\mathcal{Z}$ originating the Blaschke product B is empty, then we define $B(s) = 1, \forall s \in \mathbb{C}$.

and $\mathcal{Z}_z$ is the set of ORHP poles of P and M. Denote the Blaschke products corresponding to the sets in (8.3) by B_z, B_y and $B_{1/z}$. It is then easy to see that we can factor P and M as

$$\begin{aligned} P(s) &= \tilde{P}(s)B_P(s)/B_z(s) \,, \\ M(s) &= \tilde{M}(s)B_M(s)/B_z(s) \,, \end{aligned} \tag{8.6}$$

where $\tilde{P}$ and $\tilde{M}$ have no zeros or poles in the ORHP. We then obtain the following integral constraints.

Theorem 8.2.1 (Poisson Integral for P). Assume that F in (7.5) is a BEE. Let $q = \sigma_q + j\omega_q$, $\sigma_q > 0$, be a zero of G_y that is not also zero of G_z (i.e., $q \in \mathcal{Z}_y$ in (8.3)). Then, assuming that P is proper,

$$\int_{-\infty}^{\infty} \log|P(j\omega)| \frac{\sigma_q}{\sigma_q^2 + (\omega_q - \omega)^2}\, d\omega = \pi \log\left|B_P^{-1}(q)\right| - \pi \log\left|B_z^{-1}(q)\right| . \tag{8.7}$$

Proof. The proof follows by using factorization (8.6) and applying the Poisson integral formula (Theorem A.6.1 in Appendix A) to $\log \tilde{P}$. □

Theorem 8.2.2 (Poisson Integral for M). Suppose that F in (7.5) is a BEE. Let $p = \sigma_p + j\omega_p$, $\sigma_p > 0$, be a pole of G_z (i.e., $p \in \mathcal{Z}_{1/z}$ in (8.3)). Then, assuming that M is proper,

$$\int_{-\infty}^{\infty} \log|M(j\omega)| \frac{\sigma_p}{\sigma_p^2 + (\omega_p - \omega)^2}\, d\omega = \pi \log\left|B_M^{-1}(p)\right| - \pi \log\left|B_z^{-1}(p)\right| . \tag{8.8}$$

Proof. Same as the proof of Theorem 8.2.1, this time using $\tilde{M}$. □

The results in Theorems 8.2.1 and 8.2.2 are affected by the particular choice of filter, since the Blaschke product B_z on the RHSs of (8.7) and (8.8) depends on F. In particular, we see that the integral constraints on P and M are *relaxed* by nonminimum phase zeros of G_z that are not shared with FG_y. An explanation for this may be given if we think of a nonminimum phase zero as a "mild time-delay".[3] Then, a delay in the signal to be estimated, with respect to the measured signals, would allow for more time to compute a better estimate — in the sense discussed in §7.2.1 of Chapter 7. An extreme case of this phenomenon can be found in the process of smoothing, where the signal to be estimated is artificially delayed. See Chapters 10 and 11 for more discussion.

[3]This intuitively follows from the fact that a time-delay, analyzed by its Padé approximations (e.g., Middleton & Goodwin 1990), may be seen as containing an arbitrarily large number of nonminimum phase zeros.

Constraints that are completely independent of the estimator parameters are readily obtained from these results under an additional condition, as we see in the following corollary.

Corollary 8.2.3. Suppose that P and M in (7.5) are proper and that F is a BEE. Consider the sets defined in (8.3), and assume further that if there exists a complex number s_0 in $\mathbb{C}^+$ such that $G_z(s_0) = 0$ then also $G_y(s_0) = 0$. Then,

(i) if $q = \sigma_q + j\omega_q \in \mathcal{Z}_y$, we have that

$$\int_{-\infty}^{\infty} \log|P(j\omega)| \frac{\sigma_q}{\sigma_q^2 + (\omega_q - \omega)^2}\, d\omega \geq \pi \log\left|B_{1/z}^{-1}(q)\right|; \tag{8.9}$$

(ii) if $p = \sigma_p + j\omega_p \in \mathcal{Z}_{1/z}$, we have that

$$\int_{-\infty}^{\infty} \log|M(j\omega)| \frac{\sigma_p}{\sigma_p^2 + (\omega_p - \omega)^2}\, d\omega \geq \pi \log\left|B_y^{-1}(p)\right|. \tag{8.10}$$

Proof. First notice that the assumption that every nonminimum phase zero of G_z is also a zero of G_y implies that the set $\mathcal{Z}_z$ is empty, so B_z disappears from the RHSs of (8.7) and (8.8). Then, since $\mathcal{Z}_P$ in (8.1) includes the set $\mathcal{Z}_{1/z}$ defined in (8.3), the Blaschke product of the ORHP zeros of P, B_P, has $B_{1/z}$ as a factor and hence $\log|B_P^{-1}(q)| \geq \log|B_{1/z}^{-1}(q)|$. Thus, inequality (8.9) follows from (8.7). In a similar way, inequality (8.10) follows from (8.8) on noting that $\mathcal{Z}_M \supset \mathcal{Z}_y$. □

From the above results it can be inferred that reduction of $|P(j\omega)|$ or $|M(j\omega)|$ in one frequency range must be compensated by an increase in another frequency range. The presence of ORHP poles and zeros in those transfer functions that link the process input v to the output y and target signal z, lead to more stringent requirements. This will be discussed in more detail in §8.3.

Bode's integral theorems of feedback control (cf. §3.1.2 in Chapter 3) also have filtering analogs, as shown next.

Theorem 8.2.4 (Bode Integral for P). Suppose that P in (7.5) is proper and that F is a BEE. Let $\mathcal{Z}_P = \{p_i : i = 1, \ldots, n_p\}$ be the set of zeros of P in the ORHP and let $\mathcal{Z}_z$ in (8.3) be given by $\mathcal{Z}_z = \{\zeta_i : i = 1, \ldots, n_\zeta\}$. Then, if $P(j\infty) \neq 0$

$$\int_0^{\infty} \log\left|\frac{P(j\omega)}{P(j\infty)}\right| d\omega = \frac{\pi}{2} \lim_{s\to\infty} \frac{s[P(s) - P(\infty)]}{P(\infty)} + \pi \sum_{i=1}^{n_p} p_i - \pi \sum_{i=1}^{n_\zeta} \zeta_i. \tag{8.11}$$

Proof. Similar to the proof of (3.1.4) in Chapter 3. Note that here the contour of Figure 3.3 will have indentations into the right half plane that avoid the branch cuts of log P corresponding to both the ORHP zeros *and poles* of P. □

Theorem 8.2.5 (Bode Integral for M). Assume that M in (7.5) is proper and that F is a BEE. Let $\mathcal{Z}_M = \{q_i : i = 1, \ldots, n_q\}$ be the set of zeros of M in the ORHP and let $\mathcal{Z}_z$ in (8.3) be given by $\mathcal{Z}_z = \{\zeta_i : i = 1, \ldots, n_\zeta\}$. Then, if $M(0) \neq 0$

$$\int_0^\infty \log \left| \frac{M(j\omega)}{M(0)} \right| \frac{d\omega}{\omega^2} = \frac{\pi}{2} \frac{1}{M(0)} \lim_{s\to 0} \frac{dM(s)}{ds} + \pi \sum_{i=1}^{n_q} \frac{1}{q_i} - \pi \sum_{i=1}^{n_\zeta} \frac{1}{\zeta_i}. \quad (8.12)$$

Proof. Same as the proof of Theorem 8.2.4. □

As we see in (8.11) and (8.12), nonminimum phase zeros of G_z that are not zeros of FG_y have an effect on the Bode integral constraints that is consistent with that discussed above for Poisson integral constraints, i.e., they relax the constraints. On the other hand, we see in (8.12) — as well as in (8.8) — that the presence of nonminimum phase zeros of G_y that are not zeros of G_z can only make the constraints more severe, since $\mathcal{Z}_M \supset \mathcal{Z}_y$. This can be interpreted as these nonminimum phase zeros "blocking", this time, the measured signal from which the estimates are constructed. It is not surprising, then, that more severe limits on the achievable performance should be expected. As we will see in Chapters 10 and 11, the problem of prediction exemplifies an extreme case of this latter situation.

In the following section we discuss design implications of the constraints given by the Poisson integrals on P and M.

8.3 Design Interpretations

The objective of this section is to illustrate how the constraints imposed by the integral relations given in the §8.2 can be translated into restrictions on the shape of the frequency responses of the filtering sensitivities.

To begin with, note that, assuming G_z minimum phase, a direct application of (8.9) and (8.10) gives lower bounds on the $\mathcal{H}_\infty$ norms of P and M, i.e.,

$$\begin{aligned} \|P\|_\infty &\geq \max_{q\in\mathcal{Z}_y} \left| B_{1/z}^{-1}(q) \right|, \\ \|M\|_\infty &\geq \max_{p\in\mathcal{Z}_{1/z}} \left| B_y^{-1}(p) \right|. \end{aligned} \quad (8.13)$$

This implies, for example, that both $\|P\|_\infty > 1$ and $\|M\|_\infty > 1$ if G_z is unstable and minimum phase and G_y is nonminimum phase. Moreover, these bounds hold for all possible designs. More information, however, can be obtained if we assume that certain natural design goals are imposed.

From the discussion in §7.2.1 of Chapter 7, it is clear that the relative effect of the process disturbance on the estimation error will be reduced if $|P(j\omega)|$ is small at the frequencies where the power of the process noise v is concentrated. A similar conclusion holds for $|M(j\omega)|$ and the influence of the measurement noise w on the estimation error. In general, as discussed in §7.2.1, the spectrum of v is typically concentrated at low frequencies, while the spectrum of w is typically more significant at high frequencies. This suggests that a reasonable approach to design would be to make $|P(j\omega)|$ small at low frequencies and $|M(j\omega)|$ small at high frequencies. Thus, as previously discussed, a value of $|M(j\omega)|$ close to one, in the range of frequencies where the spectrum of v is large, will avoid a deleterious impact of the process disturbance on the estimate.

The foregoing discussion indicates that typical requirements on the filtering sensitivities are captured, for example, by the following specifications.

$$|P(j\omega)| < \alpha_1 \quad \text{for } \omega \in [0, \omega_1] , \tag{8.14}$$

$$|M(j\omega)| < \alpha_2 \quad \text{for } \omega \in [\omega_2, \infty] , \tag{8.15}$$

where α_1 and α_2 are small positive numbers. Notice that (8.15) can also be set in terms of P, taking into account the complementarity relation (7.6), i.e.,

$$|P(j\omega)| < 1 + \alpha_2 \quad \text{for } \omega \in [\omega_2, \infty] . \tag{8.16}$$

By the same token, an alternative way of expressing (8.14) is

$$|M(j\omega)| < 1 + \alpha_1 \quad \text{for } \omega \in [0, \omega_1] . \tag{8.17}$$

Figure 8.1 expresses graphically these requirements on the frequency response of P. These specifications are similar to the design objectives in feedback control, with P and M playing the role of the sensitivity S and complementary sensitivity T, respectively. In the feedback control problem, open-loop ORHP poles and zeros inhibit achieving these objectives, as we have seen in §3.3.2 of Chapter 3. We will see below that corresponding limitations apply to filtering problems.

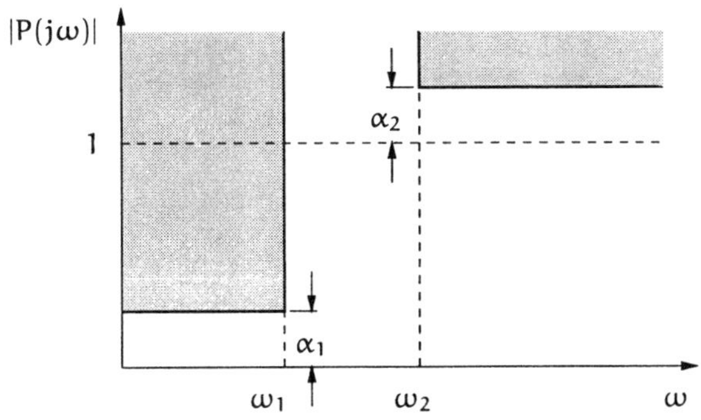

FIGURE 8.1. Frequency specifications for P.

Once design requirements such as those stated in (8.14) and (8.15) have been established, the results in §8.2 can be used to foresee their effect on achievable filter performance in other frequency ranges. For example, whenever $|P(j\omega)|$ is small, $\log|P(j\omega)|$ is negative. It follows from (8.9), for example, that if specification (8.14) holds, $\log|P(j\omega)|$ must be positive in some other frequency range.

More quantitative information regarding these peaks is given below. Let $\Theta_{s_0}(\omega_1)$ be the weighted length of the interval $[-\omega_1, \omega_1]$, as defined in (3.33) in Chapter 3. We then have the following corollary.

Corollary 8.3.1. Suppose that P and M in (7.5) are proper, G_z is minimum phase, and F is a BEE. Suppose further that (8.14) and (8.15) hold. Then,

(i) for each $q = \sigma_q + j\omega_q \in \mathcal{Z}_y$, we have that

$$\|P\|_\infty \geq \left(\frac{1}{\alpha_1}\right)^{\frac{\Theta_q(\omega_1)}{\Theta_q(\omega_2)-\Theta_q(\omega_1)}} \left(\frac{1}{1+\alpha_2}\right)^{\frac{\pi-\Theta_q(\omega_2)}{\Theta_q(\omega_2)-\Theta_q(\omega_1)}} \times \left|B_{1/z}^{-1}(q)\right|^{\frac{\pi}{\Theta_q(\omega_2)-\Theta_q(\omega_1)}}; \quad (8.18)$$

(ii) for each $p = \sigma_p + j\omega_p \in \mathcal{Z}_{1/z}$, we have that

$$\|M\|_\infty \geq \left(\frac{1}{\alpha_2}\right)^{\frac{\pi-\Theta_p(\omega_2)}{\Theta_p(\omega_2)-\Theta_p(\omega_1)}} \left(\frac{1}{1+\alpha_1}\right)^{\frac{\Theta_p(\omega_1)}{\Theta_p(\omega_2)-\Theta_p(\omega_1)}} \times \left|B_y^{-1}(p)\right|^{\frac{\pi}{\Theta_p(\omega_2)-\Theta_p(\omega_1)}}. \quad (8.19)$$

Proof. We use specification (8.15) on M to get the bound (8.16) on P. It then follows from (8.14), (8.16), and (8.9) that

$$\Theta_q(\omega_1)\log|\alpha_1| + [\Theta_q(\omega_2) - \Theta_q(\omega_1)]\log\|P\|_\infty + [\pi - \Theta_q(\omega_2)]\log|1+\alpha_2| \geq \pi\log\left|B_{1/z}^{-1}(q)\right|$$

Hence

$$\|P\|_\infty^{[\Theta_q(\omega_2)-\Theta_q(\omega_1)]} \geq \left(\frac{1}{\alpha_1}\right)^{\Theta_q(\omega_1)} \left(\frac{1}{1+\alpha_2}\right)^{\pi-\Theta_q(\omega_2)} \left|B_{1/z}^{-1}(q)\right|^{\pi}.$$

Inequality (8.18) then follows. In a similar way, (8.19) can be obtained using (8.15), (8.17) and (8.10). □

To give an idea of how severe the constraints given by (8.18) and (8.19) may be, we consider the following numerical example.

Example 8.3.1. Consider the filtering setting of Figure 7.1 in Chapter 7. Assume that the plant is as given in Figure 7.2. Further say that G_y has one real nonminimum phase zero at $q = 0.5$, and that G_z is stable and minimum phase. Suppose that we require specifications (8.14) and (8.16) to hold for the filtering sensitivity for this system. Now we propose the following experiment: fix the high-frequency specifications (8.16) with $\omega_2 = 10$ and $\alpha_2 = 0.5$, and vary the low-frequency specifications, i.e., the bandwidth ω_1 and level of reduction, α_1. Under these conditions, we analyze the lower bound b_P, defined as the RHS of (8.18), i.e.,

$$b_P \triangleq \left(\frac{1}{\alpha_1}\right)^{\frac{\Theta_q(\omega_1)}{\Theta_q(\omega_2)-\Theta_q(\omega_1)}} \left(\frac{1}{1+\alpha_2}\right)^{\frac{\pi-\Theta_q(\omega_2)}{\Theta_q(\omega_2)-\Theta_q(\omega_1)}} \times \left|B_{1/z}^{-1}(q)\right|^{\frac{\pi}{\Theta_q(\omega_2)-\Theta_q(\omega_1)}} . \quad (8.20)$$

Figure 8.2 shows the behavior of b_P vs. ω_1 for values of α_1 ranging from 0.6 to 0.1. Notice that, since G_z was assumed stable, there is no contribution of the Blaschke product $B_{1/z}$ on the RHS of (8.18). As we see in this figure, the presence of a nonminimum phase zero in G_y introduces restrictions on the minimum achievable value of $\|P\|_\infty$. More specifically, if $|P(j\omega)|$ is required to be "small" over a range of frequencies where the nonminimum phase zero of G_y has nonneglectable phase-lag (in rough lines, when q is within $[0, \omega_1)$), then $|P(j\omega)|$ will necessarily be "large" at frequencies outside this range. Hence, there is a clear trade-off on the amount of sensitivity reduction that can be obtained in the ranges $[0, \omega_1)$ and $[\omega_1, \omega_2)$.

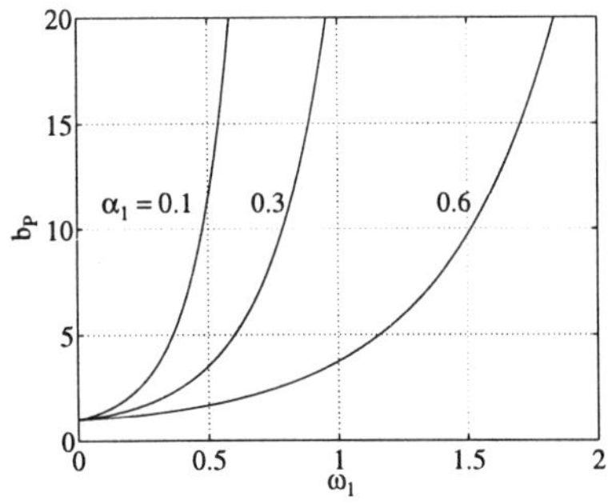

FIGURE 8.2. Lower bound b_P vs. ω_1 for different values of α_1, $\alpha_2 = 0.5$, $\omega_2 = 10$, and a nonminimum phase zero $q = 0.5$.

This trade-off is worsened if, in addition, G_z has an unstable pole, since then the Blaschke product $B_{1/z}$ in (8.20) may have a considerable contribution, depending on the position of the pole relative to the zero of G_y. Figure 8.3 again displays b_P vs. ω_1, this time for a fixed value of $\alpha_1 = 0.5$, and three different values of a single real unstable pole of G_z, $p = 2, 5, 20$.

As we see, the closer the pole and the zero are, the worse the trade-off induced on the values of $|P(j\omega)|$.

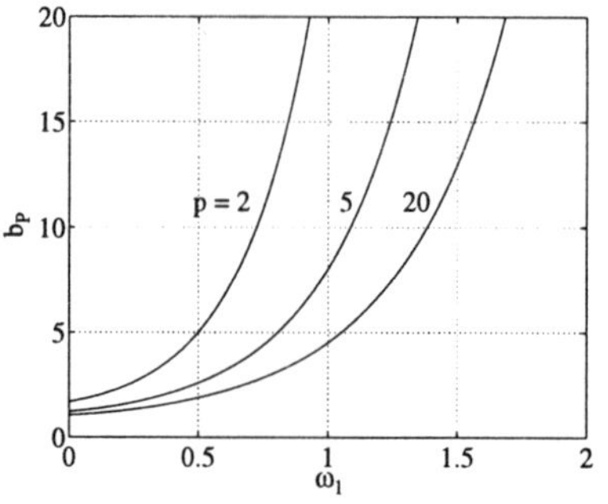

FIGURE 8.3. Lower bound b_P vs. ω_1 for different values of the pole p, $\alpha_1 = 0.6$, $\alpha_2 = 0.5$, $\omega_2 = 10$, and a nonminimum phase zero $q = 0.5$.

◦

Example 8.3.1 illustrates the effects of a real nonminimum phase zero of G_y on the achievable shapes of $|P(j\omega)|$. A practical lesson of this example is that nonminimum phase zeros closer to the $j\omega$-axis will impose tougher restrictions on P. Similar conclusions apply for complex zeros.

It is not difficult to see, after comparing expressions (8.18) and (8.19), that unstable poles of G_z play a similar role with M, although this time poles *farther* from the $j\omega$-axis are the most troublesome. We summarize these conclusions as follows.

(i) ORHP zeros of G_y constrain the magnitude of P. Zeros closer to the $j\omega$-axis impose more severe constraints.

(ii) ORHP poles of G_z constrain the magnitude of M. Poles farther from the $j\omega$-axis impose more severe constraints.

(iii) The combination of ORHP zeros and poles of the plant worsen the constraints on both sensitivities. More severe constraints should be expected if there is a pole close to a zero in the ORHP.

In connection with practical design specifications as those shown in Figure 8.1, these constraints imply that large peaks in the sensitivities may occur. For example, this will occur if reduction of P is desired over a frequency range much larger than any nonminimum phase zero of M. Consequently, high sensitivity to disturbances should be expected at an intermediate range of frequencies.

It also follows from Corollary 8.3.1 and the definition of the Blaschke product, that nonminimum phase zeros of G_z weaken the constraints on both sensitivities.

In the following section, we consider more examples for the case that particular design methodologies are used for the filter design.

8.4 Examples: Kalman Filter

We consider now a particular design methodology, namely, Kalman filter design, to illustrate how the bounds derived in the previous section can be used to assess and compare the performance achieved with different designs. The first example studies a Kalman filter design for an open-loop unstable, nonminimum phase system (i.e., one having ORHP zeros in G_y), which is affected by process noise with spectrum concentrated at low frequencies.

Example 8.4.1 (Kalman Filter Design). Consider a system of the form shown in Figure 7.1 in Chapter 7, where G, of the form (7.1) in Chapter 7, is given by the following matrices

$$A = \begin{bmatrix} 0 & 1 \\ 3 & -2 \end{bmatrix}, \quad B = \begin{bmatrix} 0 \\ 1 \end{bmatrix}, \quad C_1 = \begin{bmatrix} 2 & 1 \end{bmatrix}, \quad C_2 = \begin{bmatrix} -2 & 1 \end{bmatrix}. \tag{8.21}$$

These data give the following transfer functions for the partition of the plant shown in Figure 7.2,

$$G_z(s) = \frac{s+2}{(s+3)(s-1)}, \quad \text{and} \quad G_y(s) = \frac{s-2}{(s+3)(s-1)}. \tag{8.22}$$

From (8.22), (8.4) and (8.5), we have that P and M have nonminimum phase zeros, namely, $p = 1 \in \mathcal{Z}_P$, and $q = 2 \in \mathcal{Z}_M$. Hence, $|B^{-1}_{1/z}(q)| = |B^{-1}_y(p)| = 3$ in (8.18) and (8.19).

Arbitrarily choosing $\omega_1 = 1$, $\alpha_2 = 0.5$ and $\omega_2 = 100$ for specifications (8.16)-(8.17), leads to the bounds shown in Figure 8.4. In this figure, b_P denotes the lower bound on $\|P\|_\infty$ given by (8.20), and b_M denotes the lower bound on $\|M\|_\infty$ given by the RHS of (8.19) in Corollary 8.3.1.

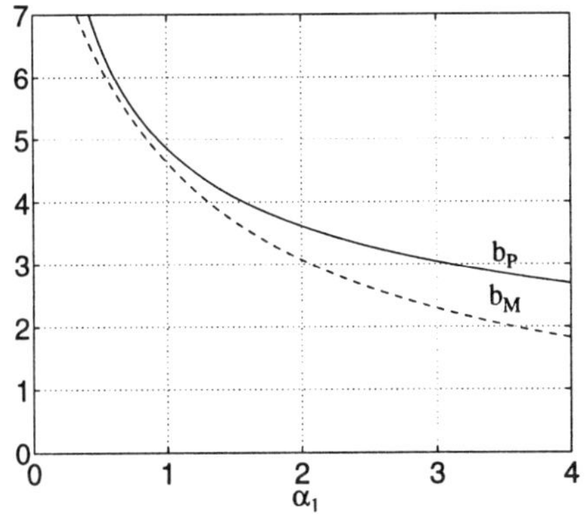

FIGURE 8.4. b_P and b_M for $\omega_1 = 1$, $\alpha_m = 0.5$ and $\omega_2 = 100$.

The curves in Figure 8.4 hold for *any* linear filter for the system (7.2) in Chapter 7 with data (8.21). As an illustration for a particular estimator, we use a standard Kalman filter (see for example Anderson & Moore 1979). Within this paradigm, we will analize the effect of changing the design for different levels of sensitivity attenuation, i.e., different values of α_1 and ω_1 in Figure 8.1.

The estimate provided by the Kalman filter achieves the optimal estimation error in the minimum variance sense, provided that the system is affected by process and measurement noises that are white and have zero mean. The filter is computed by solving an algebraic Riccati equation parametrized by the system matrices and by the noises incremental covariances.

In order to build a mechanism to adjust α_1 and ω_1, we add the following hypotheses:

(i) the measurement noise, w, is white with known incremental variance R, and

(ii) the process input, v, is generated by a low-pass model driven by white noise,

$$\dot{v} = -av + av_0, \tag{8.23}$$

where a is a positive number that serves as a free parameter, and v_0 is a white noise with known incremental variance Q.

The Kalman filter is then designed for a composite system that includes the noise model (8.23), and the number a and the ratio Q/R can be used to adjust the values of α_1 and ω_1 achieved in various designs.

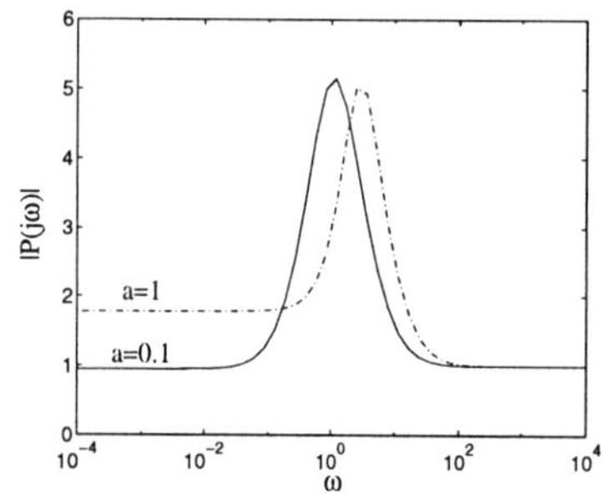

FIGURE 8.5. $|P|$ resulting from the Kalman filter for $Q = 100$, $R = 1$ and two values of a.

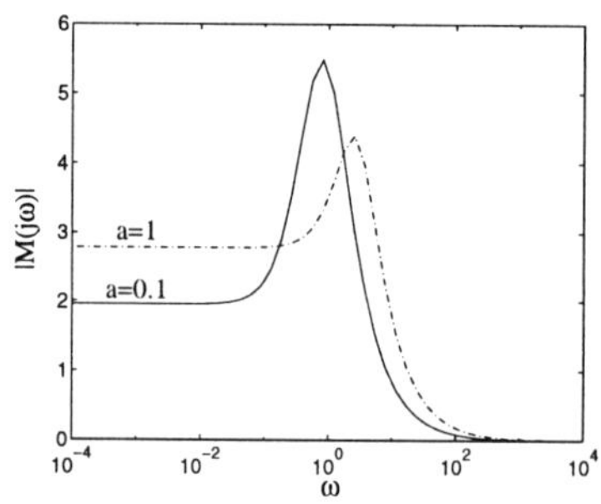

FIGURE 8.6. $|M|$ resulting from the Kalman filter for $Q = 100$, $R = 1$ and two values of a.

Some typical curves are shown in Figures 8.5 and 8.6, which illustrate the achieved shapes of $|P|$ and $|M|$ for fixed Q and R and two different values of a. On the other hand, Figures 8.7 and 8.8 show the achieved shapes of $|P|$ and $|M|$ for a fixed value $a = 1$ and two different ratios Q/R.

From various designs, we have extracted those cases for which ω_1 could be reasonably taken as 1, and have compared the corresponding value of

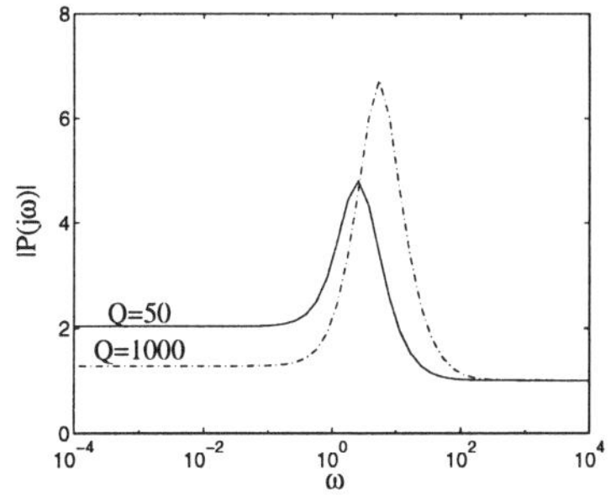

FIGURE 8.7. $|P|$ resulting from the Kalman filter for $a = 1$, $R = 1$ and two values of Q.

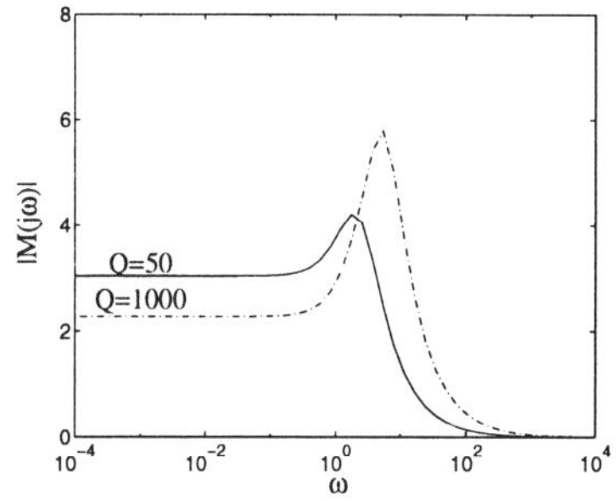

FIGURE 8.8. $|M|$ resulting from the Kalman filter for $a = 1, R = 1$ and two values of Q.

α_1 and the peak sensitivities $\|P\|_\infty$ and $\|M\|_\infty$. These are shown in Figures 8.9 and 8.10, respectively. For the sake of comparison, we also show in these figures the bounds b_P and b_M plotted earlier in Figure 8.4.

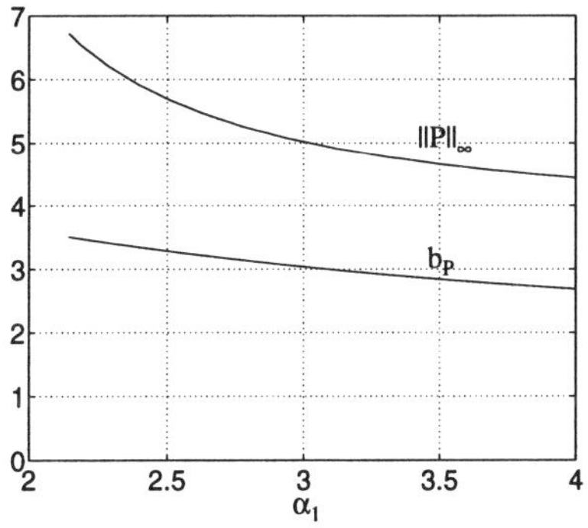

FIGURE 8.9. Achieved $\mathcal{H}_\infty$ norm for P and predicted bound b_P.

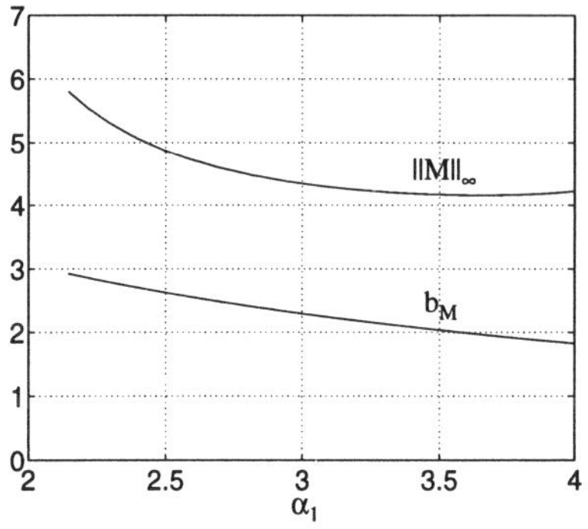

FIGURE 8.10. Achieved $\mathcal{H}_\infty$ norm for M and predicted bound b_M.

As can be seen from these last figures, the bounds b_P and b_M are rather conservative for this example; the peaks found in $|P(j\omega)|$ and $|M(j\omega)|$ approximately double the bounds predicted. ∘

Example 8.4.1 shows that the design trade-offs arising from the essential limitations imposed by ORHP poles and zeros of the plant can in fact be significantly worse than those predicted using the analysis of §8.3. The strength of this analysis lies on its generality, and the degree of its conservativeness is linked to the particular design methodology employed.

The following example compares the filtering sensitivities resulting from the previous Kalman Filter design with those achieved by a design that minimizes the $\mathcal{H}_\infty$ norm of weighted versions of P and M.

Example 8.4.2 (Kalman Filter versus $\mathcal{H}_\infty$ Designs). Consider again the plant given by (7.1) in Chapter 7 and the data (8.21). A filter design that guarantees bounds on the $\mathcal{H}_\infty$ norms of the filtering sensitivities can be achieved using a parametrization of all BEEs developed in Seron & Goodwin (1995). Under this parametrization P can be written as

$$P = T_1 - LT_2, \tag{8.24}$$

where T_1 and T_2 are proper and stable transfer functions computed from the data given in (8.21), and L is an arbitrary proper stable transfer function, which serves as a free parameter. An $\mathcal{H}_\infty$ *mixed sensitivity problem* (Kwakernaak 1985) is then obtained if we compute L to minimize

$$\left\| \begin{matrix} W_1(T_1 - LT_2) \\ W_2LT_2 \end{matrix} \right\|_\infty , \tag{8.25}$$

where the weights W_1 and W_2 serve to give an adequate shape to the solution on the $j\omega$-axis. Ideally, $|W_1(j\omega)|$ represents the inverse of the desired shape for $|P(j\omega)|$, while W_2 serves to impose the condition that M should roll off at high frequencies. Here, we have selected

$$W_1(s) = \frac{s + 1/a_1}{k(s+1)} \quad \text{and} \quad W_2(s) = \frac{0.5s + 1}{10^4}, \tag{8.26}$$

which are plotted in Figure 8.11 for $k = 10$ and $a_1 = 0.2$.

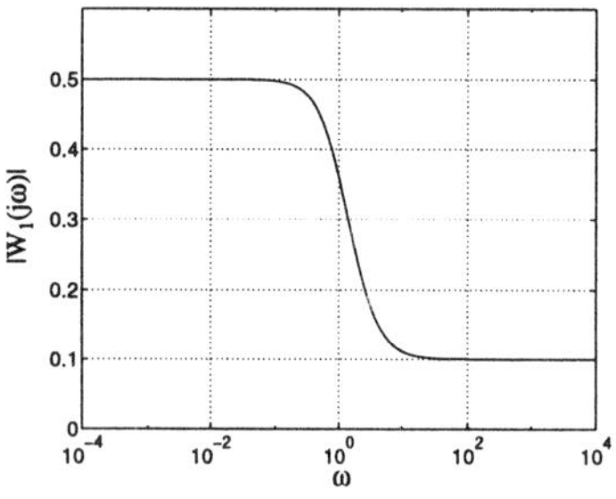

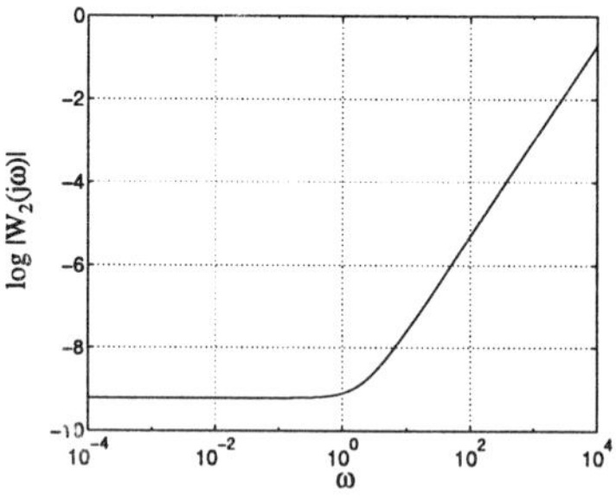

FIGURE 8.11. Weighs W_1, for $k = 10$ and $a_1 = 0.2$ (left), and W_2 (right).

By changing the values of a_1 in W_1 we can obtain different designs satisfying specification (8.14). The selection of W_1 as in Figure 8.11 was made to approximately match the shapes obtained by the Kalman filter in Example 8.4.1; in particular, note that we are requiring P to have a value close to $1/|W_1(0)| = 2$ at low frequencies. We point out that the $\mathcal{H}_\infty$ method allows for a lower reduction level simply by changing the parameters of W_1. Moreover, it turns out that this methodology can yield smaller peaks in the achieved sensitivities than those obtained with the Kalman filter of the previous example. Figures 8.12 and 8.13 compare the absolute bounds

b_P and b_M for this system with the $\mathcal{H}_\infty$ norms of P and M obtained with both methodologies for different values of α_1. The rest of the specifications for these curves are $\omega_1 = 1$, $\omega_2 = 100$, and $\alpha_2 = 0.5$.

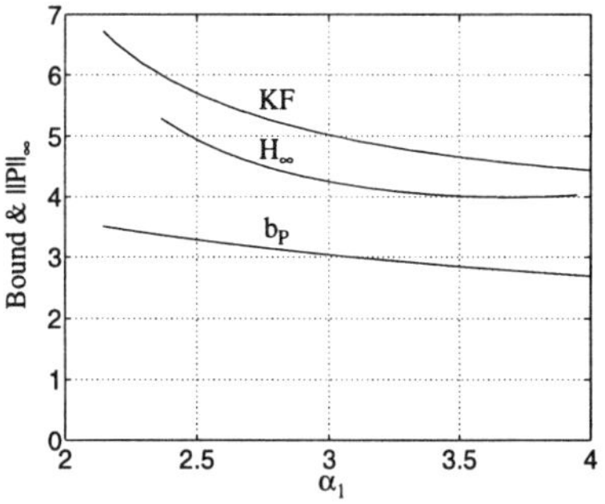

FIGURE 8.12. Achieved $\mathcal{H}_\infty$ norm for P, Kalman filter and $\mathcal{H}_\infty$ designs, and predicted bound b_P.

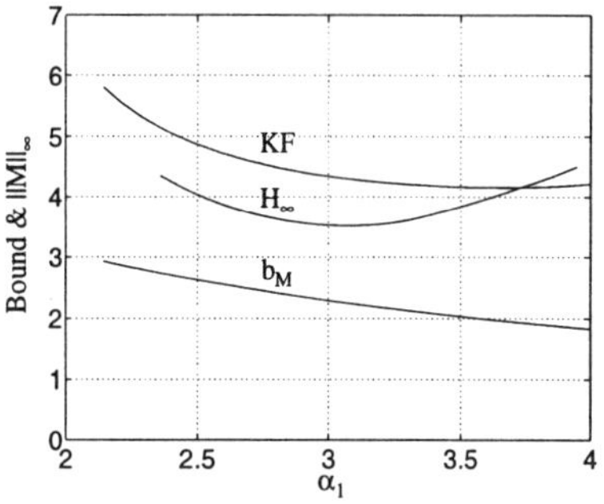

FIGURE 8.13. Achieved $\mathcal{H}_\infty$ norm for M, Kalman filter and $\mathcal{H}_\infty$ designs, and predicted bound b_M.

As can be seen from Figures 8.12 and 8.13, and as anticipated, the conservativeness of the predicted bounds depends on the particular design technique chosen. In this example, the $\mathcal{H}_\infty$ method has a water-bed effect milder than that of the Kalman filter. Sensitivities corresponding to both designs are shown in Figures 8.14 and 8.15 for $\alpha_1 = 2.53$. The parameters for W_1 in (8.26) are $k = 10$ and $a_1 = 0.2$; the parameters for the Kalman filter are taken as $a = 1$, $R = 1$, and $Q = 263$.

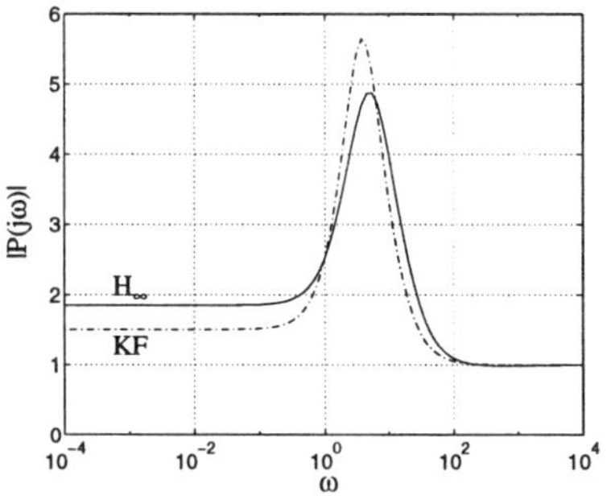

FIGURE 8.14. $|P|$ resulting from the Kalman filter and the $\mathcal{H}_\infty$ norm minimization designs.

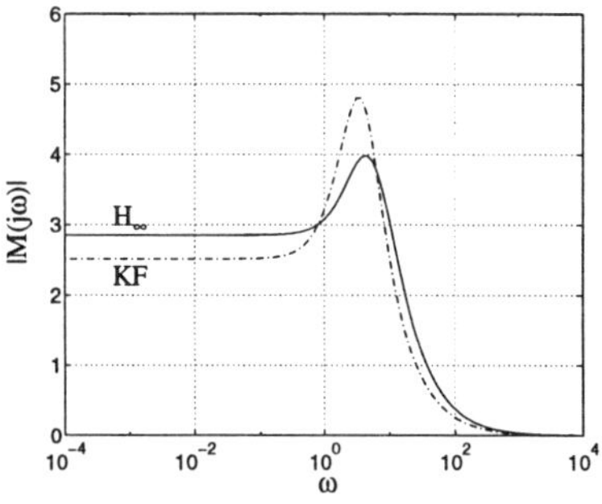

FIGURE 8.15. $|M|$ resulting from the Kalman filter and the $\mathcal{H}_\infty$ norm minimization designs.

We see from this example that the results given in this chapter can be effectively used to compare different filter design methods. ∘

8.5 Example: Inverted Pendulum

Consider the inverted pendulum analyzed in §1.3.3 of Chapter 1, and revisited in §3.3.3 of Chapter 3. It is well known that this system is observable from the carriage position measurement; however, practical evidence suggests that it is difficult to estimate the angle θ from the available data.

We will use the framework of Chapters 7 and 8 to study the sensitivity limitations that apply to the problem of angle estimation from the carriage position measurement.

Choosing $z = \theta$, the linearized model for this system (e.g., Middleton 1991) has the form (7.1) in Chapter 7, where

$$A = \begin{bmatrix} 0 & 1 & 0 & 0 \\ 0 & 0 & -g\, m_1/m_2 & 0 \\ 0 & 0 & 0 & 1 \\ 0 & 0 & (1+m_1/m_2)\, g/\ell & 0 \end{bmatrix}, \qquad B = \begin{bmatrix} 0 \\ 1/m_2 \\ 0 \\ -1/(m_2\ell) \end{bmatrix},$$

$$C_1 = \begin{bmatrix} 0 & 0 & 1 & 0 \end{bmatrix}, \qquad C_2 = \begin{bmatrix} 1 & 0 & 0 & 0 \end{bmatrix}.$$

Here m_1 is the mass at the end of the pendulum, m_2 is the carriage mass, g is the gravitational constant, and ℓ is the pendulum length. These data give the following transfer functions for the partitioning of the plant shown in Figure 7.2 of Chapter 7,

$$G_z = \frac{-1}{m_2\ell\,(s+p)(s-p)}, \quad \text{and} \quad G_y = \frac{(s+q)(s-q)}{m_2\, s^2(s+p)(s-p)}, \tag{8.27}$$

where $q = \sqrt{g/\ell}$ and $p = q\sqrt{1+m_1/m_2}$. Say we take $q = 1$. Choosing the specification parameters in (8.14) and (8.15) as $\alpha_1 = 0.25$, $\omega_1 = 0.01$, $\alpha_2 = 0.5$, and $\omega_2 = 100$, Corollary 8.3.1 gives the following lower bounds on the peak sensitivities

$$\|S\|_\infty \geq 6, \qquad \text{and} \qquad \|T\|_\infty \geq 6. \tag{8.28}$$

We next design a filter for the above system following the $\mathcal{H}_\infty$ mixed sensitivity minimization procedure outlined in Example 8.4.2. In particular, we choose the weights in (8.25) as

$$W_1(s) = 0.1\frac{s+2}{s+0.01},$$

and W_2 as given in (8.26). Figure 8.16 shows the values of $\log|P(j\omega)|$ and $\log|M(j\omega)|$ achieved by this design, for the mass ratio $m_1/m_2 = 0.2, 0.4, 1$. For $m_1/m_2 = 1$, in particular, the peak sensitivities actually achieved are around two times the values of the lower bounds in (8.28). The reasons for this conservatism include the fact that the sectionally constant shapes assumed by Corollary 8.3.1 are actually unrealistic from a practical design point of view.

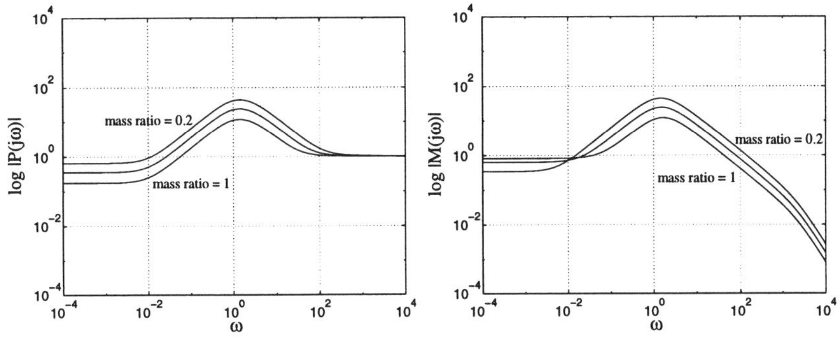

FIGURE 8.16. $\log|P(j\omega)|$ and $\log|M(j\omega)|$ for the inverted pendulum.

We note that, for the mass ratio $m_1/m_2 = 0.1$, and the same specification parameters as in (8.14) and (8.15), Corollary 8.3.1 gives the lower bounds in (8.28) equal to 44. Recalling that P maps relative input disturbances to relative estimation errors (see §7.2.1 in Chapter 7), this validates the claim made in §1.1 of Chapter 1 that relative input errors of the order of 1% will appear as angle relative estimation errors of the order of 50%.

8.6 Summary

This chapter has discussed a new concept in the area of filtering, namely that of sensitivity of the estimation error to process and measurement noise. These sensitivities always satisfy a complementarity constraint, which can be used to establish integral bounds on the magnitude of their frequency responses. These bounds give new insights into the fundamental limits that apply to *all* linear, bounded error filters.

The utility of the results has been vindicated by specific examples. In particular, §8.3 and §8.4 have shown that, for a given signal from a system to be estimated, fundamental design constraints apply to all filtering problems *irrespective* of the method used to obtain a specific design. Thus, one can use the design constraints to judge, a priori, whether or not, a desired level of performance is achievable. Moreover, the constraints can only be modified if the intrinsic nature of the problem is changed at source, e.g., by modifying the location of sensors, which, inter-alia, changes the numerator polynomials and hence the bounds in the sensitivity constraints.

Notes and References

This chapter was mainly based on Goodwin et al. (1995), Seron & Goodwin (1995) and Seron (1995).

9
MIMO Filtering

This chapter investigates sensitivity limitations in multivariable linear filtering. Similar to those obtained in Chapter 4 for the control problem, multivariable integral constraints hold for the MIMO version of the filtering sensitivities introduced in Chapter 7. Also similar to the control case, there are different ways to extend the SISO integrals to a MIMO setting. The approach followed in this chapter emphasizes the trade-offs that arise when the system is required to satisfy, besides frequency conditions, structural specifications on the multivariable sensitivities.

In the MIMO case, frequency restrictions similar to those arising in SISO systems apply to linear combinations of the scalar entries of the filtering sensitivities P and M. Thus, the penalties imposed by ORHP zeros and poles are somehow "shared" by the scalar entries of the sensitivities and, in this sense, the limitations may "relax" with respect to SISO systems. However, if, in addition to frequency specifications, a particular structure is required for these functions (e.g., diagonal, diagonal-dominant, triangular, etc.), then, as we will see, extra costs may arise.

The results presented in this chapter can be used to obtain performance limitations in a variety of filtering applications where multivariable structural features are of particular interest. To illustrate, we explore the design trade-offs arising in problems of detection and isolation of system faults.

9.1 Interpolation Constraints

We consider again the general setting introduced in Chapter 7. As in the previous chapter, the standing assumption here is that of BEE (see §7.3). We summarize the conditions required in this chapter for the functions F, G_z, and G_y defined in §7.1, in the following assumption.

Assumption 9.1.

(i) F is a BEE, i.e., F and $G_z - FG_y$ are stable.

(ii) G_z is right invertible and minimum phase.

(iii) $G_y G_z^{-1}$ is proper (here G_z^{-1} is a right inverse of G_z).

∘

Since the plant G is stabilizable, then stability of F and $G_z - FG_y$ is a necessary and sufficient condition for the filter to be a BEE, as defined in §7.3 in Chapter 8. It can be inferred from condition (i) that unstable modes shared by G_z and G_y are canceled[1] by a zero of the difference $G_z - FG_y$, while those unstable poles of G_y that do not appear in G_z must be necessarily canceled by nonminimum phase zeros of F.

In condition (ii), recall that for G_z to be right invertible it is necessary that there be at least as many process inputs as signals to be estimated. The assumption that G_z has no CRHP zeros is made to simplify the analysis.

Condition (iii) is assumed to guarantee that the sensitivities are proper.

It was shown in Chapter 8 that, in the scalar case, ORHP poles of the system transfer function G_z are zeros of the filtering sensitivity, P, whilst ORHP zeros of the system transfer function G_y that are not zeros of G_z, are zeros of the filtering complementary sensitivity, M. The following lemma generalizes these results to the MIMO case, emphasizing the multivariable character of the zeros.

Lemma 9.1.1 (Interpolation Constraints). Suppose that G and F satisfy the hypotheses stated in Assumption 9.1. Then the following conditions are satisfied by P and M:

(i) P and M are stable and proper.

(ii) Let $p \in \overline{\mathbb{C}^+}$ be a pole of G_z. Then, there exists a nonzero vector $\Phi \in \mathbb{C}^n$ such that

$$P(p)\Phi = 0, \qquad \text{and} \qquad M(p)\Phi = \Phi. \tag{9.1}$$

(iii) Let $q \in \overline{\mathbb{C}^+}$ be a zero of G_y. Then, there exists a nonzero vector $\Psi \in \mathbb{C}^n$ such that

$$P(q)\Psi = \Psi, \qquad \text{and} \qquad M(q)\Psi = 0. \tag{9.2}$$

[1] In frequency and direction.

(iv) Let $q \in \overline{\mathbb{C}^+}$ be a zero of $G_z - FG_y$. Then there exists a nonzero vector $\Psi \in \mathbb{C}^n$ such that

$$\Psi^* P(q) = 0, \qquad \text{and} \qquad \Psi^* M(q) = \Psi^*. \tag{9.3}$$

(v) Let $q \in \overline{\mathbb{C}^+}$ be a zero of F, and suppose further that the product $G_y G_z^{-1}$ does not have a pole at q. Then there exists a nonzero vector $\Psi \in \mathbb{C}^n$ such that

$$\Psi^* P(q) = \Psi^*, \qquad \text{and} \qquad \Psi^* M(q) = 0. \tag{9.4}$$

Proof. Stability of P and M follows from the fact that F is a BEE and G_z is minimum phase; their properness follows from condition (iii). Case (ii) follows in a straightforward manner from (7.5) by noting that the term $G_z - FG_y$ is stable by Assumption 9.1. For case (iii), we have that there exists a nonzero vector Ψ_1 such that $G_y(q)\Psi_1 = 0$. Since G_z is minimum phase, then there exists a nonzero vector Ψ such that $G_z^{-1}(q)\Psi = \Psi_1$. Hence, since F is assumed stable, (9.2) follows. Case (iv) is immediate on noting that G_z is minimum phase. Finally, for case (v) notice that if q is not a pole of $G_y G_z^{-1}$, then necessarily there exists a vector Ψ such that $\Psi^* F(q) G_y(q) G_z^{-1}(q) = 0$, and (9.4) follows. □

Remark 9.1.1. Conditions (iv) and (v) depend on the particular choice of filter and they are stated for completeness. However, our main focus here is on results that hold irrespective of the particular filter — this is true of the other cases in Lemma 9.1.1. Yet, as we will see in §9.4, there are filtering applications where zeros of F are directly determined by the unstable poles of the plant. ∘

Remark 9.1.2. Notice that since the plant is detectable from y, the transfer function G_y has all the unstable poles of G_z. In consequence, $G_y G_z^{-1}$ — and hence M — has no zeros at any unstable pole of G. ∘

9.2 Poisson Integral Constraints

The interpolation constraints derived in the previous section fix the values of the filtering sensitivity functions at points in the CRHP determined by poles and zeros of the plant and filter. In the SISO case addressed in Chapter 8, the Poisson integral was used to translate these constraints into equivalent integral relations that must be satisfied by P and M along the $j\omega$-axis. This section extends the results of Chapter 8 to MIMO filtering problems using the technique suggested by Gómez & Goodwin (1995) and developed for the multivariable control case in Chapter 4.

We first introduce some preliminary notation. Recalling that P and M are $n \times n$ square transfer matrices, we denote by P_{ik}, and M_{ik} their respective elements in the i-row and k-column. If Φ is a vector in $\mathbb{C}^n$, we denote its elements by ϕ_i, $i = 1, \dots, n$, i.e., $\Phi = [\phi_1^*, \phi_2^*, \dots, \phi_n^*]^*$. Also, we introduce the index set $\mathcal{I}_\Phi \triangleq \{i \in \mathbb{N} : \phi_i \neq 0\}$ as the set of indices of the nonzero elements of Φ.

As before, given a set $\mathcal{Z}_k = \{s_i, i = 1, \dots, n_k\}$ of complex numbers in the ORHP, we define its *Blaschke product* as the function

$$B_k(s) = \prod_{i=1}^{n_k} \frac{s - s_i}{s + \bar{s}_i} .$$

If $\mathcal{Z}_k$ is empty, we define $B_k(s) = 1, \forall s$. We then have the following result.

Theorem 9.2.1 (Poisson Integral for P). Let $q = \sigma_q + j\omega_q$, $\sigma_q > 0$, be a zero of G_y, and let $\Psi \in \mathbb{C}^n$, $\Psi \neq 0$, be the corresponding zero direction of M, as described in Lemma 9.1.1 (iii). Then, under Assumption 9.1, for each index k in $\mathcal{I}_\Psi$,

$$\frac{1}{\pi}\int_{-\infty}^{\infty} \log\left|\sum_{i=1}^{n} P_{ki}(j\omega)\frac{\psi_i}{\psi_k}\right| \frac{\sigma_q}{\sigma_q^2 + (\omega_q - \omega)^2} d\omega \geq 0 . \tag{9.5}$$

Proof. The proof follows the same general line as that in Chapter 4 — Gómez & Goodwin (1995) — and is an application of the Poisson integral formula, as in Freudenberg & Looze (1985), to the elements of the vector function $P\Psi : \mathbb{C} \to \mathbb{C}^n$, which are proper, stable, scalar rational functions. Pick one of these elements, say $\rho_k(s) \triangleq \sum_{i=1}^n P_{ki}(s)\psi_i$, where k is in $\mathcal{I}_\Psi$, and let B_k be the Blaschke product of the zeros of ρ_k in the ORHP. Then, the function $\tilde{\rho}(s) \triangleq B_k^{-1}(s)\rho_k(s)$ is proper, stable, and minimum phase, which implies that $\log \tilde{\rho}(s)$ is analytic in the ORHP and satisfies the conditions of the Poisson integral formula (Theorem A.6.1 in Appendix A). Evaluating this at $s = q$, and using the fact that $|\tilde{\rho_k}(j\omega)| = |\rho_k(j\omega)|$ yields

$$\begin{aligned}\frac{1}{\pi}\int_{-\infty}^{\infty} \log\left|\sum_{i=1}^{n} P_{ki}(j\omega)\psi_i\right| \frac{\sigma_q}{\sigma_q^2+(\omega_q-\omega)^2} d\omega &= \log\left|B_k^{-1}(q)\sum_{i=1}^{n} P_{ki}(q)\psi_i\right| \\ &= \log|B_k^{-1}(q)| + \log|\psi_k| ,\end{aligned} \tag{9.6}$$

where the last step follows from the interpolation condition $P(q)\Psi = \Psi$ (notice that $\psi_k \neq 0$ by assumption). Since

$$\frac{1}{\pi}\int_{-\infty}^{\infty} \frac{\sigma_q}{\sigma_q^2 + (\omega_q - \omega)^2} d\omega = 1 ,$$

subtracting $\log|\psi_k|$ from both sides of (9.6) yields

$$\frac{1}{\pi}\int_{-\infty}^{\infty} \log\left|\sum_{i=1}^{n} P_{ki}(j\omega)\frac{\psi_i}{\psi_k}\right| \frac{\sigma_q}{\sigma_q^2 + (\omega_q - \omega)^2} d\omega = \log|B_k^{-1}(q)| . \tag{9.7}$$

The proof is then completed on noting that $|B_k^{-1}(s)| \geq 1$ at any point s in $\mathbb{C}^+$, and so $\log|B_k^{-1}(q)| \geq 0$. □

Theorem 9.2.1 establishes that, for each nonminimum phase zero of the plant G_y, there is a set of integral constraints that limit the values of P on the $j\omega$-axis in an intrinsically vectorial fashion. More specifically, depending on the characteristics of the input directions associated to the zero, up to n integral constraints may arise on linear combinations of elements of P in the same row. If the zero direction happens to be canonical (i.e., it has only one nonzero entry) or P is designed to be diagonal, then the constraints reduce to those of the SISO case.

Recall that, in the SISO case, if the plant is nonminimum phase, then requiring that (as is often desirable) $|P(j\omega)| < 1$ over a frequency range, say Ω, implies that, necessarily, $|P(j\omega)| > 1$ at other frequencies. The severity of this trade-off depends on the relative location of the nonminimum phase zero and the frequency range Ω. More details and interpretations were given in Chapter 8.

A similar integral holds for the filtering complementary sensitivity function, as established below.

Theorem 9.2.2 (Poisson Integral for M). Let $p = \sigma_p + j\omega_p$, $\sigma_p > 0$, be a pole of G_z, and let $\Phi \in \mathbb{C}^n$, $\Phi \neq 0$, be the corresponding zero direction of P, as described in Lemma 9.1.1 (ii). Then, under Assumption 9.1, for each index k in $\mathcal{I}_\Phi$,

$$\frac{1}{\pi}\int_{-\infty}^{\infty} \log\left|\sum_{i=1}^{n} M_{ki}(j\omega)\frac{\phi_i}{\phi_k}\right| \frac{\sigma_p}{\sigma_p^2 + (\omega_p - \omega)^2}\, d\omega \geq 0\,. \tag{9.8}$$

Proof. The proof follows that of Theorem 9.2.1, this time using the interpolation constraint (ii) in Lemma 9.1.1. □

A straightforward corollary of these results emphasizes the vectorial nature of the associated trade-offs. Let $\Omega_1 \triangleq [0, \omega_1]$ denote a given range of frequencies of interest, and assume that the k-row of M satisfies the following design specifications:

$$|M_{ki}(j\omega)| \leq \epsilon_{ki} \quad \text{for } \omega \text{ in } [\omega_1, \infty],\, i = 1, \ldots, n, \tag{9.9}$$

where ϵ_{ki}, $i = 1, \ldots, n$, are small positive numbers. Let $\Theta_{s_0}(\omega_1)$ be the weighted length of the interval $[-\omega_1, \omega_1]$, as defined in (3.33) in Chapter 3. Then, we have the following corollary.

Corollary 9.2.3. Assume that all the conditions of Theorem 9.2.2 hold. Then, if the k-row of M achieves specifications (9.9), the following inequality must be satisfied,

$$\|M_{kk}\|_\infty + \sum_{\substack{i=1\\i\neq k}}^{n} \left|\frac{\phi_i}{\phi_k}\right| \|M_{ki}\|_\infty \geq \left(\frac{1}{\epsilon_{kk} + \sum_{\substack{i=1\\i\neq k}}^{n} \left|\frac{\phi_i}{\phi_k}\right| \epsilon_{ki}} \right)^{\frac{\pi - \Theta_p(\omega_1)}{\Theta_p(\omega_1)}} . \tag{9.10}$$

Proof. The proof follows immediately from Theorem 9.2.2 by using specifications (9.9) on M (see Corollary 4.3.4 in Chapter 4 for more details). □

As in the SISO case, the corollary shows that the integral relation (9.8) implies lower bounds on the infinity norm of elements of M. Notice that the exponent on the RHS of (9.10) is a positive number and its base is likely to be larger than one if the ϵ_{ki}'s are small enough; hence, the more demanding the specifications, the larger these lower bounds are.

If the zero direction is not canonical, an important difference in the MIMO case is that the lower bounds apply to a combination of norms of elements, which somehow relaxes the constraint over the SISO case (where there is only one element). Also, the lower bounds are smaller than in the SISO case due to the presence of extra positive terms in the denominator of the RHS of (9.10). In consequence, if M is required to be diagonal (as often occurs in a number of applications), the constraints on the values of M on the $j\omega$-axis worsen since the off-diagonal elements disappear. We analyze this in more detail in the following section.

9.3 The Cost of Diagonalization

As was concluded from Corollary 9.2.3, when the zero direction is not canonical the cost that it induces on each row of P or M can be shared among various elements of the row. On the other hand, as discussed in Chapter 4 for the control case, a diagonally decoupled system loses this potential to alleviate the cost, since all the price is paid by the diagonal elements alone.

By combining the ideas of Gómez & Goodwin and the parametrization of all diagonalizing post-compensators given by Kinnaert & Peng (1995), it is possible to make a precise statement about the cost of diagonal decoupling. For convenience, we first recall the result of Kinnaert & Peng. Given an $\ell \times n$ stable, proper, left-invertible transfer function H, it can be factorized as

$$H = Q \begin{bmatrix} \bar{H} \\ 0 \end{bmatrix} , \tag{9.11}$$

where Q is an $\ell \times \ell$ bistable and biproper transfer function and $\bar{H}$ is an upper triangular $\ell \times \ell$ matrix (Dion & Commault 1988). The matrix $[\bar{H}^* \ 0]^*$ is a column Hermite form of H over the ring of stable proper rational functions and it is unique up to units of this ring. With the aid of this factorization, Kinnaert & Peng (1995) derived a parametrization of all stable, proper postcompensators F such that FH is a stable, proper diagonal matrix with nonzero diagonal elements. Using this parametrization for F, the product FH has the form

$$FH = \bar{Q}R \triangleq \bar{Q}\,\mathrm{diag}[R_1, R_2, \ldots, R_n] , \tag{9.12}$$

where $\bar{Q}$ is an arbitrary $n \times n$ stable, proper, diagonal matrix with nonzero diagonal elements, and the R_k, $k = 1, \ldots n$, are the functions

$$R_k(s) \triangleq \frac{\prod_{i=1}^{n_k}(s - p_i)}{(s + \sigma_k)^{n_k + r_k}} . \tag{9.13}$$

In (9.13), $p_i, i = 1, \ldots n_k$, are the unstable poles of the k-row of $\bar{H}^{-1}$ in (9.11), σ_k is an arbitrary positive real number, and r_k is determined so that

$$\lim_{s \to \infty} R_k(s)\,(\text{k-row of } \bar{H}^{-1}) = \text{a nonzero row vector.}$$

Notice that the functions R_k contain information about the zeros of H, and their numerators depend on H alone.

Assume next that we are given a filtering problem where the design objective is to achieve a diagonal complementary sensitivity M. This is desirable, for example, if the filter is to perform detection and isolation of system faults, as we will discuss in §9.4. We then have the following result.

Theorem 9.3.1 (Cost of Diagonalization). Let Assumption 9.1 hold. Let $p = \sigma_p + j\omega_p$, $\sigma_p > 0$, be a pole of G_z, and let $\Phi \in \mathbb{C}^n$, $\Phi \neq 0$, be the corresponding zero direction of P, as described in Lemma 9.1.1 (ii). Assume that $G_y G_z^{-1}$ is a stable, proper and left invertible transfer function, with Hermite form

$$G_y G_z^{-1} = Q \begin{bmatrix} \bar{G} \\ 0 \end{bmatrix} . \tag{9.14}$$

Suppose that the filter F has been selected such that M in (7.5) is a stable, proper diagonal matrix with nonzero diagonal elements. Then, for each index k in $\mathcal{J}_\Phi$,

$$\frac{1}{\pi}\int_{-\infty}^{\infty} \log|M_{kk}(j\omega)| \frac{\sigma_p}{\sigma_p^2 + (\omega_p - \omega)^2} d\omega = \log|B_k^{-1}(p)| , \tag{9.15}$$

where B_k is the Blaschke product of the unstable poles of the k-row of $\bar{G}^{-1}$, and where $\bar{G}$ is defined in (9.14).

Proof. Note that the assumptions imply that Theorem 9.2.2 holds specialized to the diagonal case, i.e.,

$$\frac{1}{\pi}\int_{-\infty}^{\infty} \log|M_{kk}(j\omega)| \frac{\sigma_p}{\sigma_p^2 + (\omega_p - \omega)^2} d\omega \geq 0 \,.$$

The inequality above is turned into an equality similar to (9.7) by adding the Blaschke product of ORHP zeros of ρ_k (defined in the proof of Theorem 9.2.1). Since M has the form (9.12) with $H = G_y G_z^{-1}$, this Blaschke product is independent of the filter and equal to the unstable poles of the k-row of $\bar{G}^{-1}$, with $\bar{G}$ defined in (9.14). This completes the proof. □

The following example illustrates the result.

Example 9.3.1. Consider a plant having the following transfer functions.

$$G_y(s) = \begin{bmatrix} \frac{1}{s+1} & \frac{s-1}{s+1} & 0 \\ 0 & 0 & \frac{4-s}{(s-3)(s+2)} \\ \frac{4}{s^2+3s+5} & 0 & \frac{4}{s^2+3s+5} \end{bmatrix},$$
$$G_z^{-1}(s) = \begin{bmatrix} 0 & \frac{s-3}{s+2} \\ \frac{s+1}{s+3} & \frac{s+1/3}{s+2} \\ 0 & \frac{3-s}{s+2} \end{bmatrix}. \qquad (9.16)$$

Note that G_z^{-1} has a zero at $p = 3$ with input direction

$$\Phi = \begin{bmatrix} -1 \\ 1 \end{bmatrix},$$

which implies that integrals of the form (9.8) hold that constrain the frequency response of the rows of the complementary sensitivity M. These constraints are:

$$\frac{1}{\pi}\int_{-\infty}^{\infty} \log|M_{k1}(j\omega) - M_{k2}(j\omega)| \frac{3}{\omega^2+9} d\omega \geq 0, \quad k = 1,2. \qquad (9.17)$$

Next we compute the product $G_y G_z^{-1}$, i.e.,

$$G_y(s) G_z^{-1}(s) = \begin{bmatrix} \frac{s-1}{s+3} & \frac{s-5/3}{s+1} \\ 0 & \frac{s-4}{(s+2)^2} \\ 0 & 0 \end{bmatrix},$$

which is in the Hermite form (9.14) with $Q = I$. If the filter F is selected to achieve diagonalization of M, then necessarily M has the form (9.12) with $H = G_y G_z^{-1}$, i.e.,

$$M(s) = \bar{Q}(s) \begin{bmatrix} \frac{(s-1)(s-4)}{(s+\sigma_1)^3} & 0 \\ 0 & \frac{s-4}{(s+\sigma_2)^2} \end{bmatrix}.$$

In this case, Theorem 9.3.1 holds, giving the following integral constraints on the diagonal elements of M:

$$\begin{aligned} \frac{1}{\pi}\int_{-\infty}^{\infty} \log|M_{11}(j\omega)| \frac{3}{\omega^2+9}\, d\omega &= 2.6391, \\ \frac{1}{\pi}\int_{-\infty}^{\infty} \log|M_{22}(j\omega)| \frac{3}{\omega^2+9}\, d\omega &= 1.9459. \end{aligned} \tag{9.18}$$

Now we see that *even without imposing specifications on the values of* M *on the* $j\omega$*-axis*, there are nontrivial bounds arising from the diagonalization. Indeed, on the one hand it follows from (9.17) that

$$\begin{aligned} \|M_{11} - M_{12}\|_\infty &\geq 1, \\ \|M_{21} - M_{22}\|_\infty &\geq 1. \end{aligned}$$

On the other hand, it follows from (9.18) — after diagonalizing, i.e., $M_{12} = 0 = M_{21}$ — that

$$\begin{aligned} \|M_{11}\|_\infty &\geq 14, \\ \|M_{22}\|_\infty &\geq 7. \end{aligned} \tag{9.19}$$

Therefore, a diagonal M is obtained at the cost of large peaks in its diagonal entries. These peaks might be highly undesirable if they happen to occur at frequencies where M is required to be small. Moreover, it is not difficult to see that if frequency specifications on the entries of M are imposed in addition, the design trade-offs arising from (9.19) will be even worse. ◦

Theorem 9.3.1 has immediate application in filtering problems where diagonal sensitivities are desirable. In the next section, we illustrate our results by studying the problem of detection and isolation of faults in the context of multivariable filtering.

9.4 Application to Fault Detection

Fault detection and isolation (FDI) finds application in complex multiple component systems where, for reasons of safety or economics, tolerance to component failure is required. An approach to achieving fault tolerance[2]

[2] Avoiding the expense of hardware redundancy.

consists in exploiting the redundancy inherent in the system model, which is known as *model-based* FDI.

A common method for model-based FDI is to design a filter that generates a *residual*, or fault-sensitive signal, that is distinguishable from zero when a component of the system fails but remains close to zero otherwise. If the residual has different properties for different component faults, then it is also possible to isolate the faulty component. A filter that generates a residual that allows both detection and isolation of faults is called an FDI filter. A number of techniques are available for constructing FDI filters; see for example Massoumnia, Verghese & Willsky (1989) and the references therein.

Two issues are important in FDI filter design. Firstly, the residual should be sensitive to faults at those frequencies where the energy of the fault is likely to be concentrated. Secondly, the residual should be insensitive to other process disturbances and noise. It is well understood that these are conflicting objectives, since both sensitivity to faults and insensitivity to disturbances cannot be achieved at the same frequency. However, the implications of ORHP poles and zeros of G for achieving design objectives over a range of frequencies are less well appreciated.

In this section we show that, by approaching the FDI problem in the context of multivariable filtering, the Poisson integral constraints of §9.2 are useful to quantify the limitations imposed on FDI filters by ORHP poles and zeros of G. Furthermore, in connection with §9.3, these constraints are shown to be more severe due to a structural requirement of diagonalization imposed by the condition of isolation.

We recast the FDI problem as a filtering problem within the general framework of Figure 7.1 in Chapter 7 by taking z as the fault to be detected, and $\hat{z}$ to be the residual. By choosing the structure of G, it is possible to model a wide variety of faults. For example, the schemes of Figure 9.1 represent the important cases of *actuator* and *sensor* faults. A fault in the i-th actuator (or sensor) is modeled by $z_i \neq 0$. Nothing is assumed about the mode of fault, i.e., z is an arbitrary function of time.

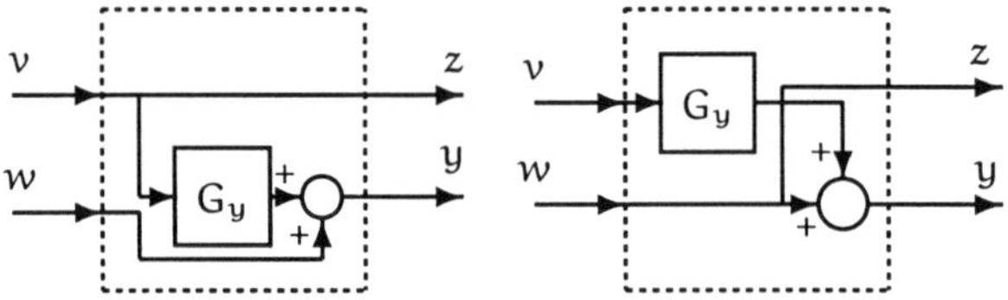

FIGURE 9.1. Plant models for actuator (left) and sensor (right) faults.

In this context, M maps the fault z to the residual $\hat{z}$, and P maps the fault to the "detection error" $\tilde{z}$. Thus, M and P quantify the quality of detection.

We will say that F is an FDI filter if

(i) (detection) for all initial conditions in G and F, and in the absence of disturbances, i.e., $v, w = 0$, then $\hat{z}(t) \to 0$; and

(ii) (isolation) if $z_i \neq 0$, then $\hat{z}_i \neq 0$ and $\hat{z}_j = 0,\ j \neq i$.

Condition (i) is equivalent to the requirement that F is a BEE. Condition (ii) is equivalent to the requirement that M is diagonal. This is not the only structure that allows isolation (Massoumnia et al. 1989), but is often studied because (a) it renders the problem of isolation trivial, and (b) it permits isolation of simultaneous faults.

The filtering sensitivity functions corresponding to the problems of actuator and sensor FDI have the forms given by Table 9.1.

	P	M
Actuator FDI	$I - FG_y$	FG_y
Sensor FDI	$I - F$	F

TABLE 9.1. Filtering sensitivity functions for actuator and sensor FDI.

We illustrate the ideas for the case of sensor faults; similar analyses can be performed for the case of actuator faults. Notice that, at first sight, the problem of sensor FDI may appear to be unconstrained, since P and M only depend on the filter, which is the variable of design. Yet, even for detection alone, F must incorporate information of the unstable dynamics of the plant, since the filter is a BEE. To see this, let $G_y = \tilde{D}^{-1}\tilde{N} = ND^{-1}$, where $(\tilde{D}, \tilde{N})$, and (N, D) are left and right coprime factorizations of G_y (Vidyasagar 1985). Then, using a parametrization of all BEEs (see e.g., Seron & Goodwin 1995), it is easy to show that F necessarily has the form $F = \tilde{Q}\tilde{D}$, where $\tilde{Q}$ is any stable, proper transfer function of appropriate dimensions. Therefore, M in the case of sensor FDI has zeros at the unstable poles of the plant. In addition, if isolation is required, then the filter must not only be a BEE, but also diagonalize the filtering complementary sensitivity M. As we have seen in §9.3, this has the associated cost — quantified by (9.15) — of further worsening the design limitations arising from unstable poles of the plant with noncanonical directions.

For sensor FDI, M not only measures the quality of detection as mentioned above, but it also maps measurement noise w and the relative effect of the input disturbance v to both $\hat{z}$ and $\tilde{z}$, the residual and detection error respectively. Hence P and M may be used to specify the properties of the filter to adequately detect faults and reject disturbances. These objectives can sometimes be conflicting, as is the case for input disturbances and low frequency faults, but can successfully be achieved, for example, in the case of low frequency faults and high frequency disturbances. In the latter case, a suitable shape for the diagonal elements of M will require that

$|M_{kk}(j\omega)|$ be close to one at low frequencies[3] and close to zero at high frequencies where the power spectrum of the disturbance concentrates its energy. These objectives can be translated into the following design specifications:

(i) $|M_{kk}(j\omega)| \approx 1$ on $[0, \omega_1]$,

(ii) $|M_{kk}(j\omega)| < \epsilon_k < 1$ on $[\omega_2, \infty]$,

with $\omega_1 < \omega_2$. Using these requirements and Theorem 9.3.1 it is easy to see that the following lower bound holds for the peak of M_{kk} between ω_1 and ω_2.

Corollary 9.4.1. Assume that the conditions of Theorem 9.3.1 hold. Then the following inequality is satisfied:

$$\sup_{\omega \in [\omega_1, \omega_2]} |M_{kk}(j\omega)| \geq \left(\frac{1}{\epsilon_k}\right)^{\frac{\pi - \Theta_p(\omega_2)}{\Theta_p(\omega_2) - \Theta_p(\omega_1)}} . \tag{9.20}$$

Proof. Immediate from Theorem 9.3.1 and the given specifications. □

We can argue from the above corollary that a large peak may occur if ϵ_k is too small, or if ω_1 is too close to ω_2. As just discussed, this may have a deleterious impact on detection if there are input disturbances with frequency content within $[\omega_1, \omega_2]$.

Corollary 9.4.1 shows the trade-offs in sensor FDI due to the constraints imposed by unstable poles of the plant G_y. Similar trade-offs occur in the case of actuator FDI in connection with nonminimum phase zeros of G_y. These results may also be extended to the case of general component FDI, where there are trade-offs due to both ORHP zeros and poles of the plant.

9.5 Summary

In this chapter we have presented integral constraints for multivariable filtering problems. In contrast to the SISO case, a new dimension arises in the analysis of MIMO filtering limitations since the structural properties of these functions have nonneglectable impact. In particular, problems in which diagonalization is required may introduce additional limitations. As one possible application, we have considered problems of fault detection and isolation in a multivariable filtering setting.

[3]For the k-residual to be sensitive to the k-fault, we require $\tilde{z}_k$ to be small, and therefore $|M_{kk}(j\omega)|$ to be close to one at frequencies where the fault is assumed to occur.

Notes and References

The results in this chapter follow Braslavsky, Seron, Goodwin & Grainger (1996). This latter work is the multivariable extension along the lines of Gómez & Goodwin (1995) of the results in Seron & Goodwin (1995).

Other results on limitations for multivariable filtering are given in Weller (1996), where integral constraints on the maximum singular values of the sensitivity functions are obtained. This latter work follows the ideas of Chen (1995).

10

Extensions to SISO Prediction

In Chapter 7 we defined filtering sensitivities, P and M, and showed that they satisfy a complementarity constraint. Furthermore, for the class of BEEs, we derived, in Chapters 8 and 9, interpolation and integral constraints that these sensitivities must satisfy. As seen, these constraints quantify fundamental limits on the filter achievable performance.

Equivalent sensitivities and constraints are obtained for the closely related estimation problem of BEE-based linear *prediction*, which is the subject of the present chapter. As we will see, the additional cost associated with the process of prediction is clearly quantifiable using the results summarized below by Theorems 10.5.1 and 10.5.2. In particular, we will discuss the influence of the prediction horizon on the achievable performance, as measured by the prediction sensitivity functions. The main conclusions are illustrated by examples where we analyze limitations in predictors based on Kalman filters.

10.1 General Prediction Problem

The problem of prediction consists of estimating a signal z at time $t + \tau$, where $\tau > 0$ is the *prediction horizon*, given observations up to time t. Similar to the filtering case, we will assume here that z is a linear combination of states of a LTI system, of which a different combination of states is available for measurement.

It is always possible to construct predictors from a given filter. To this end, suppose that we want to predict the partial state z of the system

$$\begin{aligned} \dot{x} &= Ax + Bv \,, \qquad x(0) = x_0 \,, \\ z &= C_1 x + D_1 v \,, \\ y &= C_2 x + D_2 v + w \,, \end{aligned} \tag{10.1}$$

where the pair (A, C_2) is detectable. For simplicity, we will also assume that the pair (A, B) is stabilizable. As in the filtering case, we partition the plant (10.1) as shown in Figure 10.1.

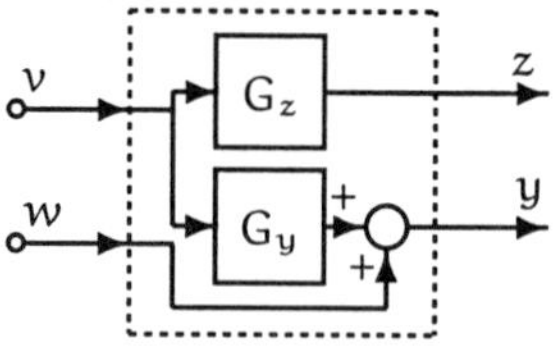

FIGURE 10.1. Structure of the plant.

Suppose that we have already designed a full-state filter having the (transfer function) form

$$\hat{X}(s) = F(s)Y(s) \,. \tag{10.2}$$

We will then consider predictors for z of the form

$$\hat{z}(t + \tau | t) = C_1 e^{A\tau} \hat{x}(t) \,, \tag{10.3}$$

where $\hat{x}(t)$ is the estimate given by the filter (10.2) at time t, and A is the evolution matrix of the system (10.1). The notation $\hat{z}(t + \tau | t)$ indicates a prediction of z at time $t + \tau$ given information up to time t.

To motivate the predictor given by (10.3), note from the system equations (10.1) that

$$z(t + \tau) = C_1 e^{A\tau} x(t) + \int_t^{t+\tau} C_1 e^{A(t+\tau-\sigma)} B v(\sigma) d\sigma \,. \tag{10.4}$$

Thus, future states are a particular linear combination of the current state x plus a function of the process input v over the interval $[t, t + \tau]$. If we assume that at time t we have no information about $v(\cdot)$ over the interval $[t, t + \tau]$, a natural τ-predictor is constructed as in (10.3).

Example 10.1.1. Under the same assumptions that guarantee the optimality in the least-squares sense of the Kalman filter (Kalman & Bucy 1961), Kailath (1968) showed that the least-squares optimal predictor for the system (10.1) has the form (10.3), where $\hat{x}(t)$ is the state estimate given by the Kalman filter. ∘

More generally, (10.3) is a sensible choice as a predictor on the assumption that we have already chosen $\hat{x}(t)$ to be reasonable in some sense.

We define the prediction error to be the difference between the future value of z, i.e., $z(t+\tau)$, and the output of the predictor, namely

$$\tilde{z}(t+\tau|t) \triangleq z(t+\tau) - \hat{z}(t+\tau|t),$$

and, using (10.3)

$$\tilde{z}(t+\tau|t) = z(t+\tau) - C_1 e^{A\tau}\hat{x}(t) . \tag{10.5}$$

In the case of prediction, it is convenient to define a *modified* Laplace transform. For a function $h(t)$ we define it to be

$$H(s) \triangleq \int_{-\tau}^{\infty} e^{-st} h(t)dt .$$

We do this to avoid an extra term in the transform of the shifted state $z(t+\tau)$, which corresponds to the transform of the state $z(t)$ truncated to the interval $[0, \tau]$. This additional term is not affine in the transforms of the input signals (see Appendix C), and hence would prevent us from obtaining an expression for the mapping from input $v(t)$ to predictor error $\tilde{z}(t+\tau|t)$, $H_{\tilde{z}v}$, as a multiplication operator.

Taking the (modified) Laplace transform of the prediction error in (10.5), we have,

$$\tilde{Z}(s) = e^{s\tau} Z(s) - C_1 e^{A\tau}\hat{X}(s) . \tag{10.6}$$

We introduce the transfer function

$$F_p(s) \triangleq C_1 e^{A\tau} F(s) , \tag{10.7}$$

which maps y to $\hat{z}$. Using (10.7) and (10.2) in (10.6), we have

$$\tilde{Z}(s) = e^{s\tau} Z(s) - F_p(s) Y(s) . \tag{10.8}$$

The prediction loop depicting the above equation is shown in Figure 10.2.

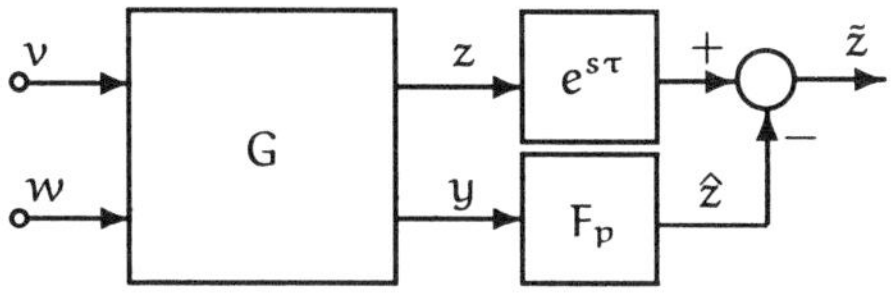

FIGURE 10.2. General configuration for prediction.

As in the filtering case, the definition of the prediction sensitivities uses the mappings $H_{\hat{z}v}$ and $H_{\tilde{z}v}$. In the case of prediction, note that we consider

$H_{\hat{z}v}$ as mapping input $v(t)$ to predictor estimate $\hat{z}(t+\tau|t)$, and $H_{\tilde{z}v}$ as mapping input $v(t)$ to predictor error $\tilde{z}(t+\tau|t)$. The operator $H_{\tilde{z}v}$ can then be derived from (10.8) as

$$H_{\tilde{z}v}(s) = e^{s\tau}H_{zv}(s) - F_p(s)H_{yv}(s) \, . \tag{10.9}$$

Similarly, from (10.3), (10.2) and (10.7), we have

$$H_{\hat{z}v}(s) = F_p(s)H_{yv}(s) \, . \tag{10.10}$$

In the following section we use the above mappings in the definition of sensitivity functions for the problem of prediction.

10.2 Sensitivity Functions

Similarly to the filtering case, we define the *prediction sensitivity* and *complementary sensitivity* functions, denoted by P and M, respectively, as[1]

$$\begin{aligned} P(s) &\triangleq e^{-s\tau}H_{\tilde{z}v}(s)H_{zv}^{-1}(s) \, , \\ M(s) &\triangleq e^{-s\tau}H_{\hat{z}v}(s)\, H_{zv}^{-1}(s) \, , \end{aligned} \tag{10.11}$$

where $H_{\tilde{z}v}$ and $H_{\hat{z}v}$ are the transfer functions given in (10.9) and (10.10), respectively.

Therefore, we obtain the *complementarity constraint for prediction* stated below.

Theorem 10.2.1. The prediction sensitivity and complementary sensitivity defined in (10.11) satisfy:

$$P(s) + M(s) = I, \tag{10.12}$$

at any finite complex frequency s that is not a pole of P and M.

Proof. Using (10.10) in (10.9) yields

$$H_{\tilde{z}v} + H_{\hat{z}v} = e^{s\tau}H_{zv} \, . \tag{10.13}$$

The result then follows on using the definitions (10.11). □

Using the partition of Figure 10.1, we have that $H_{zv} = G_z$ and $H_{yv} = G_y$, and thus P and M are alternatively expressed by

$$\begin{aligned} P(s) &= [G_z(s) - e^{-s\tau}F_p(s)G_y(s)]G_z^{-1}(s) \, , \\ M(s) &= e^{-s\tau}F_p(s)G_y(s)G_z^{-1}(s) \, . \end{aligned} \tag{10.14}$$

In the following section, we establish a result concerning predictors derived from BEEs, which essentially shows that the BEE-derived predictor inherits properties from the originating filter.

[1] In Appendix C, we show the expression of the prediction sensitivities using the usual Laplace transform.

10.3 BEE Derived Predictors

For convenience, we recall from Chapter 7 that a stable filter F in (10.2) is a BEE if and only if the transfer function (7.18), i.e.,

$$\tilde{G}_0(s) = [I - F(s)C_2](sI - A)^{-1} , \tag{10.15}$$

is stable. This, in turn, implies that the filtering error transfer function $H_{\tilde{z}\nu}$, from process input to filtering error, is stable.

We will show here that the prediction error transfer function $H_{\tilde{z}\nu}$ in (10.9) does not have finite CRHP poles whenever the generating filter, $\hat{\mathfrak{X}}$, is a BEE. Note that, due to the presence of the the entire function[2] $e^{s\tau}$ in the expression of $H_{\tilde{z}\nu}$ for prediction, we no longer refer to the *stability* of the transfer function, but rather to its *analyticity* for each finite complex number s in the CRHP.

Lemma 10.3.1 (Analyticity of $H_{\tilde{z}\nu}$). Suppose that (10.2) is a BEE, and consider the predictor given by (10.2), (10.3), for the system (10.1). Then the transfer function $H_{\tilde{z}\nu}$ in (10.9) is analytic for each finite complex number s in the CRHP.

Proof. From (10.1) and Figure 10.1, we have

$$\begin{aligned} H_{z\nu}(s) &= C_1(sI - A)^{-1}B + D_1 , \\ H_{y\nu}(s) &= C_2(sI - A)^{-1}B + D_2 . \end{aligned} \tag{10.16}$$

Using (10.16) in (10.9), we can write

$$H_{\tilde{z}\nu}(s) = \tilde{H}(s) + [e^{s\tau}D_1 - e^{A\tau}F(s)D_2] ,$$

where

$$\tilde{H}(s) \triangleq e^{s\tau}C_1[I - e^{-(sI-A)\tau}F(s)C_2](sI - A)^{-1}B . \tag{10.17}$$

Since F is stable, the term between square brackets in the expression of $H_{\tilde{z}\nu}$ is analytic for each finite complex number s in the CRHP. We hence focus on showing that the same is true for $\tilde{H}$.

For each finite complex number s, we can expand $e^{-(sI-A)\tau}$ in a power series as (cf. (A.64) in Appendix A)

$$e^{-(sI-A)\tau} = I + \sum_{k=1}^{\infty} \frac{(-\tau)^k}{k!}(sI - A)^k .$$

[2]We recall that an entire function is a complex valued function of the complex variable s that is analytic for all finite s.

Using the above expression and (10.15), $\tilde{H}$ in (10.17) can be written as

$$\tilde{H}(s) = e^{s\tau}C_1\tilde{G}_0(s)B - e^{s\tau}C_1\sum_{k=1}^{\infty}\frac{(-\tau)^k}{k!}(sI-A)^kF(s)C_2(sI-A)^{-1}B\,. \tag{10.18}$$

The first term in (10.18) is analytic for each finite s in the CRHP. This is because $\tilde{G}_0(s)$ belongs to a BEE by construction, and hence it is a stable transfer function.

As for the second term in (10.18), the only possibility of unboundedness in the finite CRHP comes from its value at the unstable poles of $(sI-A)^{-1}$, i.e., unstable eigenvalues of the matrix A. Let p be one such eigenvalue. We will show that the second term in (10.18) is actually bounded when evaluated at $s = p$.[3]

Since the generating filter F is a BEE, we know from Chapter 8 that the transfer matrix (10.15) is stable. Pre-multiplying (10.15) from the left by $(sI-A)$, does not introduce extra unstable poles. Hence,

$$(sI-A)[I-F(s)C_2](sI-A)^{-1} = I-(sI-A)F(s)C_2(sI-A)^{-1}$$

is analytic in the finite CRHP. Thus, $s = p$ cannot be a pole of the above expression. Evaluating its RHS at $s = p$, we have that the term $(pI-A)F(p)C_2(pI-A)^{-1}$ is bounded. Since p is finite, it then follows that the value at $s = p$ of the sum in (10.18), namely

$$\sum_{k=1}^{\infty}\frac{(-\tau)^k}{k!}(pI-A)^kF(p)C_2(pI-A)^{-1}B, \tag{10.19}$$

is bounded, showing that the second term in (10.18) is also bounded for each finite s in the CRHP. This completes the proof that $H_{\tilde{z}v}$ is analytic in the finite CRHP. □

The above result will be used to establish interpolation constraints in the following section.

10.4 Interpolation Constraints

In this section we show that the prediction sensitivities satisfy similar interpolation constraints to the ones that affect the filtering sensitivities, plus

[3]Note that $C_2(sI-A)^{-1}B$ may not have a pole at $s = p$ if (A, B) is not assumed to be stabilizable. In this case, the second term in (10.18) is trivially bounded. The analysis that follows, however, is in fact independent of B, and hence shows boundedness of the second term for $B = I$.

additional ones introduced by the prediction process. In the remainder of the analysis for the case of prediction, we will assume that the originating filter is a BEE and we focus on *scalar* systems.

Recall first expressions (10.14) of P and M using the partition of Figure 10.1. As was the case for the filtering sensitivities, P and M are not necessarily analytic in the finite ORHP, since G_z may have ORHP zeros that are not canceled in the division. The following result gives necessary and sufficient conditions for P and M to be analytic functions in the (finite) ORHP.

Lemma 10.4.1. Consider the prediction sensitivities given in (10.14) and assume that F in (10.7) is a BEE. Then P and M are analytic in the (finite) ORHP if and only if one of the following conditions hold:

(i) G_z is minimum phase, or

(ii) every ORHP zero of G_z is also a zero of the product $F_p G_y$.

Proof. Immediate from (10.14). □

The following lemma establishes the interpolation constraints that P and M must satisfy at the ORHP poles and zeros of G_y and G_z.

Lemma 10.4.2 (Interpolation Constraints). Assume that the originating filter F in (10.7) is a BEE. Then P and M must satisfy the following conditions.

(i) If $p \in \mathbb{C}^+$ is a pole of G_z, then

$$P(p) = 0, \quad \text{and} \quad M(p) = 1.$$

(ii) If $q \in \mathbb{C}^+$ is a zero of G_y that is not also zero of G_z, then

$$P(q) = 1, \quad \text{and} \quad M(q) = 0.$$

(iii) If $q \in \mathbb{C}^+$ is a zero of G_z that is not also zero of $F_p G_y$, then $P(s)$ and $M(s)$ have a pole at $s = q$.

Proof. If $p \in \mathbb{C}^+$ is a pole of $G_z = C_1(sI - A)^{-1}B + D_1$, then p is an eigenvalue of A. It follows from Lemma 10.3.1 that $H_{\tilde{z}v}(s)$ is analytic at $s = p$. Case (i) then follows from (10.11), since $H_{zv} = G_z$. Case (ii) is immediate from (10.14) since F — and hence F_p in (10.7) — is stable. Finally, case (iii) follows from Lemma 10.4.1. □

As in §8.1 of Chapter 8, we denote by $\mathcal{Z}_P$ and $\mathcal{Z}_M$ the sets of ORHP zeros of P and M respectively, repeated according to their multiplicities. Then, Lemma 10.4.2 identifies subsets of $\mathcal{Z}_P$ and $\mathcal{Z}_M$. It is easy to see that $\mathcal{Z}_M$ is completed by those zeros of F_p that are not also zeros of G_z. As for P, it may have (possibly infinitely many) other ORHP zeros, which are a result of the prediction action. Specifically, the function M in (10.14) is analytic in a neighborhood of infinity, and, due to the factor $e^{-s\tau}$, it has an essential singularity at infinity (see Example A.8.3 in Appendix A). It then follows from the Great Picard Theorem (Conway 1973) that, in each neighborhood of infinity, M assumes the value 1 (and hence P assumes the value zero) an infinite number of times.

The infinite sequence of zeros of P can be further studied by means of Lemma A.11.1 in Appendix A. Indeed, note that the numerator of P in (10.14) has the form of f in (A.87), for g_1 equal to minus the numerator of $F_p G_y / G_z$ and g_2 equal to the denominator of $F_p G_y / G_z$. Then Lemma A.11.1 applies with

$$\delta = \mathrm{RD}\, \frac{F_p G_y}{G_z}, \tag{10.20}$$

where RD H denotes the relative degree of the transfer function H. If $\delta = 0$, then η is given by

$$\eta = -\lim_{s \to \infty} \frac{G_z(s)}{F_p(s) G_y(s)}. \tag{10.21}$$

In §10.5, we restrict our attention to those cases for which $\delta \geq 0$, where δ is defined in (10.20). Using Lemma A.11.1 we conclude that P has infinitely many zeros in the ORHP when $\delta = 0$ and $|\eta| < 1$ in (10.21). When $\delta > 0$, Lemma A.11.1 indicates that the high frequency zeros of P have negative real part. However, P may still have "low frequency" zeros of positive real part, as shown in the following example.

Example 10.4.1. Let P in (10.14) be given by

$$P = 1 - e^{-s\tau} \frac{e^{a\tau} b}{e_1 s + e_2},$$

where e_1, e_2, b, a are positive real constants. Assume further that $b > e_2$. The above sensitivity corresponds to prediction of a scalar plant having an unstable pole at $s = 1/a$.

The numerator of P is

$$f(s) = e_1 s + e_2 - b e^{a\tau} e^{-s\tau}.$$

It was shown in Example A.11.1 of Appendix A that f will have ORHP zeros inside a semicircular contour of radius $R < (b e^{a\tau} - e_2)/(e_1 + \tau)$ in the ORHP. Moreover, the number of those zeros increases with the prediction horizon τ.
∘

In summary, $\mathcal{Z}_P$ in (8.1), in contrast to $\mathcal{Z}_M$, may have an infinite number of elements, which are zeros of P introduced by the prediction process.

Although some information about these zeros can be obtained in specific cases (see §10.6.2), their exact location is, in general, difficult to compute.

In the following section, we will use the interpolation constraints provided by Lemma 10.4.2 to derive integral relations on the frequency responses of the prediction sensitivities.

10.5 Integral Constraints

Before stating the Poisson integral theorems for prediction, we introduce some notation. Let the sets (8.1) be given by

$$\begin{aligned}\mathcal{Z}_P &= \{p_i : i = 1, \dots, n_p\}\,,\\ \mathcal{Z}_M &= \{q_i : i = 1, \dots, n_q\}\,,\end{aligned}$$

and let the corresponding Blaschke products be[4]

$$B_P(s) = \prod_{i=1}^{n_p} \frac{s - p_i}{s + \overline{p}_i}\,, \qquad \text{and} \qquad B_M(s) = \prod_{i=1}^{n_q} \frac{s - q_i}{s + \overline{q}_i}\,. \tag{10.22}$$

Note that, according to the discussion in the previous section, n_p, i.e., the number of ORHP zeros of P, may be infinity.

Following the filtering case in Chapter 8, we introduce the sets of zeros:

$$\begin{aligned}\mathcal{Z}_z &\triangleq \{s \in \mathbb{C}^+ : G_z(s) = 0 \quad \text{and} \quad F_p(s)G_y(s) \neq 0\}\,,\\ \mathcal{Z}_y &\triangleq \{s \in \mathbb{C}^+ : G_y(s) = 0 \quad \text{and} \quad G_z(s) \neq 0\}\,,\\ \mathcal{Z}_{1/z} &\triangleq \{s \in \mathbb{C}^+ : G_z^{-1}(s) = 0\}\,,\end{aligned} \tag{10.23}$$

and denote the corresponding Blaschke products by B_z, B_y and $B_{1/z}$. Note that $\mathcal{Z}_z$ is the set of ORHP poles of P and M.

In order to restrict the behavior at infinity of the sensitivities, we require the following

Assumption 10.1. $\delta = \mathrm{RD}\,\dfrac{F_p G_y}{G_z} \geq 0.$ ∘

Under Assumption 10.1 the function $B_z P$ is analytic and bounded in the ORHP. It follows from De Branges (1968, Theorem 8, p. 20) that P in (10.14) can be factorized in the form

$$P = \tilde{P} B_P / B_z\,, \tag{10.24}$$

---

[4]Recall that, if the set $\mathcal{Z}$ originating the Blaschke product B is empty, we define $B(s) = 1$, $\forall s \in \mathbb{C}$.

where $\tilde{P}$ has no zeros or poles in the ORHP. The convergence of the Blaschke products in (10.24) also follows from De Branges (1968, Theorem 8, p. 20), even when $n_p = \infty$.

Similarly, M in (10.14) can be factorized as

$$M = e^{-s\tau} \tilde{M} B_M / B_z \,, \tag{10.25}$$

where $\tilde{M}$ has no zeros or poles in the ORHP.

The following results use the factorizations (10.24) and (10.25) to obtain integral constraints on the functions $\log|P|$ and $\log|M|$.

Theorem 10.5.1 (Poisson Integral for P). Suppose that F in (10.7) is a BEE and consider the prediction sensitivity, P, defined in (10.14). Let $q = \sigma_q + j\omega_q$, $\sigma_q > 0$, be a zero of G_y that is not also zero of G_z (i.e., $q \in \mathcal{Z}_y$ in (10.23)). Then, under Assumption 10.1,

$$\int_{-\infty}^{\infty} \log|P(j\omega)| \frac{\sigma_q}{\sigma_q^2 + (\omega_q - \omega)^2} \, d\omega = \pi \log\left|B_P^{-1}(q)\right| - \pi \log\left|B_z^{-1}(q)\right| . \tag{10.26}$$

Proof. $\tilde{P}$ in (10.24) has no zeros or poles in the ORHP. Also, under Assumption 10.1, $B_z P$ is bounded in the ORHP, and thus it is not difficult to see that the same is true for $\tilde{P}$. Hence, we can apply the Poisson integral formula to $\log \tilde{P}$ as in Theorem 3.3.1 of Chapter 3. □

Theorem 10.5.2 (Poisson Integral for M). Suppose that F in (10.7) is a BEE and consider the prediction complementary sensitivity, M, defined in (10.14). Let $p = \sigma_p + j\omega_p$, $\sigma_p > 0$, be a pole of G_z (i.e., $p \in \mathcal{Z}_{1/z}$ in (10.23)). Then, under Assumption 10.1,

$$\int_{-\infty}^{\infty} \log|M(j\omega)| \frac{\sigma_p}{\sigma_p^2 + (\omega_p - \omega)^2} \, d\omega = \pi\sigma_p\tau + \pi \log\left|B_M^{-1}(p)\right| - \pi \log\left|B_z^{-1}(p)\right| . \tag{10.27}$$

Proof. $\tilde{M}$ in (10.25) has no zeros or poles in the ORHP and, since Assumption 10.1 holds, it is bounded in the ORHP. We can then apply the Poisson integral formula to $\log \tilde{M}$ as in Theorem 3.3.1. □

The results in Theorems 10.5.1 and 10.5.2 are affected by the particular choice of predictor. However, we readily obtain the following corollary, which presents a constraint that is independent of the estimator parameters.

Corollary 10.5.3. Consider the prediction sensitivities, P and M, given in (10.14), and suppose that F in (10.7) is a BEE. Consider the sets of zeros defined in (10.23), and assume further that every ORHP zero of G_z is also a zero of G_y (i.e., $\mathcal{Z}_z = \emptyset$). Then, under Assumption 10.1,

(i) if $q = \sigma_q + j\omega_q \in \mathcal{Z}_y$, we have that

$$\int_{-\infty}^{\infty} \log|P(j\omega)| \frac{\sigma_q}{\sigma_q^2 + (\omega_q - \omega)^2}\, d\omega \geq \pi \log\left|B_{1/z}^{-1}(q)\right|; \qquad (10.28)$$

(ii) if $p = \sigma_p + j\omega_p \in \mathcal{Z}_{1/z}$, we have that if, in addition, $\mathcal{Z}_z = \emptyset$, then

$$\int_{-\infty}^{\infty} \log|M(j\omega)| \frac{\sigma_p}{\sigma_p^2 + (\omega_p - \omega)^2}\, d\omega \geq \pi\sigma_p\tau + \pi \log\left|B_y^{-1}(p)\right|.$$

Proof. Same as the proof of Corollary 8.2.3. □

Note that the inequality (10.28) would become an equality if we included a term corresponding to the Blaschke product of the zeros introduced by the prediction process, which are in general not known.

Clearly from the above results, the problem of prediction is subject to similar constraints in performance — arising from ORHP poles and zeros of the plant — as those present in filtering. As a matter of fact, the restrictions in prediction are inherently more severe, as can be directly seen, for example, from the integral constraint on M, which worsens as the prediction horizon τ expands. Indeed, a worse constraint means unavoidable higher sensitivity to noise, which is naturally expected, since the forecast of information, i.e., prediction, generally becomes more difficult in noisy environments (Anderson & Moore 1979, p. 11).

10.6 Effect of the Prediction Horizon

As a way of interpreting Theorems 10.5.1 and 10.5.2, we provide here a more detailed discussion of the influence of the prediction horizon on the frequency responses of the prediction sensitivities and their corresponding integral constraints. In order to keep the analysis simple, we *assume that* G_z *is minimum phase,* i.e., the prediction sensitivities are analytic in the ORHP. We consider two cases separately: (i) large values of τ, which can be studied directly from the expressions of the sensitivities; and (ii) intermediate values of τ, which we study using Theorems 10.5.1 and 10.5.2.

10.6.1 *Large Values of* τ

Consider the expressions of P and M given in (10.14), evaluated on the imaginary axis. Some preliminary conclusions may be drawn for large values of the prediction horizon by just analysing the expressions of the prediction sensitivities. Since the analysis varies if the plant — more precisely, G_z — is stable or not, we treat these two cases separately.

Unstable G_z

If the system matrix A is unstable and at least one unstable eigenvalue is observable from $z = C_1x + D_1v$ (i.e., G_z is unstable), then, from the definition of F_p in (10.7), $C_1e^{A\tau}$ will increase exponentially with τ. Thus, for a sufficiently large prediction horizon, the second term in the expression of P in (10.14) will become dominant, and hence the magnitude of both sensitivities on the imaginary axis will tend to be equally large, which means that there will be equally high sensitivity to noise in both the estimate and the prediction error.

Stable G_z

If the system matrix A is stable, then $|M(j\omega)|$ will become negligible (i.e., $|M(j\omega)| \to 0$) when τ goes beyond several multiples of the dominant time constant of the system[5] observable from z. Correspondingly, the magnitude of $P(j\omega)$ will tend to one at all frequencies. This is the opposite to the ideal desired situation, and may be attributed to the fact that, after such a large τ, no more information about the system is carried by the filter estimate that gives origin to the predictor, and hence there is no point in predicting beyond this time.

10.6.2 *Intermediate Values of* τ

For intermediate values of the prediction horizon, we can obtain further information from the integral constraints (10.26) and (10.27). Once more, we analyze both unstable and stable cases separately, assuming always that G_y has nonminimum phase zeros.

Unstable G_z

The first situation that we consider is where both P and M have zeros independent of the prediction process. This is the case when G_z is unstable and G_y is nonminimum phase (recall that we assume that G_z is minimum phase). Note that both sensitivities are constrained already in the equivalent filtering problem, as seen from Theorems 8.2.1 and 8.2.2 in Chapter 8.

Under these conditions, the term $\sigma_p\tau$ on the RHS of (10.27), (where σ_p is the real part of each unstable pole of G_z), will be present. Hence, independent of the existence of additional ORHP zeros of P originating from the prediction process, the value of the weighted integral of M, in equation (10.27), increases directly proportional to the prediction horizon τ, giving an additional cost to that for filtering under the same conditions.

---

[5]The dominant time constant of a stable system can be taken as the inverse of the real part of the eigenvalue with smallest magnitude for the real part.

The corresponding situation for P is not as straightforward, but we can still draw some conclusions. From §10.4, we know that P has infinitely many zeros, and the asymptotic location of these zeros depends on the relative degree δ given in (10.20). When $\delta > 0$, the sequence of zeros of P converge to the OLHP. However, as shown in Example 10.4.1 (see also Example 10.6.1 below), the first zeros of the sequence are likely to lie in the ORHP, and even increase in number with τ. Thus, the term depending on the inverse of the Blaschke product $B_P(q)$ (where q is an ORHP zero of G_y) on the RHS of (10.26) may be increased with respect to the corresponding filtering term.

When $\delta = 0$, the asymptotic locus of the sequence of prediction zeros of P depends on the value of η in (10.21), i.e., of

$$\eta = -\lim_{s\to\infty} \frac{G_z(s)}{C_1 e^{A\tau} F(s) G_y(s)}, \tag{10.29}$$

where we have replaced F_p using (10.7). Indeed, the sequence of zeros will converge to the ORHP if $|\eta| < 1$ in (10.29). Note that η has the vector $C_1 e^{A\tau}$ in its denominator. Since some unstable eigenvalues of A are observable from z (since G_z is unstable), a similar analysis as before tells us that the denominator of $|\eta|$ will grow exponentially with τ. It follows that there exists a sufficiently large value of τ for which $|\eta|$ will become smaller than 1, and then the term depending on the inverse of the Blaschke product $B_P(q)$ (where q is an ORHP zero of G_y) on the RHS of (10.26) will be increased with respect to the corresponding filtering term. This produces an increase in the value of the weighted integral of P, though this time not directly proportional to τ. The following example illustrates some of these points.

Example 10.6.1. We consider again the Kalman filter design given in Example 8.4.1 of Chapter 8. Following the idea described in §10.1, here we derive a predictor from the Kalman filter obtained with weights $Q = 100$ and $R = 1$. We take the parameter a in (8.23) as $a = 1$.

In contrast with the filtering case, the prediction sensitivity has an infinite number of zeros that may constraint its magnitude on the $j\omega$-axis. As seen in Lemma A.11.1 in Appendix A, the asymptotic location of these zeros depends on the sign of the number δ in (10.20). In this case we have that δ is positive, and hence it follows that the high frequency zeros of P converge to the OLHP. However, P has low frequency ORHP zeros whose number *increase* with the prediction horizon. Figures 10.3 and 10.4 show plots of $\log|P(s)|$ for this example in the case of filtering ($\tau = 0$) and prediction with $\tau = 0.41$, respectively. Notice that the negative peaks in the plots indicate the position of the zeros of P. As we see from these figures, while there is a single ORHP zero at $s = 1$ in the filtering case, there are five (one real and two complex pairs) in the case of prediction.

Figure 10.5 shows the number of zeros of P in the ORHP as a function of the prediction horizon τ on the interval $[0, 1]$. Clearly, the number of ORHP zeros of P is an increasing function of the prediction horizon. This fact will

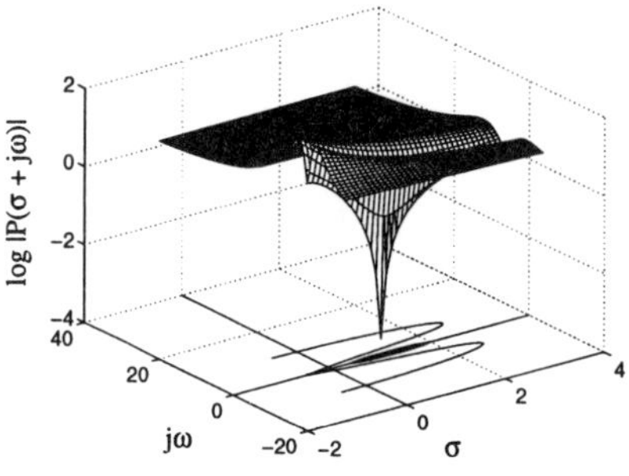

FIGURE 10.3. ORHP zeros of the filtering sensitivity for the system of Example 10.6.1, $\tau = 0$.

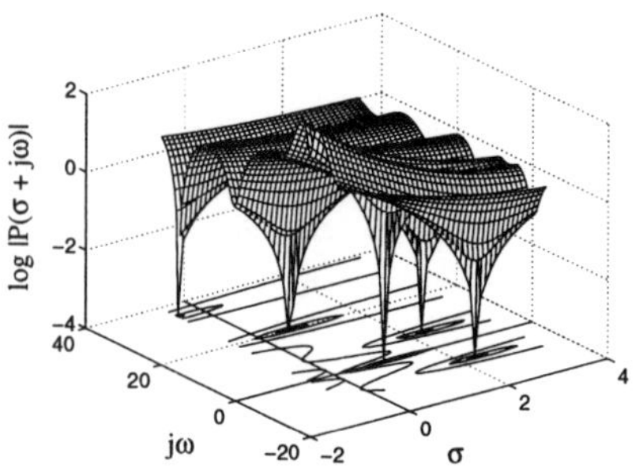

FIGURE 10.4. ORHP zeros of the prediction sensitivity for the system of Example 10.6.1, $\tau = 0.41$.

imply that the weighted integral, on the LHS of (10.26), also increases with the prediction horizon.

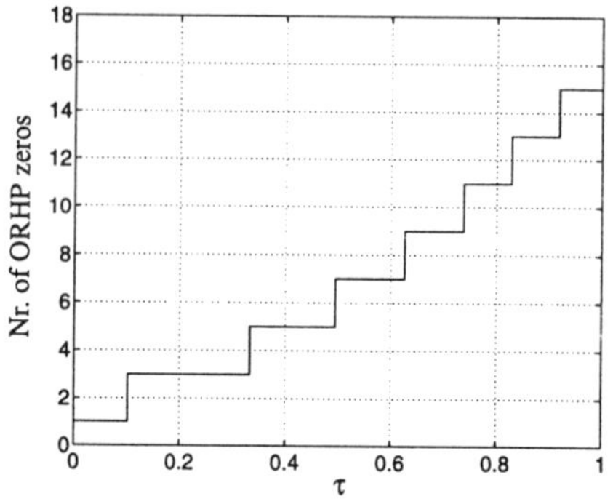

FIGURE 10.5. ORHP zeros of P versus the prediction horizon τ.

As for M, we know from (10.27) and the previous discussion, that its weighted integral increases with the prediction horizon as well.

The frequency responses of $|P|$ and $|M|$ are shown in Figures 10.6 and 10.7, respectively, for three different values of the horizon τ. Note that the filtering sensitivities correspond to $\tau = 0$ in those figures. Also note that, as τ increases, so does the peak in the sensitivity functions, thus confirming the additional cost associated with prediction.

It is interesting to see that, although the magnitude of the prediction sensitivities exceeds that of the filtering sensitivities over almost every frequency range, their shapes approximately follow those of the filtering sensitivities. This phenomenon is in total accordance with what we discussed before: the same weighted integrals on both sensitivities, which already constrained the filtering problem, now achieve larger values for prediction. ◦

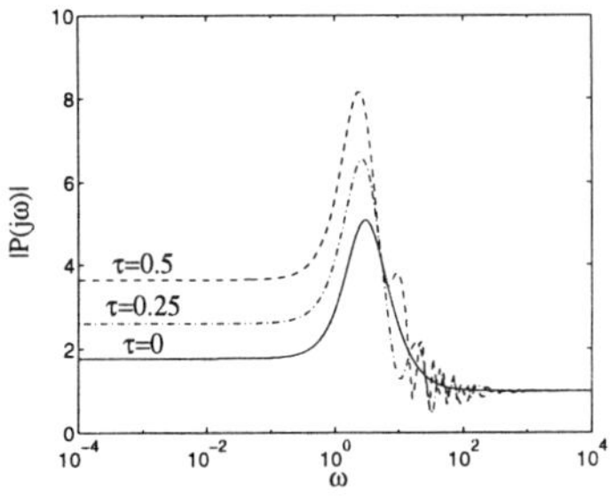

FIGURE 10.6. |P| achieved by the Kalman filter-based predictor for the unstable plant of Example 10.6.1.

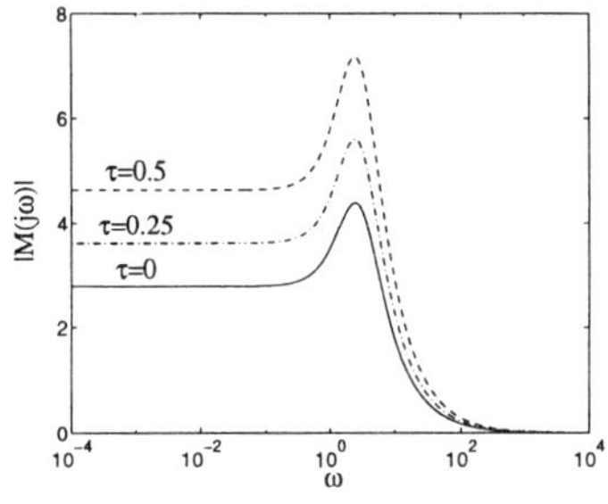

FIGURE 10.7. |M| achieved by the Kalman filter-based predictor for the unstable plant of Example 10.6.1.

Summarizing, for the case of an unstable and nonminimum phase plant, — i.e., when both sensitivities are already constrained in the filtering case — *the prediction process worsens the constraints imposed by ORHP poles and zeros of the plant with respect to the corresponding filtering case*. Indeed, prediction adds positive terms to the RHSs of (10.27) and (in general) of (10.26) too. This means that *the same weighted integrals* on the LHS (with the prediction sensitivities replacing the filtering sensitivities, of course) will *achieve higher values for prediction*. Additional integral constraints on M also appear when there are new ORHP zeros of P introduced by prediction.

Stable G_z

We analize now the effect of τ when the plant is stable and G_y has ORHP zeros. In this case only M has zeros independent of the prediction horizon, and therefore, only P is already constrained in the corresponding filtering problem. Concerning the sequence of zeros of P for $\tau \neq 0$, when $\delta > 0$, the zeros converge to the OLHP and, for a stable plant, it is not clear that the first zeros of the sequence would lie in the ORHP. In this case, Theorems 10.5.1 and 10.5.2 do not indicate additional costs associated with prediction. The following example illustrates this.

Example 10.6.2 (Stable Plant. Case $\delta > 0$). Consider the plant given by the following state-space model,

$$A = \begin{bmatrix} -2 & -3 \\ 1 & 0 \end{bmatrix}, \quad B = \begin{bmatrix} 1 \\ 0 \end{bmatrix}, \quad C_1 = \begin{bmatrix} 1 & 2 \end{bmatrix}, \quad C_2 = \begin{bmatrix} 1 & -2 \end{bmatrix}. \tag{10.30}$$

For this stable plant, we construct a predictor based on a Kalman filter obtained with the weights $Q = 100$ and $R = 1$, and the same noise model of Example 8.4.1 with the parameter a in (8.23) taken as $a = 1$.

Since G_y has a zero at $q = 2$, then the weighted integral of P in (10.26) can be evaluated with the RHS equal to 1. This, however, is the same case as that of filtering for the same problem. The achieved $|P(j\omega)|$ is shown in

Figure 10.9 for increasing values of the prediction horizon. It is seen from this figure that, although the value of the weighted integral in (10.26) is the same as that for filtering, the shape of $|P(j\omega)|$ varies with τ.

For this example, δ in (10.21) is equal to one. This says that the sequence of infinite zeros of P converges to the OLHP. Further, no low frequency ORHP zero was found for several values of τ smaller than the dominant time constant of the system. Then, the weighted integral of M in (10.27) cannot even be stated. Thus, for this example, no additional cost associated with prediction is given by Theorems 10.5.1 and 10.5.2. Note that, the achieved $|M(j\omega)|$, shown in Figure 10.8, is seen to decrease to zero as τ increases, whilst the achieved $|P(j\omega)|$ tends to 1.

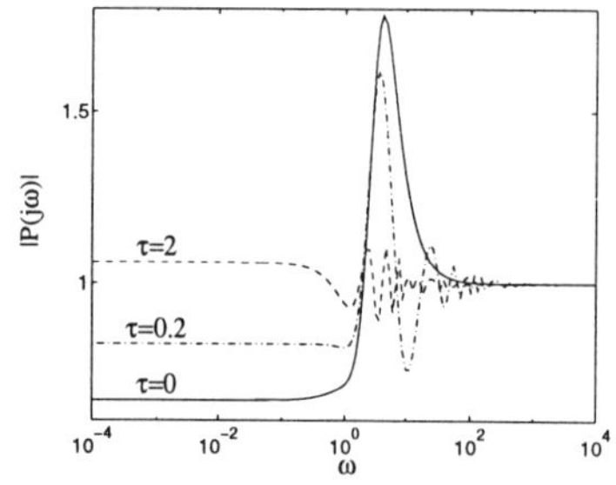

FIGURE 10.8. $|P|$ resulting from the Kalman filter-based predictor for the stable plant of Example 10.6.2.

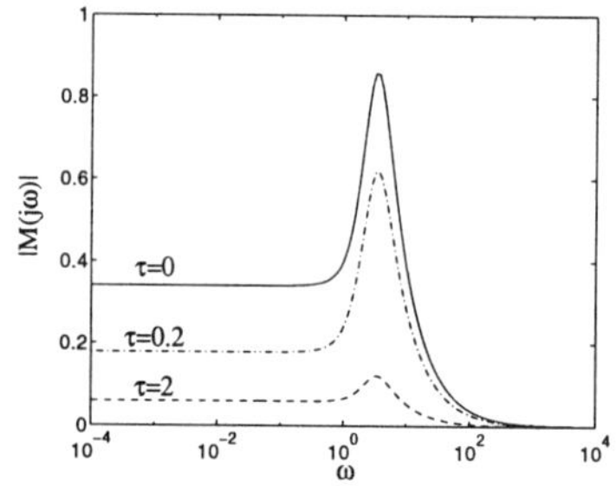

FIGURE 10.9. $|M|$ achieved by the Kalman filter-based predictor for the stable plant of Example 10.6.2.

$\circ$

In the case of a stable plant, the interesting situation is then when $\delta = 0$. Assume further that for a prediction horizon much smaller than the dominant time constant of the system observable from z, $|\eta|$ in (10.29) is smaller than 1. For such a τ, then, P has an infinite sequence of zeros, p_i say, converging to the ORHP. Hence, the term depending on the inverse of the Blaschke product $B_P(q)$ (where q is an ORHP zero of G_y) on the RHS of (10.26) will be increased with respect to the corresponding filtering term. Similarly, the integral constraint (10.27) holds for each zero p_i, with the terms $\mathrm{Re}(p_i)\tau$ and $\log|B_y^{-1}(p_i)|$, on its RHS. The values of both RHSs in (10.26) and (10.27) are thus clear functions of τ. But this stops when τ is large enough so that $C_1 e^{A\tau}$ in the denominator of η in (10.29) turns the value of $|\eta|$ larger than 1. Thus, unless P has low frequency zeros in the ORHP, the RHSs of both weighted integrals become independent of τ. In particular, the integral constraint (10.27) on M disappears.

Example 10.6.3 (Stable Plant. Case $\delta = 0$.). Consider a Kalman filter-based predictor ($Q = 100$, $R = 1$ and $\alpha = 1$) for the system given by A, B and C_2 in (10.30), but with $C_1 = [0 \quad 1]$. In this case, $\delta = 0$ in (10.20). Moreover, it was found that, for τ smaller than approximately 0.77, $|\eta|$ in (10.29) was smaller than one. Thus, for $0 < \tau < 0.77$, P has an infinite sequence of zeros, p_i say, converging to the ORHP. Hence, the term depending on the inverse of the Blaschke product $B_P(q)$, for $q = 2$ (i.e., the zero of G_y), on the RHS of (10.26) will be increased with respect to the corresponding filtering term. Similarly, the integral constraint (10.27) holds for each zero p_i, with the terms $\mathrm{Re}(p_i)\tau$ and $\log|B_y^{-1}(p_i)|$, on its RHS.

Note that the worst constraints for M come from the first zeros of the infinite sequence of zeros of P, which are the closest to the nonminimum phase zero of M at $q = 2$. This is because the magnitude of the inverse of a Blashke product achieves it maximum at the complex point equal to the zero that generates it, and then decreases monotonically to 1. It was found by plotting that, for $\tau = 0.25$ the first pair of prediction zeros of P where approximately $p_{1,2} = 1.9 \pm j14.2$. For $\tau = 0.5$ the first pair of prediction zeros of P where approximately $p_{1,2} = 0.6 \pm j7.8$. Then, the RHS of (10.27) is larger for $\tau = 0.25$ than for $\tau = 0.5$, particularly due to the term $\mathrm{Re}(p_i)\tau$. This effect is seen from Figure 10.11, where the magnitude of $|M(j\omega)|$ decreases as τ increases.

Figure 10.10 shows the achieved shapes of $|P(j\omega)|$ for three values of τ in the above-mentioned range. Note that peaks greater than one continue to appear. However, since it is expected that the level of low frequency reduction achieved by the filtering sensitivity ($\tau = 0$) should be worsened by the process of prediction, then the peaks outside this low frequency range do not exceed those achieved for filtering.

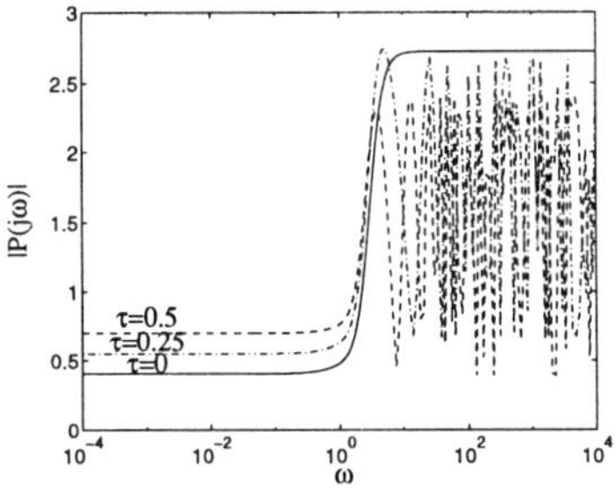

FIGURE 10.10. |P| resulting from the Kalman filter-based predictor for the stable plant of Example 10.6.3.

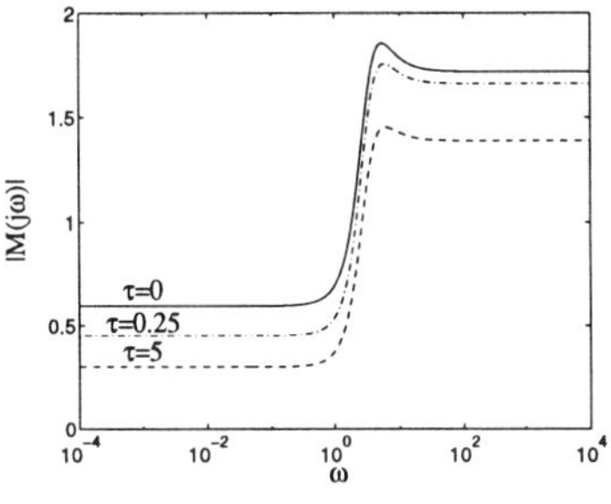

FIGURE 10.11. |M| resulting from the Kalman filter-based predictor for the stable plant of Example 10.6.3.

○

Summarizing, for the case of stable and nonminimum phase plants, *only the filtering sensitivity* P is constrained. In the case of $\delta = 0$, and for a certain range of values of the prediction horizon, the prediction process *adds a term to the RHSs of* (10.26) with respect to the corresponding value for filtering. This means that *the same weighted integral* on the LHS of (10.26) will *achieve higher values for prediction*. Also, *the complementary sensitivity* M *will start to be constrained*. Moreover, infinitely many integral constraints (10.27) can be stated on M, one for each of the prediction zeros.

10.7 Summary

This chapter has extended the sensitivity analysis developed in Chapter 8 for the problem of filtering with bounded error estimators to the case of prediction.

The results show that there is an additional quantifiable cost associated with prediction, summarized in Theorems 10.5.1 and 10.5.2. The exact values of the weighted integrals of P and M are, in general, not computable. However, we have shown via different situations that the integral constraints can be used to study the effect of the prediction horizon on the frequency responses of the sensitivities.

Notes and References

The sensitivity results of this chapter are based on Seron (1995). Additional material was taken from Kailath (1981) and Kailath (1968).

11
Extensions to SISO Smoothing

Following similar developments to those in the previous chapter, a theory of design limitations can also be extended to problems of *fixed-lag smoothing*. In this chapter, we define appropriate complementary sensitivities and obtain fundamental limitations that apply to scalar smoothers derived from BEEs.

As opposed to the case of prediction, which induces additional costs to that encountered in the corresponding filtering problem, smoothing accounts for an improvement in performance proportional to the extent of the smoothing lag. This is certainly unsurprising, since more information is taken into account in order to produce a *smoothed estimate*, and so smoothers will generally be expected to perform better than filters (Anderson & Moore 1979, Chapter 7). Here, we cast in a sensitivity setting this well-known improvement phenomenon due to smoothing.

11.1 General Smoothing Problem

The fixed lag smoothing problem consists of estimating a signal z at time $t - \tau$, where $\tau > 0$ is the *smoothing lag*, given measurements up to time t. Hence, smoothing may be seen as the counterpart of the problem of prediction dealt with in the preceding chapter.

Many solutions to the continuous time smoothing problem appeared in the literature during the 1960's. Kailath & Frost (1968) showed that many of the already existing formulae could be derived using the innovations approach to least-squares estimation. In their work it was also established that the least-squares smoothing solution is completely determined by the results for the least-squares filtering problem. This result is valid for a general second order (finite-variance) signal process in white noise. The class of smoothers that we consider here is inspired by the general smoothing formula of Kailath & Frost (see also Kailath 1981).

As before, we assume that the signal z is the partial state of the system

$$\begin{aligned} \dot{x} &= Ax + Bv\,, \qquad x(0) = x_0\,, \\ z &= C_1 x + D_1 v\,, \\ y &= C_2 x + D_2 v + w\,, \end{aligned} \tag{11.1}$$

where the pair (A, C_2) is detectable and the pair (A, B) is stabilizable. Suppose that we have constructed the following full-state filter for (11.1):

$$\begin{aligned} \dot{\xi} &= \hat{A}\xi + K_y y, \qquad \xi(0) = \xi_0\,, \\ \hat{x} &= \hat{C}\xi\,, \end{aligned} \tag{11.2}$$

and let its Laplace transform be

$$\hat{X}(s) = F(s)Y(s)\,. \tag{11.3}$$

We define the associated innovations process as (see §7.2.1 in Chapter 7)

$$\iota \triangleq y - C_2\hat{x}\,, \tag{11.4}$$

where y is the measured output of (11.1). The innovations process $\iota(t)$ may be regarded as the new information that is brought into the system at time t (Kailath 1968).

In the present problem of smoothing, we want to estimate $z(t-\tau)$ using all the data up to time t. A natural way to do this is to add a linear function of the innovations process over the interval $[t-\tau, t]$ to the filtered estimate at time $t-\tau$. Motivated by the structure of the optimal least-squares smoother (Kailath & Frost 1968), we will then consider smoothers of the form

$$\hat{z}(t-\tau|t) = C_1\hat{x}(t-\tau) + \int_{t-\tau}^{t} C_1 V_1 e^{\hat{A}'(\sigma-t+\tau)} V_2 \iota(\sigma)\,d\sigma\,, \tag{11.5}$$

where $\hat{A}'$ is the transpose of the state transition matrix of the filter (11.2), and V_1, V_2 are constant matrices of appropriate dimensions. The notation $\hat{z}(t-\tau|t)$ indicates estimation at time $t-\tau$ given information up to time t. Note that, for consistency with the definition of the filter (11.2), the matrix V_1 must be of the form $V_1 = \hat{C}\bar{V}_1$, for some matrix $\bar{V}_1$.

Smoothers of the form (11.5) include, for example, the least-squares smoother of Kailath & Frost (1968).

Example 11.1.1. Assume that, in (11.1), $D_1 = D_2 = 0$, and v and w are uncorrelated white noises with incremental covariances equal to Q and R, respectively. Further, assume that the initial condition $x(0)$ is a zero-mean random variable uncorrelated with the process noise v.

Let $\hat{x}(t)$ be the state estimate given by the Kalman filter for the system (11.1) and let $\Phi \geq 0$ be a stabilizing solution to the Riccati equation

$$A\Phi + \Phi A' + BQB' - \Phi C_2' R^{-1} C_2 \Phi = 0. \tag{11.6}$$

Using the innovations technique, Kailath & Frost (1968) showed that a fixed lag ($\tau > 0$) least-squares smoother for the full-state x can be derived from the Kalman filter as follows

$$\hat{x}(t-\tau|t) = \hat{x}(t-\tau) + \int_{t-\tau}^{t} \Phi e^{\hat{A}'(\sigma - t + \tau)} C_2' R^{-1} \iota(\sigma) d\sigma\,, \tag{11.7}$$

i.e., in the general expression (11.5), the special choices $V_1 = \Phi$ and $V_2 = C_2' R^{-1}$ are made in this particular design methodology. ∘

We define the smoothing error to be the difference between the past value of z, i.e., $z(t-\tau)$, and the output of the smoother, namely

$$\tilde{z}(t-\tau|t) \triangleq z(t-\tau) - \hat{z}(t-\tau|t)\,.$$

Using (11.5), we have

$$\tilde{z}(t-\tau|t) = z(t-\tau) - C_1\hat{x}(t-\tau) - \int_{t-\tau}^{t} C_1 V_1 e^{\hat{A}'(\sigma-t+\tau)} V_2 \iota(\sigma) d\sigma\,. \tag{11.8}$$

As before, we introduce the input-output operators $H_{\hat{z}v}$, mapping input $v(t)$ to smoother estimate $\hat{z}(t-\tau|t)$, and $H_{\tilde{z}v}$, mapping input $v(t)$ to smoothing error $\tilde{z}(t-\tau|t)$. In order to define the smoothing sensitivity functions, we require the Laplace transforms of these operators. We start deriving the transform of the integral in (11.8). Making the change of variable $\alpha = t - \sigma$ in this integral, we obtain, assuming that $\iota(t) = 0,\ t < 0$,

$$\begin{aligned}\int_{t-\tau}^{t} C_1 V_1 e^{\hat{A}'(\sigma-t+\tau)} V_2 \iota(\sigma) d\sigma &= \int_{0}^{\tau} C_1 V_1 e^{\hat{A}'(\tau-\alpha)} V_2 \iota(t-\alpha) d\alpha \\ &= \int_{0}^{t} C_1 V_1 e^{\hat{A}'(\tau-\alpha)} V_2 \iota(t-\alpha) d\alpha \\ &\triangleq \int_{0}^{t} H_s(\alpha)\, \iota(t-\alpha)\, d\alpha \\ &= (H_s * \iota)(t)\,,\end{aligned} \tag{11.9}$$

where the symbol "$*$" denotes real convolution and where the function H_s is defined as (Anderson & Doan 1977)

$$H_s(t) = \begin{cases} C_1 V_1 e^{\hat{A}'(\tau - t)} V_2 & \text{if } 0 \le t \le \tau, \\ 0 & \text{otherwise.} \end{cases}$$

It thus follows that the Laplace transform of (11.9) is $H_s(s)\mathcal{I}(s)$, where $\mathcal{I}$ denotes the Laplace transform of the innovations, and where

$$H_s(s) \triangleq C_1 V_1 (sI + \hat{A}')^{-1} \left[e^{\tau \hat{A}'} - e^{-s\tau} I \right] V_2 \,. \tag{11.10}$$

The transforms of the smoother (11.5), and of the smoothing error (11.8), are then given by

$$\begin{aligned} \hat{Z}(s) &= C_1 e^{-s\tau} \hat{X}(s) + H_s(s)\mathcal{I}(s), \\ \tilde{Z}(s) &= e^{-s\tau} Z(s) - C_1 e^{-s\tau} \hat{X}(s) - H_s(s)\mathcal{I}(s) \,. \end{aligned} \tag{11.11}$$

Next, note from (11.4) and (11.3) that the transform of the innovations is given by

$$\mathcal{I}(s) = [I - C_2 F(s)] Y(s) \,. \tag{11.12}$$

Hence, using (11.3) and (11.12), the equations in (11.11) can be written as

$$\begin{aligned} \hat{Z}(s) &= F_s(s) Y(s) \,, \\ \tilde{Z}(s) &= e^{-s\tau} Z(s) - F_s(s) Y(s) \,, \end{aligned} \tag{11.13}$$

where we have used the definition

$$F_s(s) \triangleq e^{-s\tau} C_1 F(s) + H_s(s)[I - C_2 F(s)] \,, \tag{11.14}$$

which is the transfer function that maps y to $\hat{z}$. The smoothing loop depicting equations (11.13) is shown in Figure 11.1.

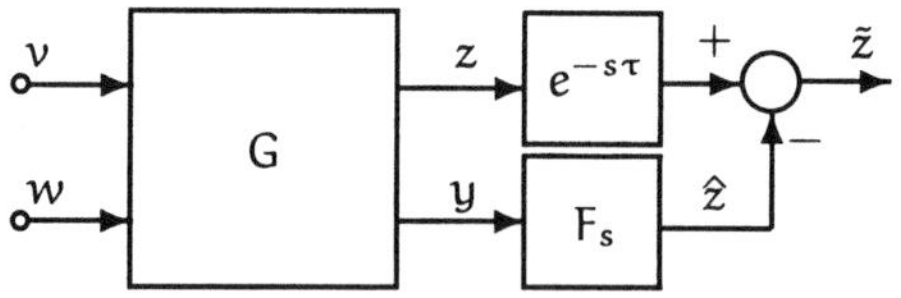

FIGURE 11.1. General configuration for smoothing.

The operators $H_{\hat{z}v}$ and $H_{\tilde{z}v}$ can now be derived from (11.13) as

$$H_{\hat{z}v}(s) = F_s(s) H_{yv}(s) \,, \tag{11.15}$$

$$H_{\tilde{z}v}(s) = e^{-s\tau} H_{zv}(s) - F_s(s) H_{yv}(s) \,. \tag{11.16}$$

In the following section we use the above mappings in the definition of sensitivity functions for fixed-lag smoothing.

11.2 Sensitivity Functions

As for the filtering and prediction problems, we define the *smoothing sensitivity* and *complementary sensitivity* functions, denoted also by P and M, respectively, as

$$P(s) \triangleq e^{s\tau} H_{\tilde{z}v}(s) H_{zv}^{-1}(s),$$
$$M(s) \triangleq e^{s\tau} H_{\hat{z}v}(s) H_{zv}^{-1}(s),$$
(11.17)

where $H_{\tilde{z}v}$ and $H_{\hat{z}v}$ are the transfer functions given in (11.16) and (11.15), respectively.

Therefore, we obtain the *complementarity constraint for smoothing* stated below.

Theorem 11.2.1. The smoothing sensitivity and complementary sensitivity defined in (11.17) satisfy:

$$P(s) + M(s) = I,$$
(11.18)

at any finite complex frequency s that is not a pole of P and M.

Proof. Using (11.15) in (11.16) yields

$$H_{\tilde{z}v} + H_{\hat{z}v} = e^{-s\tau} H_{zv}.$$
(11.19)

The result then follows on using the definitions (11.17). □

In the following section, we establish a result concerning smoothers derived from BEEs.

11.3 BEE Derived Smoothers

We will show here that the smoothing error transfer function $H_{\tilde{z}v}$ in (11.16) does not have finite CRHP poles whenever the generating filter, F in (11.3), is a BEE. Recall from Chapter 7 that a stable full-state filter F is a BEE if and only if the transfer function (7.18) is stable.

We first show that the smoother transfer function F_s in (11.14) is analytic in the finite CRHP if the generating filter F is stable.

Lemma 11.3.1. If F is stable then $F_s(s)$ in (11.14) is analytic at each finite complex number s in the CRHP.

Proof. Since F in (11.14) is stable, it only remains to show that H_s is analytic in the finite CRHP. In fact, H_s is an entire function, i.e., analytic for all finite complex frequency s. This is because the factor $[e^{\tau \hat{A}'} - e^{-s\tau} I]$ in the expression for H_s in (11.10) cancels the unstable poles of $(sI + \hat{A}')^{-1}$. Indeed,

$$\begin{aligned} e^{\tau\hat{A}'} - e^{-s\tau}I &= e^{-s\tau}[e^{\tau(sI+\hat{A}')} - I] \\ &= e^{-s\tau}\sum_{k=1}^{\infty}\frac{(-\tau)^k}{k!}(sI+\hat{A}')^k , \end{aligned}$$

and so

$$\begin{aligned} H_s &= C_1V_1(sI+\hat{A}')^{-1}[e^{\tau\hat{A}'} - e^{-s\tau}I]V_2 \\ &= e^{-s\tau}C_1V_1\sum_{k=1}^{\infty}\frac{(-\tau)^k}{k!}(sI+\hat{A}')^{k-1} \end{aligned}$$

is analytic in the whole finite complex plane. Hence the result follows. □

We then have the following result on analyticity of $H_{\tilde{z}\nu}$.

Lemma 11.3.2. Consider the smoother given by (11.3), (11.5), for the system (11.1). Suppose that (11.3) is a BEE. Then the transfer function $H_{\tilde{z}\nu}$ in (11.16) is analytic for each finite complex number s in the CRHP.

Proof. Replacing F_s from (11.14) into (11.16), we can write

$$H_{\tilde{z}\nu}(s) = e^{-s\tau}\tilde{H}_1(s) - H_s(s)\tilde{H}_2(s) , \tag{11.20}$$

where

$$\begin{aligned} \tilde{H}_1 &\triangleq H_{z\nu} - C_1FH_{y\nu} , \\ \tilde{H}_2 &\triangleq (I - C_2F)H_{y\nu} . \end{aligned}$$

Using the expressions for $H_{z\nu}$ and $H_{y\nu}$ given in (10.16), yields

$$\begin{aligned} \tilde{H}_1(s) &= C_1[I - F(s)C_2](sI - A)^{-1}B + D_1 - C_1F(s)D_2 , \\ \tilde{H}_2(s) &= C_2[I - F(s)C_2](sI - A)^{-1}B + [I - C_2F(s)]D_2 , \end{aligned}$$

which are stable transfer functions since F is a BEE (i.e., F and $[I - F(s)C_2](sI - A)^{-1}$ are stable). Since H_s is entire (as shown in the proof of Lemma 11.3.1), inspection of (11.20) then shows that $H_{\tilde{z}\nu}$ in is analytic for each finite complex number s in the CRHP. This completes the proof. □

The above results will be used to establish interpolation constraints in the following section.

11.4 Interpolation Constraints

In this section we extend the result of §10.4 in Chapter 10 to the case of smoothing. In the remainder of the analysis of the smoothing problem, we will assume that the originating filter is a BEE and we focus on *scalar* systems.

We first express P and M in (10.11) using the partition of Figure 10.1. Since $H_{z\nu} = G_z$ and $H_{y\nu} = G_y$, we have that

$$\begin{aligned} P(s) &= [G_z(s) - e^{s\tau}F_s(s)G_y(s)]G_z^{-1}(s) , \\ M(s) &= e^{s\tau}F_s(s)G_y(s)G_z^{-1}(s) . \end{aligned} \tag{11.21}$$

As was the case for the filtering and prediction problems, P and M are not necessarily analytic in the finite ORHP. The following result gives necessary and sufficient conditions for P and M to be analytic functions in the (finite) ORHP.

Lemma 11.4.1. Consider the smoothing sensitivities given in (11.21) and assume that F in (11.14) is a BEE. Then P and M are analytic in the (finite) ORHP if and only if one of the following conditions hold:

(i) G_z is minimum phase, or

(ii) every ORHP zero of G_z is also a zero of the product $F_s G_y$.

Proof. Immediate from (11.21) and Lemma 11.3.1. □

The following lemma establishes the interpolation constraints that P and M must satisfy at the ORHP poles and zeros of G_y and G_z.

Lemma 11.4.2 (Interpolation Constraints). Assume that the originating filter F in (11.14) is a BEE. Then P and M must satisfy the following conditions.

(i) If $p \in \mathbb{C}^+$ is a pole of G_z, then

$$P(p) = 0, \qquad \text{and} \qquad M(p) = 1.$$

(ii) If $q \in \mathbb{C}^+$ is a zero of G_y that is not also zero of G_z, then

$$P(q) = 1, \qquad \text{and} \qquad M(q) = 0.$$

(iii) If $q \in \mathbb{C}^+$ is a zero of G_z that is not also zero of $F_s G_y$, then $P(s)$ and $M(s)$ have a pole at $s = q$.

Proof. Case (i) follows from (11.17), on noting that $H_{\tilde{z}\nu}(s)$ is analytic in the ORHP by Lemma 11.3.2 and hence it cannot cancel any unstable pole of G_z. Case (ii) is immediate from (11.21) since F_s is stable by Lemma 11.3.1. Finally, case (iii) follows from Lemma 11.4.1. □

Recall from §8.1 in Chapter 8 that we denote by $\mathcal{Z}_P$ and $\mathcal{Z}_M$ the sets of ORHP zeros of P and M respectively, repeated according to their multiplicities. Then, Lemma 11.4.2 identifies subsets of $\mathcal{Z}_P$ and $\mathcal{Z}_M$. However, these zeros are in general only a proper subset of the complete set of zeros of the smoothing sensitivities. In the case of prediction, we have shown in §10.4 of Chapter 10, that P may have ORHP zeros other than the poles of G_z. Here, *both sensitivities* P *and* M may incorporate other ORHP zeros that are inherent to the process of smoothing. Indeed, it is easy to see that P and M can be written as

$$P = e^{s\tau}\frac{f_1}{d_1}, \qquad \text{and} \qquad M = e^{s\tau}\frac{f_2}{d_2},$$

where d_1 and d_2 are polynomials in s, and f_1 and f_2 have the form of (A.87) in Appendix A, i.e.,

$$f_i(s) = g_1^i(s)e^{-s\tau} + g_2^i(s), \qquad i = 1, 2.$$

Therefore, from Lemma A.11.1 in Appendix A (on zeros of entire functions), we have that both P and M have an infinite sequence of zeros converging to infinity. Furthermore, according to the relative orders of the pairs (g_1^i, g_2^i), $i = 1, 2$, (and depending on the ratio between the highest order coefficients if they have the same order), some of these zeros (possibly infinitely many) may have positive real parts. A further analysis[1] shows that P has generically infinitely many zeros converging to the ORHP, whilst the situation for M cannot, in general, be predicted. However, we remark that the complementary sensitivity M achieved by the smoother derived from the Kalman filter (see Example 11.1.1) has infinitely many zeros converging to the OLHP.

In summary, both sets of zeros $\mathcal{Z}_P$ and $\mathcal{Z}_M$ may have an infinite number of elements — other than those identified in Lemma 11.4.2 — which are zeros of P and M introduced by the smoothing process.

In the following section, we will use the information of Lemma 11.4.2 to derive Poisson integral constraints on the frequency responses of the smoothing sensitivities.

11.5 Integral Constraints

Let the Blaschke products corresponding to the sets $\mathcal{Z}_P$ and $\mathcal{Z}_M$ be given by (10.22) in Chapter 10, where now both n_p and n_q may be infinity. Also,

[1]See Seron (1995) for more details.

similar to Chapter 10, we introduce the sets of zeros:

$$\begin{aligned} \mathcal{Z}_z &\triangleq \{s \in \mathbb{C}^+ : G_z(s) = 0 \quad \text{and} \quad F_s(s)G_y(s) \neq 0\}\,, \\ \mathcal{Z}_y &\triangleq \{s \in \mathbb{C}^+ : G_y(s) = 0 \quad \text{and} \quad G_z(s) \neq 0\}\,, \\ \mathcal{Z}_{1/z} &\triangleq \{s \in \mathbb{C}^+ : G_z^{-1}(s) = 0\}\,, \end{aligned} \tag{11.22}$$

and also denote the corresponding Blaschke products by B_z, B_y and $B_{1/z}$.

In order to restrict the behavior at infinity of the sensitivities, we require the following assumption on the relative degree of the ratio G_y/G_z.

Assumption 11.1. F in (11.14) is proper and RD $\dfrac{G_y}{G_z} \geq 0$. ∘

Under Assumption 11.1, the rational factors of P and M in (11.21) are proper. It can then be seen by inspection of (11.21) and (11.14) that, under Assumption 11.1, the functions $e^{-s\tau}P$ and $e^{-s\tau}M$ will be bounded functions for $s \to \infty$ in the ORHP. It thus follows from De Branges (1968, Theorem 8, p.20) that P and M in (11.21) can be factored as

$$\begin{aligned} P &= e^{s\tau}\tilde{P}B_P/B_z\,, \\ M &= e^{s\tau}\tilde{M}B_M/B_z\,, \end{aligned} \tag{11.23}$$

where $\tilde{P}$ and $\tilde{M}$ have no zeros or poles in the ORHP. The convergence of the Blaschke products in (11.23) also follows from De Branges (1968), even when $n_p = \infty$ and/or $n_q = \infty$.

The smoothing sensitivities satisfy the following integral constraints.

Theorem 11.5.1 (Poisson Integral for P). Suppose that F in (11.14) is a BEE and consider the smoothing sensitivity, P, defined in (11.21). Let $q = \sigma_q + j\omega_q$, $\sigma_q > 0$, be a zero of G_y that is not also zero of G_z (i.e., $q \in \mathcal{Z}_y$ in (11.22)). Then, under Assumption 11.1,

$$\int_{-\infty}^{\infty} \log|P(j\omega)| \frac{\sigma_q}{\sigma_q^2 + (\omega_q - \omega)^2}\, d\omega = \pi \log\left|B_P^{-1}(q)\right| - \pi \log\left|B_z^{-1}(q)\right| - \pi\sigma_q\tau. \tag{11.24}$$

Proof. $\tilde{P}$ in (11.23) has no zeros or poles in the ORHP. Also, under Assumption 11.1, $e^{-s\tau}B_zP$ is bounded in the ORHP, and it is not difficult to see that the same is true for $\tilde{P}$. Hence, we can apply the Poisson integral formula to $\log \tilde{P}$ as in Theorem 3.3.1 of Chapter 3. □

Theorem 11.5.2 (Poisson Integral for M). Suppose that F in (11.14) is a BEE and consider the smoothing complementary sensitivity, M, defined in (11.21). Let $p = \sigma_p + j\omega_p$, $\sigma_p > 0$, be a pole of G_z (i.e., $p \in \mathcal{Z}_{1/z}$ in (11.22)). Then, under Assumption 11.1,

$$\int_{-\infty}^{\infty} \log|M(j\omega)| \frac{\sigma_p}{\sigma_p^2 + (\omega_p - \omega)^2}\, d\omega = \pi \log\left|B_M^{-1}(p)\right| - \pi \log\left|B_z^{-1}(p)\right| - \pi\sigma_p\tau. \tag{11.25}$$

Proof. Similar to the proof of Theorem 11.5.1. □

The following corollary gives integral constraints that are independent of the estimator parameters.

Corollary 11.5.3. Consider the smoothing sensitivities, P and M, given in (11.21), and suppose that F in (11.14) is a BEE. Consider the sets of zeros defined in (11.22), and assume further that every ORHP zero of G_z is also a zero of G_y (i.e., $\mathcal{Z}_z = \emptyset$). Then, under Assumption 11.1,

(i) if $q = \sigma_q + j\omega_q \in \mathcal{Z}_y$, we have that

$$\int_{-\infty}^{\infty} \log|P(j\omega)| \frac{\sigma_q}{\sigma_q^2 + (\omega_q - \omega)^2}\, d\omega \geq \pi \log\left|B_{1/z}^{-1}(q)\right| - \pi\sigma_q\tau; \tag{11.26}$$

(ii) if $p = \sigma_p + j\omega_p \in \mathcal{Z}_{1/z}$, we have that

$$\int_{-\infty}^{\infty} \log|M(j\omega)| \frac{\sigma_p}{\sigma_p^2 + (\omega_p - \omega)^2}\, d\omega \geq \pi \log\left|B_y^{-1}(p)\right| - \pi\sigma_p\tau.$$

Proof. Same as the proof of Corollary 8.2.3 in Chapter 8. □

Note that a term proportional to the smoothing lag appears with a negative sign on the RHS of the integral constraints for both sensitivities. This effect is analyzed next.

11.5.1 *Effect of the Smoothing Lag*

As we discussed for the prediction problem, the effect of smoothing will be more dramatic when the original filtering problem has both sensitivities constrained. This is the case, for example, when G_z is minimum phase but has an unstable pole, p say, and G_y has a nonminimum phase zero, q say. Under these conditions, there is an integral constrain on P of the form (11.24) where $\sigma_q = \text{Re}(q)$ is fixed independently of τ. Analogously, an integral constraint of the form (11.25) can be stated on M, with $\sigma_p = \text{Re}(p)$ fixed for all τ.

From the discussion at the end of §11.4, we can argue that P will generally have an infinite sequence of zeros converging to the ORHP. Also, we may assume that the infinite sequence of zeros of M converges to the OLHP (this is the case, for example, for the smoother derived from the Kalman filter). Then, the term $\log|B_P^{-1}(q)|$ on the RHS of (11.24) will be enlarged by the inverse Blaschke product of the infinite zeros of P, whereas the term $\log|B_M^{-1}(q)|$ on the RHS of (11.25) will probably not be larger than the corresponding term for filtering.

The improvement due to smoothing is now clear for M. Indeed, the term $-\sigma_p\tau$, for $\sigma_p = \mathrm{Re}(p)$ fixed for all τ, will continuously reduce the value of the weighted integral (11.25) as the smoothing lag increases, until the constraint finally disappears. This means that *the same weighted integral* on the LHS (with the smoothing complementary sensitivity M replacing the corresponding one for filtering) will *achieve a lower value for smoothing.*

The integral constraint (11.24) for P also has the negative linear term in τ, $-\sigma_q\tau$, but the enlarged term $\log|B_P^{-1}(q)|$ is adding to the value of the integral. Evidently, the lower bound given by the inequality in (11.26) decreases with the smoothing lag. However, since we cannot, in general, evaluate the Blaschke product, we do not have an exact estimate of how tight this bound may be.

11.6 Sensitivity Improvement of the Optimal Smoother

The improvement that smoothing represents upon filtering is a well-known fact in the filtering literature. Indeed, it has been proven for the case of optimal smoothers that the smoothing error variance is always lower than the filtering error variance (Anderson & Chirarattananon 1971).

As we have discussed in §11.5.1, there is also an improvement in sensitivity associated with smoothing, in the sense that design restrictions imposed by ORHP poles and zeros of the plant relax with the smoothing lag. However, even for a large smoothing lag, these results only show less stringent design constraints, and do not guarantee the reduction of peaks in the values of $|P|$ and $|M|$ on the $j\omega$-axis.

In this section, we give a more precise result concerning improvement in sensitivity by smoothing. We will see that for the class of smoothers based on the Kalman filter, i.e., smoothers that are optimal in the least-squares sense, the frequency responses of both smoothing sensitivities P and M have no peak values greater than one provided that the smoothing lag is sufficiently large. In fact, we will see that a smoothing lag of several times the dominant time constant of the Kalman filter, essentially achieves all the possible improvement in sensitivity reduction. We then illustrate these results with two numerical examples.

Consider then the smoother based on least squares presented in Example 11.1.1, where the plant is given by (11.1) with $D_1 = D_2 = 0$. We will assume that the solution, Φ, of the Riccati equation (11.6) is positive definite. Further, we assume that the system evolution matrix A in (11.1) has no eigenvalues on the imaginary axis. Also, we take the process noise incremental covariance $R = 1$ for simplicity of notation.

Let $\tau = \bar{\tau} \gg \tau_{max}(\hat{A})$, where $\tau_{max}(\hat{A})$ is the dominant time constant[2] of the Kalman filter. It is shown in Lemma D.0.3 in Appendix D that, for $\tau = \bar{\tau}$, the scalar smoothing sensitivities in (11.21) are approximately given by

$$\begin{aligned} M_{[\bar{\tau}]}(s) &= Q H_{\iota\nu}(-s) H_{\iota\nu}(s) \, , \\ P_{[\bar{\tau}]}(s) &= 1 - Q H_{\iota\nu}(-s) H_{\iota\nu}(s) \, , \end{aligned} \tag{11.27}$$

where $H_{\iota\nu}$ is the transfer function from the process input, ν, to the innovations process $\iota = y - C_2\hat{x}$ corresponding to the Kalman filter (see (D.2) in Appendix D), and given by

$$H_{\iota\nu} = H(sI - \hat{A})^{-1} B \, . \tag{11.28}$$

The following result shows that the modulus of $M_{[\bar{\tau}]}$ and $P_{[\bar{\tau}]}$ on the imaginary axis is bounded above by one.

Theorem 11.6.1. Consider the expressions (11.27) of the smoothing sensitivities for a smoothing lag $\tau = \bar{\tau} \gg \tau_{max}(\hat{A})$. Assume that the evolution matrix A of the system to be estimated has no eigenvalues on the imaginary axis. Then

$$\begin{aligned} \|M_{[\bar{\tau}]}\|_\infty &< 1 \, , \\ \|P_{[\bar{\tau}]}\|_\infty &= 1 \, . \end{aligned} \tag{11.29}$$

Proof. We first show that $\|H_{\iota\nu}\sqrt{Q}\|_\infty < 1$, where $H_{\iota\nu}$ is given in (11.28).

Since $\hat{A}$ has no imaginary eigenvalues, we can use the result that

$$\|H_{\iota\nu}\sqrt{Q}\|_\infty < 1$$

if and only if the Hamiltonian matrix

$$A_H \triangleq \begin{bmatrix} \hat{A} & BQB' \\ -C_2'C_2 & -\hat{A}' \end{bmatrix} \tag{11.30}$$

has no eigenvalues on the imaginary axis (Willems 1971*b*).

[2]See the footnote on page 222.

We thus compute

$$|sI - A_H| = \begin{vmatrix} sI - \hat{A} & -BQB' \\ C_2'C_2 & sI + \hat{A}' \end{vmatrix}$$

$$= \left| \begin{bmatrix} I & 0 \\ 0 & \Phi \end{bmatrix} \begin{bmatrix} sI - \hat{A} & -BQB' \\ C_2'C_2 & sI + \hat{A}' \end{bmatrix} \begin{bmatrix} I & 0 \\ 0 & \Phi^{-1} \end{bmatrix} \right|$$

$$= \begin{vmatrix} sI - \hat{A} & -BQB'\Phi^{-1} \\ \Phi C_2'C_2 & sI + \Phi\hat{A}'\Phi^{-1} \end{vmatrix}$$

$$= \begin{vmatrix} sI - A + \Phi C_2'C_2 & -BQB'\Phi^{-1} \\ \Phi C_2'C_2 & sI - A - BQB'\Phi^{-1} \end{vmatrix}$$

$$= \begin{vmatrix} sI - A & 0 \\ \Phi C_2'C_2 & sI - A - BQB'\Phi^{-1} + \Phi C_2'C_2 \end{vmatrix}$$

$$= \begin{vmatrix} sI - A & 0 \\ \Phi C_2'C_2 & sI + \Phi A'\Phi^{-1} \end{vmatrix},$$

where we have used the Riccati equation (11.6) in the form $\Phi\hat{A}'\Phi^{-1} = \Phi(A' - C_2'C_2\Phi)\Phi^{-1} = -(A + BQB'\Phi^{-1})$, and applied some elementary algebra of determinants.

It follows that A_H has no eigenvalues on the imaginary axis since the evolution matrix of the Kalman filter, $\hat{A}$, is a stability matrix, and A has no eigenvalues on the imaginary axis by assumption. Hence, $\|H_{\iota\nu}\sqrt{Q}\|_\infty < 1$ and thus, from (11.27),

$$\|M_{[\tau]}\|_\infty \leq \|H_{\iota\nu}\sqrt{Q}\|_\infty^2 < 1.$$

Now note that $0 \leq Q|H_{\iota\nu}(j\omega)|^2 < 1$ implies that $P_{[\tau]}$ in (11.27) satisfies $0 < P_{[\tau]}(j\omega) \leq 1$. Thus

$$\|P_{[\tau]}\|_\infty = 1$$

since $H_{\iota\nu}$ is strictly proper. □

Theorem 11.6.1 shows that the frequency responses of the smoothing sensitivities will experience no peaks above one if the smoothing lag is chosen several times greater than the dominant time constant of the filter. We emphasize that the result holds for unstable and nonminimum phase systems, which were shown in Chapter 8 to produce filtering sensitivities with large peaks above one.

The following examples illustrate Theorem 11.6.1 for the cases of unstable and stable plants.

Example 11.6.1 (Unstable Plant). We consider again the Kalman filter given in Example 8.4.1 of Chapter 8. Here we construct a smoother as in (11.7), derived from the Kalman filter obtained for the weights $Q = 100$ and $R = 1$. We take the parameter a in (8.23) as $a = 1$.

The frequency responses of $|P|$ and $|M|$ are shown in Figures 11.2 and 11.3, respectively, for three different values of the smoothing lag τ. Note that the filtering sensitivities correspond to $\tau = 0$.

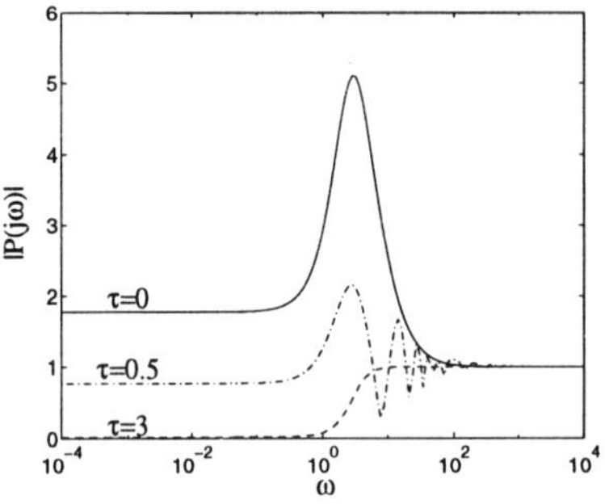

FIGURE 11.2. $|P|$ achieved by the Kalman filter-based smoother for the unstable plant of Example 11.6.1.

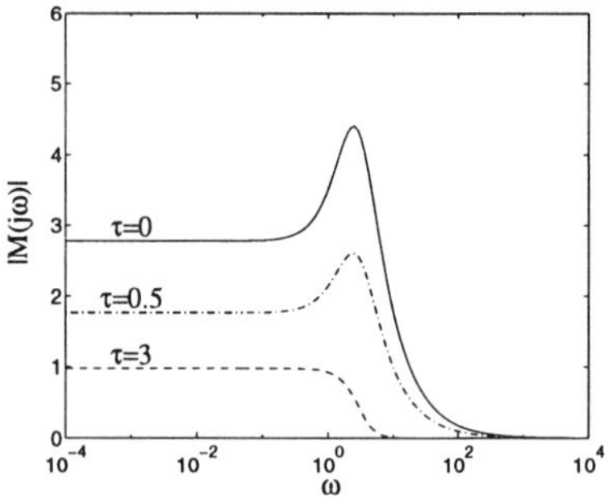

FIGURE 11.3. $|M|$ achieved by the Kalman filter-based smoother for the unstable plant of Example 11.6.1.

These plots clearly indicate that the process of smoothing reduces the constraints on the frequency responses of both sensitivities, which tend to "ideal" shapes as we increase the smoothing lag. ○

Example 11.6.2 (Stable Plant). Consider again the stable plant used in Example 8.4.2 of Chapter 8 and revisited in §10.6.2 of Chapter 10. For this plant, we construct a smoother as in (11.7). The generating Kalman filter is again designed for an aggregated plant that includes the disturbance model (8.23) with the parameter $a = 1$. The weights are chosen as $Q = 100$ and $R = 0.1$.

The plots of Figures 11.4 and 11.5 show the frequency responses of P and M for three values of τ. These figures show essentially the same effect of an improved frequency response as for the unstable system in Example 11.6.1.

Note that the poles of the generating Kalman filter for this example are $-2.1035 \pm j2.5701$ and -1.8335, which implies that $\tau = 0.8$ is approximately 0.9 times the filter dominant time constant, taken as $\tau_f \approx 1/1.8335 \approx 0.55$, whilst $\tau = 2$ is approximately 3.6 times τ_f. As seen in §11.6, taking a smoothing lag of several times (approximately 5 times according to Anderson (1969)) the dominant time constant of the filter, gives all the improvement possible using smoothing. We can see that this phenomenon of "saturation" of improvement is also present in the frequency responses of the error sensitivities.

○

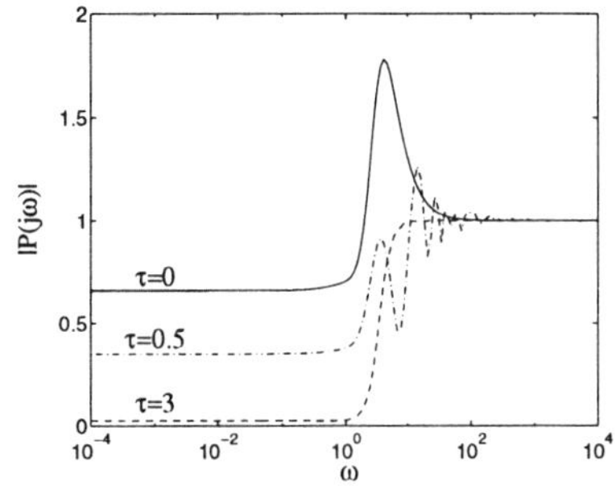

FIGURE 11.4. |P| achieved by the Kalman filter-based smoother for the stable plant of Example 11.6.2.

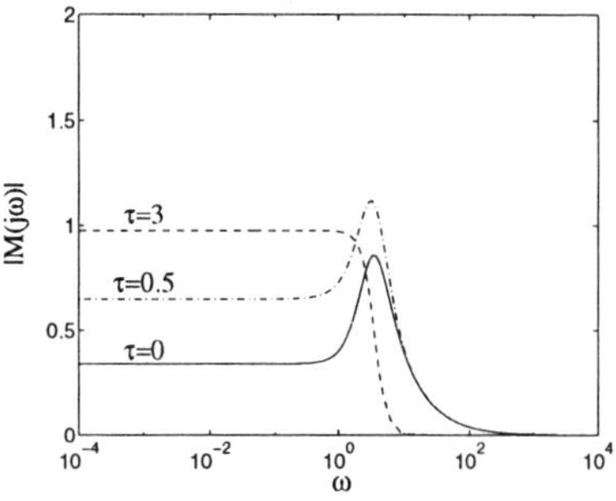

FIGURE 11.5. |M| achieved by the Kalman filter-based smoother for the stable plant of Example 11.6.2.

11.7 Summary

This chapter has extended the sensitivity analysis developed in Chapter 8 for the problem of filtering with bounded error estimators to the case of fixed-lag smoothing. The results indicate that, in contrast to the problem of prediction, the constraints in sensitivity are reduced as the smoothing lag increases.

In particular, notice that if one considers the estimation horizon to be either positive (for prediction) or negative (for smoothing), then the results are seen to be consistent. This is clear from the structural relations for filtering, given by (7.7), for prediction, given by (10.13) and (10.10), and for smoothing, given by (11.19) and (11.15). These relations are recalled in Table 11.1.

Filtering	$H_{\tilde{z}v} + FH_{yv} = H_{zv}$
Prediction	$H_{\tilde{z}v} + F_p H_{yv} = e^{s\tau} H_{zv}$
Smoothing	$H_{\tilde{z}v} + F_s H_{yv} = e^{-s\tau} H_{zv}$

TABLE 11.1. Structural relations in filtering, prediction and smoothing.

The structural relations shown in Table 11.1 are the basis for the definition of the corresponding sensitivities. It should be noticed, however, that the nature of the error transfer function $H_{\tilde{z}v}$ in prediction and smoothing is not dual.

Notes and References

The sensitivity results of this chapter are based on Seron (1995). Background information on smoothing was collected from Kailath (1981), Kailath & Frost (1968) and Anderson & Doan (1977).

Part IV

Limitations in Nonlinear Control and Filtering

12
Nonlinear Operators

In Chapters 13 and 14 we will investigate fundamental constraints that hold for the problems of nonlinear feedback control and nonlinear filtering. The general framework used is that of input-output nonlinear operators acting on linear signal spaces. Within this framework, the linear concepts of nonminimum phase zeros and unstable poles of transfer functions are easily handled using ideas of defect in the domain and range of the nonlinear operators. Some properties of this approach essential to our analysis are reviewed in this chapter.

12.1 Nonlinear Operators

Much of the research effort to date in nonlinear control theory has focused on controller design. This area has experienced substantial progress, giving rise to numerous systematic procedures for controller design applicable to different classes of nonlinear systems. For a recent survey on this topic see Coron, Praly & Teel (1995). The theory for nonlinear observers is more incipient than that for nonlinear control, though some techniques for observer construction have become available in the recent years. Most of these nonlinear controller and observer design methods use a state space description of the system.

The state space approach provides satisfactory answers when dealing with synthesis problems, yet it does not appear suitable for a theory of limitations imposed by the structure of the feedback loop. For this problem, a more natural framework is provided by the *input-output operator* approach. This approach received considerable attention during the 60's and 70's (Sandberg 1965, Zames 1966*a*, Zames 1966*b*, Willems 1971*a*, Desoer & Vidyasagar 1975, Hill & Moylan 1980). In this input-output framework, systems are represented by nonlinear operators acting on normed signal spaces. Two important measures related to these operators are the maximum *gain*[1] over the input signal space, and the maximum *incremental gain,* also called *Lipschitz constant* or *Lipschitz norm.* In the linear case, both measures collapse into one. Moreover, for linear time-invariant systems (LTI) on $\mathcal{L}_2$,[2] they coincide with the $\mathcal{H}_\infty$ norm of the transfer function. Thus, both the $\mathcal{L}_2$-gain and the Lipschitz norm are possible nonlinear extensions of the $\mathcal{H}_\infty$ norm of linear systems.

We next review properties of nonlinear operators acting on linear, Banach and Hilbert spaces.

12.1.1 Nonlinear Operators on a Linear Space

Let $\mathcal{X}$ be a linear space and let H be a nonlinear operator on $\mathcal{X}$, i.e., a mapping between its domain $\mathcal{D}(\mathsf{H}) \subset \mathcal{X}$ into $\mathcal{X}$. The *domain* and *range* of H are defined as follows:

$$\begin{aligned} \mathcal{D}(\mathsf{H}) &\triangleq \{x \in \mathcal{X} : \mathsf{H}x \in \mathcal{X}\}, \\ \mathcal{R}(\mathsf{H}) &\triangleq \{\mathsf{H}x : x \in \mathcal{D}(\mathsf{H})\}. \end{aligned} \tag{12.1}$$

If H_1 and H_2 are nonlinear operators on $\mathcal{X}$ and $\mathcal{D}(\mathsf{H}_1) \cap \mathcal{D}(\mathsf{H}_2) \neq \emptyset$, the *addition* $\mathsf{H}_1 + \mathsf{H}_2 : \mathcal{D}(\mathsf{H}_1 + \mathsf{H}_2) \to \mathcal{X}$ is defined by

$$(\mathsf{H}_1 + \mathsf{H}_2)x = \mathsf{H}_1 x + \mathsf{H}_2 x, \quad \forall x \in \mathcal{D}(\mathsf{H}_1 + \mathsf{H}_2) = \mathcal{D}(\mathsf{H}_1) \cap \mathcal{D}(\mathsf{H}_2).$$

If $\mathcal{R}(\mathsf{H}_2) \cap \mathcal{D}(\mathsf{H}_1) \neq \emptyset$, the *composition* $\mathsf{H}_1\mathsf{H}_2 : \mathcal{D}(\mathsf{H}_1\mathsf{H}_2) \to \mathcal{X}$ is defined by

$$(\mathsf{H}_1\mathsf{H}_2)x = \mathsf{H}_1(\mathsf{H}_2 x), \quad \forall x \in \mathcal{D}(\mathsf{H}_1\mathsf{H}_2) = \mathsf{H}_2^{-1}(\mathcal{D}(\mathsf{H}_1)).$$

Here, the notation $\mathsf{H}^{-1}(\mathcal{U})$, for a set $\mathcal{U} \subset \mathcal{X}$, indicates the pre-image set of $\mathcal{U}$ through H, i.e.,

$$\mathsf{H}^{-1}(\mathcal{U}) = \{x \in \mathcal{D}(\mathsf{H}) : \mathsf{H}x \in \mathcal{U}\}.$$

Similar to the notation used for the linear case, the symbol H_{ba} will be used to represent an open-loop nonlinear operator mapping signal a to signal b. Also, $\mathbf{H}_{\mathrm{ba}}$ will stand for the total nonlinear operator mapping signal a to signal b.

---

[1]I.e., the maximum ratio of the norm of the output signal to the norm of the input signal.
[2]The definition of the signal space $\mathcal{L}_2$ is recalled in §12.1.3.

12.1.2 Nonlinear Operators on a Banach Space

Let $\mathcal{X}$ be a Banach space. We say that the operator H is *stable* if its domain is $\mathcal{X}$ and we say that H is *unstable* if its domain is a strict subset of $\mathcal{X}$. This definition of operator stability/instability can be interpreted as the usual system stability/instability when $\mathcal{X}$ is taken to be a space of "physically meaningful" signals, such as $\mathcal{L}_2$ or its isomorphic equivalent $\mathcal{H}_2$.

We say that H is *nonminimum phase* if the closure of its range is a strict subset of $\mathcal{X}$.

Example 12.1.1. To motivate the concept of nonminimum phase nonlinear operator, consider the simple case of a scalar, LTI operator on $\mathcal{H}_2$, represented by a transfer function, H, say. Assume further that H is stable and proper. Then, if H has a nonminimum phase zero, the range of such an operator is the set of signals in $\mathcal{H}_2$ that have a zero at the same frequency, which is not dense in $\mathcal{H}_2$. ∘

The closure of a set $\mathcal{U} \subset \mathcal{X}$ will be denoted by $\mathrm{cl}\,\mathcal{U}$. We denote by $\mathrm{Lip}(\mathcal{D}, \mathcal{X})$ the class of all operators $H : \mathcal{D}(H) = \mathcal{D} \subset \mathcal{X} \to \mathcal{X}$ such that

$$\|H\|_L \triangleq \sup\{|Hx - Hy|/|x - y| : x, y \in \mathcal{D}, x \neq y\} < \infty, \tag{12.2}$$

where $|\cdot|$ is the norm in $\mathcal{X}$. A member H of $\mathrm{Lip}(\mathcal{D}, \mathcal{X})$ is called a Lipschitz operator and $\|H\|_L$ is the Lipschitz constant, Lipschitz gain, or incremental gain of H. When $\mathcal{D} = \mathcal{X}$, the class will be denoted by $\mathrm{Lip}(\mathcal{X})$. Note that members of $\mathrm{Lip}(\mathcal{X})$ are stable operators, since their domain is $\mathcal{X}$.

It is easy to see, that if H is a linear operator, then $\|H\|_L$ in (12.2) reduces to the operator norm induced by the norm in $\mathcal{X}$. Indeed, it is claimed in Zarantonello (1967) that the ordinary norm for linear operators naturally extends to the Lipschitz norm in the nonlinear case. To be precise, the Lipschitz norm is in fact a *semi-norm*, thus the space $(\mathrm{Lip}(\mathcal{X}), \|\cdot\|_L)$ is a semi-normed linear space.

$\mathrm{Lip}(\mathcal{X})$ is closed under operator composition. Moreover, if $H_1, H_2 \in \mathrm{Lip}(\mathcal{X})$ then the composition $H_1 H_2 \in \mathrm{Lip}(\mathcal{X})$ and the sub-multiplicative property $\|H_1 H_2\|_L \leq \|H_1\|_L \|H_2\|_L$ holds for the Lipschitz norm.

A member H of $\mathrm{Lip}(\mathcal{X})$ is said to be *invertible* in $\mathrm{Lip}(\mathcal{X})$ if there is an operator $H_i \in \mathrm{Lip}(\mathcal{X})$ such that $HH_i = H_i H = I$. We denote $H_i = H^{-1}$. Clearly, it is necessary that $\mathcal{R}(H) = \mathcal{X}$ for H to be invertible in $\mathrm{Lip}(\mathcal{X})$ and thus a nonminimum phase operator is not invertible in $\mathrm{Lip}(\mathcal{X})$. The following result gives a condition for the invertibility of members of $\mathrm{Lip}(\mathcal{X})$.

Lemma 12.1.1. Let $\mathcal{X}$ be a Banach space and let $H \in \mathrm{Lip}(\mathcal{X})$. Suppose that $\|H\|_L < 1$. Then $I - H$ is invertible in $\mathrm{Lip}(\mathcal{X})$ and $\|(I - H)^{-1}\|_L \leq (1 - \|H\|_L)^{-1}$.

Proof. See Martin Jr. (1976, Theorem 2.2, p.66). □

Lemma 12.1.1 appears to have been stated for the first time in the control literature by Zames (1966*a*, 1966*b*), although it was probably already known among mathematicians. This important result is the basis of the proof of the small gain theorem in its Lipschitz version (Sandberg 1965, Zames 1966*a*, Zames 1966*b*).

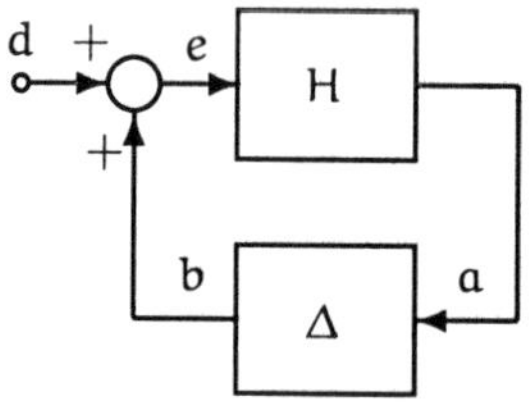

FIGURE 12.1. Basic perturbation model.

In §13.5 we will address the issue of stability robustness and refer to the basic perturbation model shown in Figure 12.1, where $H : \mathcal{D}(H) \to \mathcal{X}$ and $\Delta : \mathcal{D}(\Delta) \to \mathcal{X}$ are nonlinear operators. We say that the feedback loop of Figure 12.1 is *stable* if $e, a, b \in \mathcal{X}$ whenever $d \in \mathcal{X}$. If $H, \Delta \in \mathrm{Lip}(\mathcal{X})$, we also say that the feedback loop of Figure 12.1 is *Lipschitz stable* if the operator $\mathbf{H}_{ed} \in \mathrm{Lip}(\mathcal{X})$. It follows from Lemma 12.1.1 that, provided $H, \Delta \in \mathrm{Lip}(\mathcal{X})$, the loop in Figure 12.1 is Lipschitz stable if $\|\Delta H\|_L < 1$. Moreover,

$$\|\mathbf{H}_{ed}\|_L = \|(I - \Delta H)^{-1}\|_L \leq (1 - \|\Delta H\|_L)^{-1} \tag{12.3}$$

and

$$\|\mathbf{H}_{ad}\|_L = \|H(I - \Delta H)^{-1}\|_L \leq \|H\|(1 - \|\Delta H\|_L)^{-1}. \tag{12.4}$$

12.1.3 *Nonlinear Operators on a Hilbert Space*

This section reviews some notation and terminology that will be used in §13.4 of Chapter 13. Let $\mathcal{L}_2$ be the standard Hilbert space of real-valued measurable square-integrable functions defined on $\mathbb{R}_+$ (the positive real line) with norm $\|\cdot\|_{\mathcal{L}_2}$ and inner product $< f, g >_{\mathcal{L}_2}$. For $f \in \mathcal{L}_2$, $\mathbf{f}$ denotes its Fourier transform. Let $\Omega \subset \mathbb{R}$ be a set with nonzero measure. For $f \in \mathcal{L}_2$, we define $\|f\|_\Omega$ as

$$\|f\|_\Omega \triangleq \left(\frac{1}{2\pi} \int_\Omega |\mathbf{f}(j\omega)|^2 d\omega \right)^{1/2}.$$

Note that $\|\cdot\|_\Omega$ represents the signal-energy distributed over the frequency range Ω (Shamma 1991). A sequence $\{f_i\} \subset \mathcal{L}_2$ is said to *weakly converge* to $f_0 \in \mathcal{L}_2$ if for all $g \in \mathcal{L}_2$,

$$\lim_i < f_i, g >_{\mathcal{L}_2} = < f_0, g >_{\mathcal{L}_2}.$$

The $\mathcal{L}_2$-*weak-closure* of a set $\mathcal{U}$ is denoted wk–cl $\mathcal{U}$.

A mapping $\mathsf{H} : \mathcal{L}_2 \to \mathcal{L}_2$ is an *I/O operator* if it is unbiased (i.e., $\mathsf{H}\,0 = 0$) and causal. The domain and range of an I/O operator H are defined as in (12.1) with $\mathcal{X}$ replaced with $\mathcal{L}_2$.

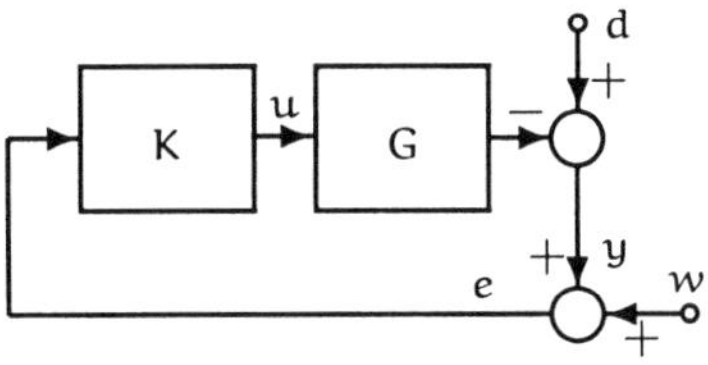

FIGURE 12.2. Feedback Control Loop.

The feedback system shown in Figure 12.2 is said to be *stable* if u and e belong to $\mathcal{L}_2$ whenever d and w belong to $\mathcal{L}_2$; in other words, the nonlinear operators that map d and w into u and e are stable nonlinear operators. In this case, the controller K is said to *stabilize* the plant G.

12.2 Nonlinear Cancelations

For the problem of nonlinear feedback control, to be discussed in Chapter 13, stability of closed-loop operators is achieved via feedback, and hence it is not necessary to address the issue of unstable "nonlinear zero-pole (pole-zero) cancelation". Therefore, the background on nonlinear operators given so far is enough to treat this problem. By way of contrast, the filtering problem is of open-loop nature[3], and therefore, if the plant is unstable, the stability of the estimation error operator (see (14.10) in Chapter 14) has to be achieved by some nonlinear extension of the linear notion of unstable cancelations.

To gain insight into what a possible definition of unstable "nonlinear zero-pole cancelation" may be, let us informally discuss the case of a scalar, LTI operator on $\mathcal{H}_2$, represented by a transfer function, H_1, say. Assume further that H_1 is stable and proper. Then, if H_1 has a nonminimum phase zero, we have discussed in Example 12.1.1 that the closure of the range of H_1 does not cover $\mathcal{H}_2$, and this has motivated the definition of nonminimum phase nonlinear operator used in §12.1.2. However, note also that H_1 has the additional property that there exist signals outside $\mathcal{H}_2$ (from its orthogonal complement, $\mathcal{H}_2^{\perp}$, in fact) that are mapped by H_1 into $\mathcal{H}_2$, (i.e., signals that have poles at the nonminimum phase zero of H_1). Now, assume that H_1 is multiplied by (composed with) another transfer function, H_2 say, that has an unstable pole at the same frequency where H_1 has a nonminimum phase zero. If instead of restricting the domain of H_2

[3]See comments at the end of §7.2.2 in Chapter 7.

(according to the definition of unstable operator given in §12.1.2) we let H_2 act on all of $\mathcal{H}_2$, then bounded signals will be mapped into unbounded signals. However, when composed with H_1, these unbounded signals are precisely those that H_1 maps into $\mathcal{H}_2$, giving a stable product $H_1 H_2$.

From the above discussion, it is clear that the framework used in §12.1 is insufficient to handle "nonlinear cancelations", since signals that are unbounded are not formally defined. The use of extended spaces, where unbounded signals lie, becomes then necessary.

12.2.1 *Nonlinear Operators on Extended Banach Spaces*

We will consider nonlinear operators defined on an extended linear space $\mathcal{X}_e \supset \mathcal{X}$, where $\mathcal{X}$ is a Banach space. The way in which $\mathcal{X}_e$ "extends" $\mathcal{X}$ is arbitrary. The domain and range of the operator $H : \mathcal{X}_e \to \mathcal{X}_e$ are defined as in (12.1) in §12.1.1. Also, the definitions of stable and unstable operators are the same used in §12.1.1.

Given a set $\mathcal{D} \subset \mathcal{X}_e$, the symbol $\overline{\mathcal{D}}^e$ will denote its complement in $\mathcal{X}_e$, i.e.,

$$\overline{\mathcal{D}}^e \triangleq \{x \in \mathcal{X}_e : x \notin \mathcal{D}\}.$$

Similarly, for a set $\mathcal{D} \subset \mathcal{X}$, $\overline{\mathcal{D}}$ will indicate its complement in $\mathcal{X}$, i.e.,

$$\overline{\mathcal{D}} \triangleq \{x \in \mathcal{X} : x \notin \mathcal{D}\}.$$

The set $\overline{\mathcal{X}}^e$, i.e., the complement of $\mathcal{X}$ in $\mathcal{X}_e$, will be considered to contain "unbounded" signals.

In this context of extended spaces, we say that $H : \mathcal{X}_e \to \mathcal{X}_e$ is *nonminimum phase* (NMP) if the following two conditions hold:

(i) (defect in range) the closure of the range of H, $\text{cl}\,\mathcal{R}(H)$, is a strict subset of $\mathcal{X}$.

(ii) (excess in domain) the *extended domain* of H, defined as the set

$$\mathcal{E}(H) \triangleq \{x \in \overline{\mathcal{X}}^e : Hx \in \mathcal{X}\}$$

is not empty.

Property (ii) implies that a nonempty set of unbounded signals are mapped into bounded images by H. Note that this corresponds to the property of LTI systems previously discussed.

It may be possible for a nonlinear operator to satisfy only one of the properties of NMP operators. If (i) holds, we will say that the operator has the *the defect-in-range property*, and that it is of D-NMP type. Alternatively, if (ii) holds, we will say that the operator has the *excess-in-domain property*, and that it is of E-NMP type. Figure 12.3 illustrates a NMP operator in the context of extended spaces, i.e., with both the defect-in-range and excess-in-domain properties, as just described.

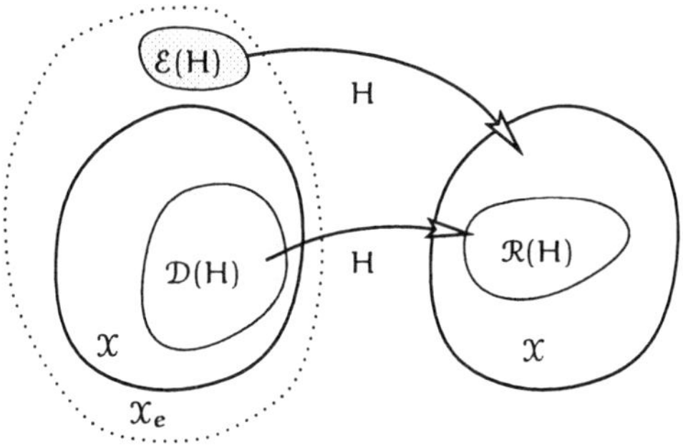

FIGURE 12.3. Set interpretation of a NMP nonlinear operator.

In order to allow for cancelations, we need to adjust the definition of domain of the composition of nonlinear operators to this setting of extended spaces. Let H_1 and H_2 be nonlinear operators on $\mathcal{X}_e$ such that $\mathcal{R}(H_2) \cap \mathcal{D}(H_1) \neq \emptyset$. Then the domain of the composition $H_1 H_2$ is defined as

$$\mathcal{D}(H_1 H_2) = H_2^{-1}(\mathcal{D}(H_1)) \cup \mathcal{A}(H_1 H_2), \tag{12.5}$$

where

$$H_2^{-1}(\mathcal{D}(H_1)) = \{x \in \mathcal{D}(H_2) : H_2 x \in \mathcal{D}(H_1)\},$$

and $\mathcal{A}(H_1 H_2)$ is the *added domain of the composition*, given by

$$\mathcal{A}(H_1 H_2) \triangleq \{x \in \overline{\mathcal{D}(H_2)} : H_2 x \in \mathcal{E}(H_1)\}. \tag{12.6}$$

Note that $\mathcal{A}(H_1 H_2) \neq \emptyset$ only if H_2 is unstable and H_1 is E-NMP. This is because then there exists a signal in the complement of the domain of H_2 that is mapped by H_2 into $\mathcal{E}(H_1)$, which is then nonempty. On the other hand, if H_1 is E-NMP, then $\mathcal{E}(H_1) \neq \emptyset$ and then $\mathcal{A}(H_1 H_2)$ *may* be nonempty if H_2 is unstable.

When the set $\mathcal{A}(H_1 H_2)$ is nonempty, we say that there is an *unstable nonlinear zero-pole cancelation* in the composition $H_1 H_2$. The nonlinear zero-pole cancelation idea is depicted in Figure 12.4.

Note that the definition of composition domain used in §12.1.1, consisted only of the first term of the set union on the RHS of (12.5), i.e., the set $H_2^{-1}(\mathcal{D}(H_1))$. This first term is different from $\mathcal{D}(H_2)$ only if H_1 is unstable. When H_1 is unstable and $H_2^{-1}(\mathcal{D}(H_1)) = \mathcal{D}(H_2)$, we say that there is an *unstable nonlinear pole-zero cancelation* in the composition $H_1 H_2$.

The concepts considered in this subsection will be used in Chapter 14 to address the issue of stability of the estimation error in nonlinear filtering.

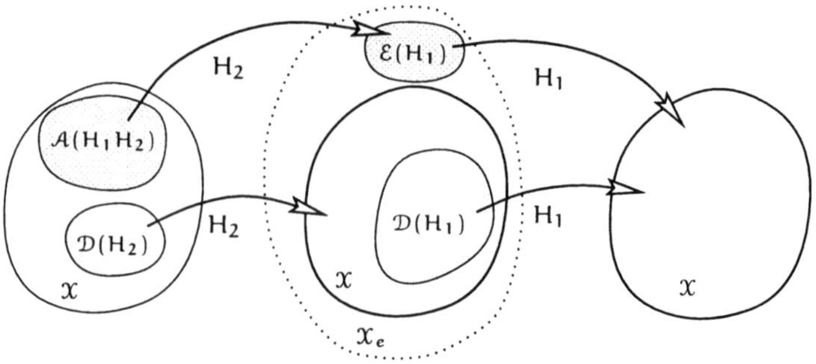

FIGURE 12.4. Set interpretation of unstable nonlinear zero-pole cancelation.

12.3 Summary

We have presented background on nonlinear operators, necessary for the developments in the next two chapters. In particular, we have described properties of nonlinear operators as mappings on Banach or Hilbert spaces, with emphasis on Lipschitz operators. A set-based framework has been provided to handle nonlinear extensions of the linear concepts of nonminimum phase systems, unstable systems, and unstable zero-pole (pole-zero) cancelations.

Notes and References

The material of §12.1 is mainly taken from Martin Jr. (1976). §12.1.2 and §12.1.3 also have input from Doyle, Georgiou & Smith (1993) and Shamma (1991). The set-based framework of §12.2 is taken from Seron (1995).

The definition of unstable operator as one having a defect in its domain is used in Doyle et al. (1993) and it is common in references dealing with the input-output approach to feedback control (e.g., Desoer & Vidyasagar 1975). The concept of non-minimum phaseness that we use here is similar to the one given in Shamma (1991).

13
Nonlinear Control

This chapter analyzes performance limitations and stability robustness of the unity feedback configuration of Figure 12.2 considered in Chapter 12. We use the material on nonlinear operator theory developed in §12.1.1 of Chapter 12.

13.1 Review of Linear Sensitivity Relations

For convenience, we review here the central result on linear sensitivity limitations that we will extend to the nonlinear case in §13.3. This result shows that the infinity norm of S is bounded below by one for nonminimum phase systems and the infinity norm of T is bounded below by one for open-loop unstable systems.

We use the notions of zeros and poles of multivariable systems introduced in §2.1.1 of Chapter 2. The symbol $|\cdot|$ denotes here the Euclidean norm in $\mathbb{C}^n$. If M is a matrix in $\mathbb{C}^{n\times n}$, $\|M\|_2$ denotes its matrix norm induced by the Euclidean vector norm.

The following result is well-known for linear systems.[1]

Theorem 13.1.1. Let L be the transfer matrix of a LTI open-loop system. Assume that the closed-loop sensitivity $S = (I+L)^{-1}$ and complementary sensitivity $T = L(I+L)^{-1}$ are stable transfer functions. Then:

[1]It can be inferred, for example, from the inequality (4.34) in and 4.

(i) If L is nonminimum phase, then $\|S\|_\infty \geq 1$.

(ii) If L is unstable, then $\|T\|_\infty \geq 1$.

Proof. The result can be derived from Chen (1995, Corollary 4.2), but we give here a simpler proof exploiting the fact that the $\mathcal{H}_\infty$ norm is the operator norm on $\mathcal{H}_2$, the space of Laplace transforms of signals in $\mathcal{L}_2$.[2]

(i) Let $L(s)$ have a zero at $s = q$ in the ORHP with output direction $\Psi \in \mathbb{C}^n$, such that $\Psi^*\Psi = 1$; i.e., $\Psi^* L(q) = 0$. Then $\Psi^* S(q) = \Psi^*$ (cf. (4.5) in Chapter 4). The infinity norm of S satisfies

$$\begin{aligned}\|S\|_\infty &= \sup_{\mathrm{Re}(s)>0} \|S(s)\|_2 \\ &= \sup_{\mathrm{Re}(s)>0} \max_{\substack{\eta\in\mathbb{C}^n \\ |\eta|=1}} |\eta^* S(s)| \\ &\geq |\Psi^* S(q)| \\ &= 1.\end{aligned}$$

(ii) Let $L(s)$ have a pole at $s = p$ in the ORHP. Then, using (4.4) in Chapter 4, we have that $\Phi^* T(p) = \Phi^*$ and the proof follows as in (i).

□

The proof of Theorem 13.1.1 uses the fact that the operator norm on an infinite dimensional space can be bounded below by the matrix norm on $\mathbb{C}^n$ with the Euclidean norm. Then, the zeros and poles in the ORHP of the open-loop system give particular interpolation values of the closed-loop operators, which provide lower bounds on the matrix induced norm. Although this property generally does not hold for nonlinear operators, we will see that nonminimum phase and unstable open-loop behavior, as defined in §12.1.2, still represent constraints on the closed-loop system.

13.2 A Complementarity Constraint

Consider the feedback interconnection of plant G and controller K as displayed in Figure 13.1.

We define the *nonlinear sensitivity* operator, $\mathbf{S}$, and the *nonlinear complementary sensitivity* operator, $\mathbf{T}$, as

$$\begin{aligned}\mathbf{S} &= \mathbf{H}_{yd}|_{(w=0)}, \\ \mathbf{T} &= -\mathbf{H}_{yw}|_{(d=0)},\end{aligned} \tag{13.1}$$

[2] In fact, $\mathcal{H}_2$ (for signals) is the Banach space of functions $x : \mathbb{C} \to \mathbb{C}^n$, which are analytic in the ORHP, and satisfy $\sup_{\sigma>0}(2\pi)^{-1}\int_{-\infty}^{\infty} |x(\sigma + j\omega)|^2 d\omega < \infty$.

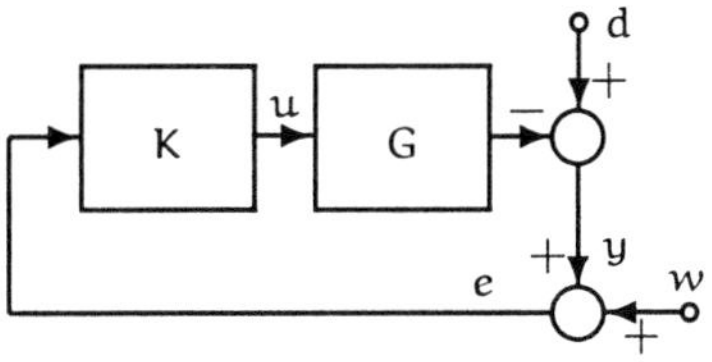

FIGURE 13.1. Feedback Control Loop.

where the notation $\mathbf{H}_{ba}|_{(c=0)}$ stands for the total map from signal a to signal b when signal c is identically zero. In words, **S** is the mapping between output disturbance and system output in the absence of the sensor noise, and **T** is the mapping between sensor noise and system output in the absence of output disturbances.

We will consider **S** and **T** as nonlinear operators on some linear space $\mathcal{X}$, i.e., $\mathbf{S} : \mathcal{D}(\mathbf{S}) \subset \mathcal{X} \to \mathcal{X}$ and $\mathbf{T} : \mathcal{D}(\mathbf{T}) \subset \mathcal{X} \to \mathcal{X}$. In fact, the domains of **S** and **T** are the same, as shown in the following result.

Lemma 13.2.1. Assume that **S** and **T** defined in (13.1) are nonlinear operators on a linear space $\mathcal{X}$. Then $\mathcal{D}(\mathbf{S}) = \mathcal{D}(\mathbf{T})$.

Proof. Consider Figure 13.1 and assume that $d = 0$ and $w \in \mathcal{D}(\mathbf{S})$. Then

$$e = \mathbf{S}w \in \mathcal{X}.$$

It follows that $\mathbf{T}w = -y = w - e \in \mathcal{X}$ since $\mathcal{X}$ is a linear space. Thus $w \in \mathcal{D}(\mathbf{T})$ and therefore $\mathcal{D}(\mathbf{S}) \subset \mathcal{D}(\mathbf{T})$. In a similar way it can be shown that $\mathcal{D}(\mathbf{T}) \subset \mathcal{D}(\mathbf{S})$. Hence $\mathcal{D}(\mathbf{S}) = \mathcal{D}(\mathbf{T})$. □

The operators **S** and **T** satisfy the following complementarity constraint.

Theorem 13.2.2. Consider the operators defined in (13.1) as nonlinear operators on a linear space. Then on $\mathcal{D}(\mathbf{S}) = \mathcal{D}(\mathbf{T})$

$$\mathbf{S} + \mathbf{T} = \mathrm{I}. \tag{13.2}$$

Proof. From the definition (13.1) and Figure 13.1, the sensitivity operator is given by

$$\mathbf{S} = (\mathsf{GK} + \mathrm{I})^{-1}, \tag{13.3}$$

and the complementarity sensitivity operator is given by

$$\mathbf{T} = \mathsf{GK}(\mathsf{GK} + \mathrm{I})^{-1} = \mathsf{GK}\mathbf{S}. \tag{13.4}$$

Clearly, on $\mathcal{D}(\mathbf{S}) = \mathcal{D}(\mathbf{T})$,

$$\mathbf{S} + \mathbf{T} = \mathbf{S} + \mathsf{GK}\mathbf{S} = (\mathsf{GK} + \mathrm{I})\,\mathbf{S} = \mathrm{I},$$

by definition of addition of operators. □

Theorem 13.2.2 clearly indicates that the complementarity constraint given by (13.2) is purely determined by the structure of the feedback control loop together with the additive nature of the disturbance inputs. Hence, this constraint is independent of whether the plant or controller are linear or nonlinear operators.

The following result is straightforward.

Corollary 13.2.3. Let $\mathcal{X}$ be a Banach space and assume that $\mathbf{S}$ and $\mathbf{T}$ in (13.1) are nonlinear operators on $\mathcal{X}$. Then $\mathbf{S}+\mathbf{T} = \mathrm{I}$ on $\mathcal{X}$ if and only if $\mathbf{S}$ and $\mathbf{T}$ are stable operators on $\mathcal{X}$.

Proof. Immediate from Theorem 13.2.2 and the definition of stable operator on page 249. □

13.3 Sensitivity Limitations

We consider in this section that the open loop $\mathrm{L} \triangleq \mathrm{GK}$ is a nonlinear operator $\mathrm{L} : \mathcal{D}(\mathrm{L}) \subset \mathcal{X} \to \mathcal{X}$, where $\mathcal{X}$ is a Banach space.

Before obtaining the main result of this section on sensitivity limitations, we need a preliminary lemma, which is a nonlinear generalization of the interpolation constraints satisfied by the linear sensitivities.

Lemma 13.3.1. Let $\mathrm{L} : \mathcal{D}(\mathrm{L}) \subset \mathcal{X} \to \mathcal{X}$ be such that the operators $\mathbf{S}$ and $\mathbf{T}$ have a nonempty domain. Then

(i) If L is nonminimum phase, then $\mathbf{T}$ is nonminimum phase.

(ii) If L is unstable and $\mathcal{D}(\mathrm{L})$ is closed, then $\mathbf{S}$ is nonminimum phase.

Proof.

(i) Assume that L is nonminimum phase, i.e., $\mathrm{cl}\,\mathcal{R}(\mathrm{L})$ is a strict subset of $\mathcal{X}$. From (13.4) it follows that $\mathcal{R}(\mathbf{T}) \subset \mathcal{R}(\mathrm{L})$ and hence $\mathrm{cl}\,\mathcal{R}(\mathbf{T}) \subset \mathrm{cl}\,\mathcal{R}(\mathrm{L})$, which means that $\mathbf{T}$ is nonminimum phase.

(ii) Assume that L is unstable, i.e., $\mathcal{D}(\mathrm{L})$ is a strict subset of $\mathcal{X}$. Since $\mathbf{T}$ maps $\mathcal{D}(\mathbf{T}) = \mathcal{D}(\mathbf{S})$ into $\mathcal{X}$, it follows from (13.4) that $\mathcal{R}(\mathbf{S}) \subset \mathcal{D}(\mathrm{L})$. Hence $\mathrm{cl}\,\mathcal{R}(\mathbf{S}) \subset \mathrm{cl}\,\mathcal{D}(\mathrm{L}) = \mathcal{D}(\mathrm{L})$ (since $\mathcal{D}(\mathrm{L})$ is closed) and thus $\mathbf{S}$ is nonminimum phase.

□

Note that, for (ii) to hold, it is essential that $\mathcal{D}(\mathbf{T}) = \mathcal{D}(\mathbf{S})$. Indeed, if $\mathbf{S}$ and $\mathbf{T}$ were any nonlinear operators satisfying $\mathbf{T} = \mathrm{L}\mathbf{S}$, then $\mathcal{D}(\mathbf{T}) = \mathbf{S}^{-1}(\mathcal{D}(\mathrm{L})) \triangleq \{x \in \mathcal{D}(\mathbf{S}) : \mathbf{S}x \in \mathcal{D}(\mathrm{L})\}$ and hence the relation $\mathcal{R}(\mathbf{S}) \subset \mathcal{D}(\mathrm{L})$ does not hold in general.

The following theorem gives lower bounds on the Lipschitz constants of the nonlinear sensitivity operators for open-loop nonminimum phase and unstable systems.

Theorem 13.3.2. Let $L : \mathcal{D}(L) \subset \mathcal{X} \to \mathcal{X}$ be such that the sensitivity operator $\mathbf{S}$ is in $\mathrm{Lip}(\mathcal{X})$. Then

(i) If L is nonminimum phase, then $\|\mathbf{S}\|_L \geq 1$.

(ii) If L is unstable, then $\|\mathbf{T}\|_L \geq 1$.

Proof. Note first that since $\mathbf{S} \in \mathrm{Lip}(\mathcal{X})$, it is a stable operator. It follows from Corollary 13.2.3 that $\mathbf{S}+\mathbf{T} = I$ on $\mathcal{X}$. Then, $\mathbf{S} \in \mathrm{Lip}(\mathcal{X}) \Rightarrow \mathbf{T} = I-\mathbf{S} \in \mathrm{Lip}(\mathcal{X})$ since $\mathrm{Lip}(\mathcal{X})$ is a linear space (Martin Jr. 1976). Next:

(i) Assume that L is nonminimum phase. If $\|\mathbf{S}\|_L < 1$, it follows from Lemma 12.1.1 that $I-\mathbf{S} = \mathbf{T}$ is invertible in $\mathrm{Lip}(\mathcal{X})$, which is not possible since $\mathbf{T}$ is nonminimum phase by Lemma 13.3.1. Hence necessarily $\|\mathbf{S}\|_L \geq 1$.

(ii) Assume that L is unstable. Since $\mathbf{S}$ is in $\mathrm{Lip}(\mathcal{X})$, it is possible to show that L is a closed operator (cf. Seron 1995, Theorem 6.4.3), which implies that $\mathcal{D}(L)$ is a closed set. Now, if $\|\mathbf{T}\|_L < 1$, then by Lemma 12.1.1, $I-\mathbf{T} = \mathbf{S}$ is invertible in $\mathrm{Lip}(\mathcal{X})$, which is not possible since $\mathbf{S}$ is nonminimum phase by Lemma 13.3.1. Hence $\|\mathbf{T}\|_L \geq 1$.

□

Theorem 13.3.2 is the nonlinear extension of Theorem 13.1.1 for the case of a closed-loop system belonging to the class of Lipschitz operators. We point out that the bounds given in Theorem 13.3.2 do not assume sensitivity reduction and hence they represent limits for any possible control design.

The assumption that the sensitivities of interest are Lipschitz operators is a rather strong requirement. A relaxation would be, for example, to assume that they have a finite $\mathcal{L}_2$-gain. We stress the fact, however, that the Lipschitz assumption brings a substantial amount of mathematical structure to the problem, since many tools from linear functional analysis have immediate counterparts in Lipschitz operator theory.

As a final remark, we note that the results on Lipschitz sensitivity limitations given in this section, also hold if the extended-space setting of §12.2 of Chapter 12 is used. Indeed, the crucial fact is that Lemma 13.3.1 on interpolation constraints is still valid since $\mathcal{D}(\mathbf{S}) = \mathcal{D}(\mathbf{T})$.

To see this, recall from (13.4) that $\mathbf{T} = GK\mathbf{S} = L\mathbf{S}$. Then, using (12.5), the domain of $\mathbf{T}$ can be computed as

$$\begin{aligned}\mathcal{D}(\mathbf{T}) &= \mathcal{D}(L\mathbf{S})\\ &= \mathbf{S}^{-1}\left(\mathcal{D}(L)\right) \cup \mathcal{A}(L\mathbf{S})\\ &= \{x \in \mathcal{D}(\mathbf{S}) : \mathbf{S}x \in \mathcal{D}(L)\} \cup \{x \in \overline{\mathcal{D}(\mathbf{S})} : \mathbf{S}x \in \mathcal{E}(L)\}.\end{aligned}$$

Since $\mathcal{D}(\mathbf{S}) = \mathcal{D}(\mathbf{T})$, it follows that $\mathcal{A}(L\mathbf{S}) = \emptyset$. Thus, the analysis reduces to the definition of composition domain used in §12.1. This means that the conclusions of Theorem 13.3.2 are valid when the operators are assumed to be defined on an extended Banach space.

Interestingly, if the open-loop system, L, is unstable, then, according to the terminology introduced in §12.2, there is an unstable nonlinear "pole-zero" cancelation between L and **S**. This cancelation, however, is achieved via feedback.

13.4 The Water-Bed Effect

An important initial result on performance limitations imposed by non-minimum phase open-loop plants, was obtained by Shamma (1991) for the problem of nonlinear sensitivity reduction. In this section we will state this result and give its counterpart for the complementary sensitivity.

Consider again the feedback loop in Figure 13.1. Let $\mathcal{D}$ be a given class of finite-energy disturbances. We define the performance measure

$$\mu_S(G, K, \Omega, \mathcal{D}) \triangleq \sup_{d \in \mathcal{D}} \|\mathbf{S}d\|_\Omega , \tag{13.5}$$

where **S** is the nonlinear sensitivity operator and $\|\cdot\|_\Omega$ denotes the signal-energy distributed over the range Ω. As mentioned in Shamma (1991), $\mu_S(G, K, \Omega, \mathcal{D})$ expresses the maximum effect of a disturbance $d \in \mathcal{D}$ on the energy of $y = \mathbf{S}d$ in the frequency interval Ω.

Shamma (1991) considered the problem of minimization of (13.5) and derived the nonlinear counterpart of the water-bed phenomenon experienced by nonminimum phase plants as stated below.

Theorem 13.4.1 (Shamma 1991). Let $\Omega \subset \mathbb{R}$ have nonzero measure and let $\mathcal{D} \subset \mathcal{L}_2$ be a bounded set of disturbances. Let $\{K_i\}$ be a sequence of I/O operators which stabilize the I/O operator G. Suppose that

$$\mathcal{D} \not\subset \text{wk-cl}\,\mathcal{R}(G).$$

Then $\mu_S(G, K_i, \Omega, \mathcal{D}) \to 0$ implies

$$\sup_i \sup_{d \in \mathcal{D}} \|\mathbf{S}_i d\|_{\mathcal{L}_2} = \infty.$$

∘

$\mathbf{S}_i$ in Theorem 13.4.1 is the sensitivity operator induced by the controller $K = K_i$ in Figure 13.1. The result shows that, if the nonlinear plant is "nonminimum phase", then an arbitrarily small value of the frequency-weighted sensitivity results in an arbitrarily large response to some admissible disturbance. The plant G in Theorem 13.4.1 is "nonminimum phase" in the sense that there exists a $\tilde{d} \in \mathcal{D}$ such that $\tilde{d} \notin \text{wk-cl}\,\mathcal{R}(G)$.

This condition may be interpreted as an inability to construct a "stable approximate inverse" of the plant (Shamma 1991). Note that, since $\mathrm{cl}\,\mathcal{R}(G) \subset \mathrm{wk\text{-}cl}\,\mathcal{R}(G)$ in general, a system may be nonminimum phase according to the definition given in §12.1.2, but the weak-closure of its range may be the whole space. Thus, the nonminimum phaseness concept given in §12.1.2 seems to be insufficient to prove a water-bed effect for nonlinear feedback systems.

We consider next the corresponding phenomenon for the complementary sensitivity operator $\mathbf{T}$ and unstable open-loop dynamics. Let $\mathcal{W}$ be an admissible class of finite-energy sensor disturbances. Define the performance measure

$$\mu_T(G, K_i, \Omega, \mathcal{W}) \triangleq \sup_{w \in \mathcal{W}} \|\mathbf{T}_i w\|_\Omega, \tag{13.6}$$

where $\mathbf{T}_i$ is the complementary sensitivity operator induced by the controller $K = K_i$ in Figure 13.1. We then have the following result.

Theorem 13.4.2. Let $\Omega \in \mathbb{R}$ have nonzero measure and let $\mathcal{N} \subset \mathcal{L}_2$ be a bounded set of sensor disturbances. Let $\{K_i\}$ be a sequence of I/O operators which stabilize the I/O operator G. Suppose that

$$\mathcal{W} \not\subset \mathrm{wk\text{-}cl}\,\mathcal{D}(GK_i).$$

Then $\mu_T(G, K_i, \Omega, \mathcal{W}) \to 0$ implies

$$\sup_i \sup_{w \in \mathcal{W}} \|\mathbf{T}_i w\|_{\mathcal{L}_2} = \infty. \tag{13.7}$$

Proof. We follow the same line of development as in Shamma (1991).

Suppose, by contradiction, that $\{K_i\}$ is such that

$$\sup_i \sup_{w \in \mathcal{W}} \|\mathbf{T}_i w\|_{\mathcal{L}_2} \le \alpha < \infty. \tag{13.8}$$

Let $w = \tilde{w} \notin \mathrm{wk\text{-}cl}\,\mathcal{D}(GK_i)$ and define $y_i \triangleq \mathbf{T}_i \tilde{w}$. From (13.8), the sequence of outputs $\{y_i\}$ is bounded. Hence, it contains a weakly convergent subsequence (Conway 1990), which we rename $\{y_i\}$. Since

$$\mu_T(G, K_i, \Omega, \mathcal{W}) = \sup_{w \in \mathcal{W}} \|\mathbf{T}_i w\|_\Omega \to 0,$$

then $\|y_i\|_\Omega \to 0$. It follows that $\{y_i\} \to 0$ weakly (Shamma 1991).

Now, since each K_i stabilizes G, we have that both y_i and $e_i \in \mathcal{L}_2$. Hence, since $y_i = GK_i e_i$, it follows that $e_i \in \mathcal{D}(GK_i)$. But

$$e_i = y_i + \tilde{w}. \tag{13.9}$$

Since $\{y_i\} \to 0$ weakly, it follows from (13.9) that $\tilde{w} \in \mathrm{wk\text{-}cl}\,\mathcal{D}(GK_i)$, which contradicts the open-loop instability assumption. Hence (13.7) follows.

□

Theorem 13.4.2 states the trade-off imposed upon the complementary sensitivity operator by unstable open-loop dynamics of the composition of the plant and controller. Loosely speaking, the instability is interpreted as the existence of a bounded admissible input that produces an unbounded output.

13.5 Sensitivity and Stability Robustness

In this section we consider sufficient conditions that guarantee robust Lipschitz stability of the nonlinear sensitivities when the open-loop system is perturbed in different ways. Indeed, we consider additive, output-multiplicative and input-divisive perturbations. Each case will be reconfigured in the basic perturbation model shown in Figure 12.1 of Chapter 12 for appropriate choices of the nonlinear operator H.

We consider the feedback loop in Figure 13.1 to be the nominal system. We denote the nominal open-loop operator by $L \triangleq GK$ and the nominal sensitivity and complementary sensitivity by $\mathbf{S}$ and $\mathbf{T}$, respectively. We use the symbols $\tilde{\mathbf{S}}$ and $\tilde{\mathbf{T}}$ for the actual sensitivity operators corresponding to the perturbed plant, which will be denoted by $\tilde{L}$.

We then have the following result.

Theorem 13.5.1. Let the nominal open-loop system $L : \mathcal{D}(L) \subset X \to \mathcal{X}$ be such that the nominal sensitivity operator $\mathbf{S}$ is in $\mathrm{Lip}(\mathcal{X})$. Then:

(i) Assume that the perturbed system is modeled as $\tilde{L} \triangleq L - \Delta$, with $\Delta \in \mathrm{Lip}(\mathcal{X})$ (additive perturbation model, Figure 13.2). If $\|\Delta\mathbf{S}\|_L < 1$ then the actual sensitivity operators are in $\mathrm{Lip}(\mathcal{X})$ and

$$\|\tilde{\mathbf{S}}\|_L \leq \|\mathbf{S}\|_L(1 - \|\Delta\mathbf{S}\|_L)^{-1}. \tag{13.10}$$

(ii) Assume that the perturbed system is modeled as $\tilde{L} \triangleq (I - \Delta)L$, with $\Delta \in \mathrm{Lip}(\mathcal{X})$ (output-multiplicative perturbation model, Figure 13.3). If $\|\Delta\mathbf{T}\|_L < 1$ then the actual sensitivity operators are in $\mathrm{Lip}(\mathcal{X})$ and

$$\|\tilde{\mathbf{S}}\|_L \leq \|\mathbf{S}\|_L(1 - \|\Delta\mathbf{T}\|_L)^{-1}. \tag{13.11}$$

(iii) Assume that the perturbed system is modeled as $\tilde{L} \triangleq L(I - \Delta)^{-1}$, with $\Delta \in \mathrm{Lip}(\mathcal{X})$ (input-divisive perturbation model, Figure 13.4). If $\|\Delta\mathbf{S}\|_L < 1$ then the actual sensitivity operators are in $\mathrm{Lip}(\mathcal{X})$ and

$$\|\tilde{\mathbf{S}}\|_L \leq 1 + \|\mathbf{T}\|_L(1 - \|\Delta\mathbf{S}\|_L)^{-1}. \tag{13.12}$$

Proof. (i) Let $\tilde{L} \triangleq L - \Delta$, $\Delta \in \mathrm{Lip}(\mathcal{X})$, as shown in the left scheme of Figure 13.2.

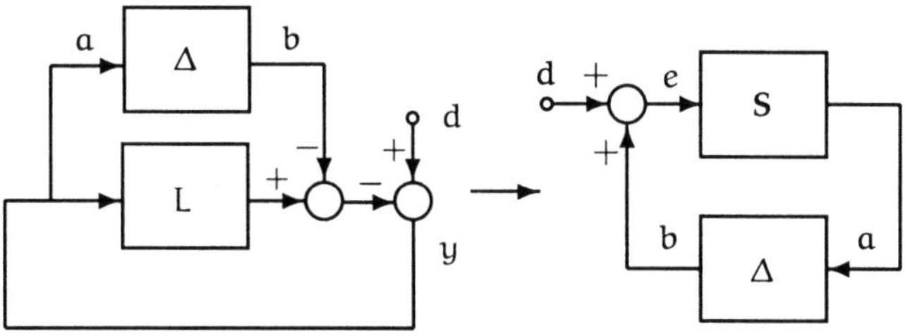

FIGURE 13.2. Additive perturbation model (left) and equivalent configuration in terms of **S** (right).

Solving for $y = a$, we have that $y = a = \mathbf{S}(d + b)$, thus the loop on the left can be drawn as the basic perturbation model on the right of Figure 13.2. If $\|\Delta\mathbf{S}\|_L < 1$ it follows from Lemma 12.1.1 that $(I - \Delta\mathbf{S})^{-1} \in \text{Lip}(\mathcal{X})$ and hence $\tilde{\mathbf{S}} = \mathbf{H}_{yd}|_{(n=0)} = \mathbf{S}(I - \Delta\mathbf{S})^{-1} \in \text{Lip}(\mathcal{X})$ with the bound given in (13.10) for its Lipschitz constant (see (12.4)).

(ii) Let $\tilde{L} \triangleq (I - \Delta)L$, $\Delta \in \text{Lip}(\mathcal{X})$, as depicted in the left scheme of Figure 13.3.

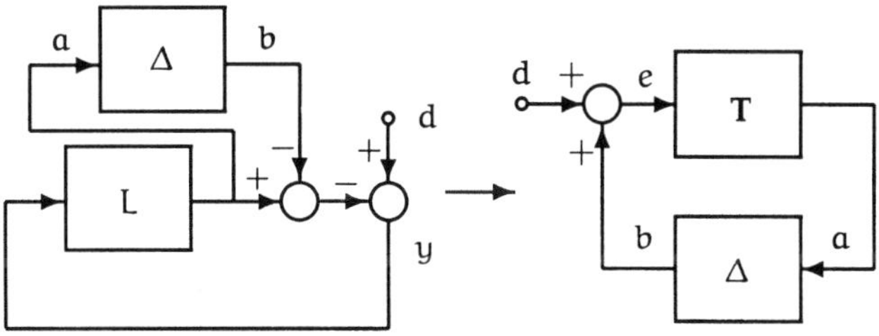

FIGURE 13.3. Multiplicative perturbation model (left) and equivalent configuration in terms of **T** (right).

Solving for y, we have that $y = \mathbf{S}(d+b) \triangleq \mathbf{S}e$, and then $a = Ly = \mathbf{T}e$. Thus the loop on the left can be drawn as the loop on the right of Figure 13.3. If $\|\Delta\mathbf{T}\|_L < 1$ Lemma 12.1.1 shows that $(I - \Delta\mathbf{T})^{-1} \in \text{Lip}(\mathcal{X})$ and hence $\tilde{\mathbf{S}} = \mathbf{H}_{yd}|_{(n=0)} = \mathbf{S}\mathbf{H}_{ed} = \mathbf{S}(I - \Delta\mathbf{T})^{-1} \in \text{Lip}(\mathcal{X})$. The bound given in (13.11) for its Lipschitz constant follows using (12.3).

(iii) Let $\tilde{L} \triangleq L(I - \Delta)^{-1}$, $\Delta \in \text{Lip}(\mathcal{X})$, as depicted in the left scheme of Figure 13.4.

Solving for a, we have that $a = \mathbf{S}(d + b) = \mathbf{S}e$ (and then $y = d - La = d - \mathbf{T}e$). Thus the loop on the left can be drawn as the loop on the right of Figure 13.4. If $\|\Delta\mathbf{S}\|_L < 1$ it follows from Lemma 12.1.1 that $(I - \Delta\mathbf{S})^{-1} \in \text{Lip}(\mathcal{X})$ and hence $\tilde{\mathbf{S}} = \mathbf{H}_{yd}|_{(n=0)} = I - \mathbf{T}\mathbf{H}_{ed} =$

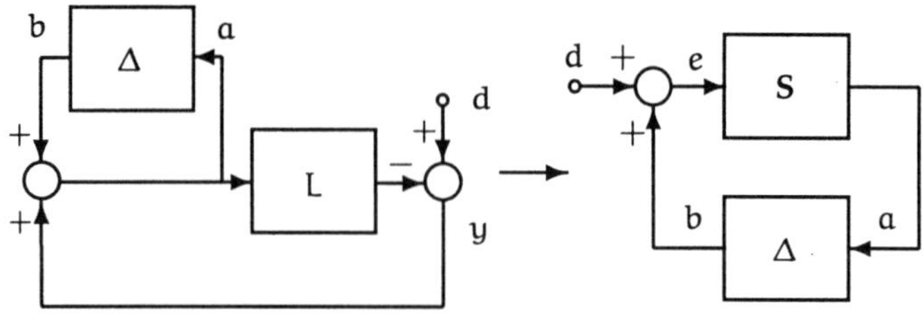

FIGURE 13.4. Divisive perturbation model (left) and equivalent configuration in terms of $\mathbf{S}$ (right).

$I - \mathbf{T}(I - \Delta\mathbf{S})^{-1} \in \mathrm{Lip}(\mathcal{X})$, with the bound given in (13.12) for its Lipschitz constant.

□

The additive perturbation model may result, for example, from an additive perturbation of the plant, i.e., $\tilde{G} = G - (\Delta_G)$, giving the values $L = GK$ and $\Delta = (\Delta_G)K$ in Figure 13.2. The output-multiplicative perturbation model comes directly from an output-multiplicative perturbation of the plant of the form $\tilde{G} = (I - \Delta)G$. The input-divisive perturbation model may represent (originally unmodeled) sensor dynamics.

It can be concluded from Theorems 13.3.2 and 13.5.1 that nonminimum phase and unstable systems allow only "small" perturbations if robust stability is to be achieved. Indeed, consider for example robust stability under additive perturbations; it then follows from Theorem 13.5.1 and the submultiplicative property of the Lipschitz norm that

$$\|\Delta\|_L \|\mathbf{S}\|_L < 1$$

is also a sufficient condition for robust stability. This, together with Theorem 13.3.2 thus imply that $\|\Delta\|_L < 1$ is required for the sufficient condition for stability of the perturbed system to hold.

13.6 Summary

This chapter has extended the known concept of sensitivity operators, as used in the case of linear feedback systems, to the case of nonlinear systems. The nonlinear sensitivity operators corresponding to the unity feedback control problem with additive disturbance inputs satisfy the structural complementarity constraint $\mathbf{S} + \mathbf{T} = I$ and "interpolation constraints" similar to their linear counterparts. Namely, if the open-loop system is nonminimum phase, then $\mathbf{T}$ is nonminimum phase, and if the open-loop system is unstable, then $\mathbf{S}$ is nonminimum phase. The nonlinear extensions of the linear concepts of nonminimum phase and unstable systems are handled as properties of the domain and range of nonlinear operators.

For the case of Lipschitz sensitivities, the complementarity and interpolation constraints can be used to establish the fact that the Lipschitz constant of sensitivity is bounded below by one for nonminimum phase open-loop systems and that the Lipschitz constant of complementary sensitivity is bounded below by one for open-loop unstable systems. These results parallel those known in linear control theory on the bounding of the H_∞ norms of S and T.

Probably the first result on performance limitations in general nonlinear feedback control was given in Shamma (1991). Considering the nonlinear sensitivity as an operator in a Hilbert space, Shamma established that, for a nonminimum phase system, the achievement of arbitrary sensitivity reduction over a frequency range necessarily implies an arbitrarily large response to some admissible disturbance. This is the nonlinear extension of the water-bed phenomenon experienced by the linear sensitivity (see §3.2 in Chapter 3). As expected, a result parallel to that of Shamma can also be derived on the trade-off induced by unstable open-loop dynamics for the problem of complementary sensitivity reduction.

The nonlinear sensitivities developed here are the appropriate tool with which to state sufficient conditions for robust stability. This shows that the role of the linear sensitivities, S and T, in measuring closed-loop robustness against unstructured perturbations can be extended to nonlinear systems.

Notes and References

Sensitivity Limitations

The results on nonlinear Lipschitz sensitivity limitations are based on Seron & Goodwin (1996), Seron (1995) and Goodwin & Seron (1995).

The nonlinear extension of the water-bed effect is due to Shamma (1991).

Stability Robustness

§13.5 is based on Seron & Goodwin (1996). Robustness tests similar to those given in §13.5 were obtained by Astolfi & Guzzella (1993) using the $\mathcal{L}_2$-gain (or "nonlinear $\mathcal{H}_\infty$ norm") *in lieu* of the Lipschitz constant. Astolfi & Guzzella, however, used a state-space approach. Other results on robustness for nonlinear feedback systems are the works of Georgiou & Smith (1995*a*, 1995*b*, 1995*c*), who use a nonlinear generalization of the gap metric.

14
Nonlinear Filtering

This chapter represents a preliminary extension to the nonlinear case of the sensitivity approach to filtering of Chapters 8 and 9. We use the background on nonlinear operators given in Chapter 12. Indeed, we first study the complementarity of the filtering sensitivities in the framework of §12.1 and then use the concept of nonlinear cancelations developed in §12.2 to address the issue of stability of the estimation error.

14.1 A Complementarity Constraint

Consider the filtering problem shown schematically in Figure 14.1.

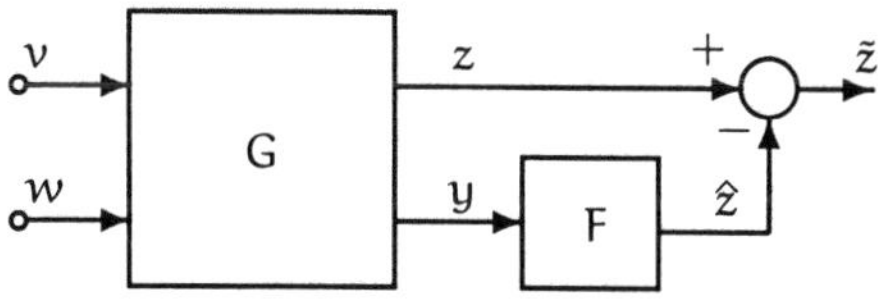

FIGURE 14.1. General filtering configuration.

For the purposes of this chapter, the "plant", depicted in Figure 14.2, is characterized by two nonlinear operators. The "output" operator, G_y, is assumed to be a nonlinear operator on a linear space, $\mathcal{X}$, i.e., $G_y : \mathcal{D}(G_y) \subset \mathcal{X} \to \mathcal{X}$. It maps the process input, v, into the noise-free output, $y - w$. The "state" operator, G_z, is the nonlinear operator $G_z : \mathcal{D}(G_z) \subset \mathcal{X} \to \mathcal{X}$, which maps ν into the signal (or internal state) to be estimated, z. Specifically,

$$\begin{aligned} y &= G_y\nu + w, \\ z &= G_z\nu, \end{aligned} \tag{14.1}$$

where w is measurement noise corrupting the system output. We will further assume that the operator G_y is "unbiased", i.e., $G_y\,0 = 0$.

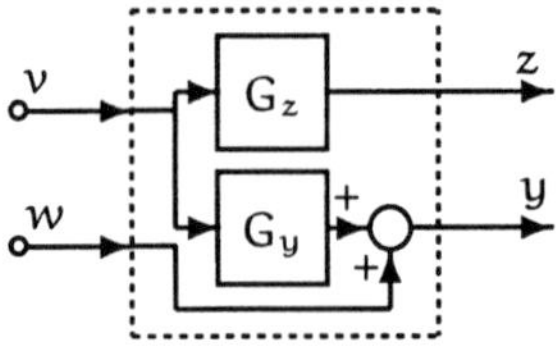

FIGURE 14.2. Structure of the plant.

A nonlinear state estimator for (14.1) is chosen as

$$\hat{z} = Fy, \tag{14.2}$$

where $\hat{z}$ is the estimate of z and the filter, F, is a nonlinear operator $F : \mathcal{D}(F) \subset \mathcal{X} \to \mathcal{X}$. F is assumed to be a stable operator, i.e., $\mathcal{D}(F) = \mathcal{X}$.

Let the estimation error be, as in Chapter 9,

$$\tilde{z} = z - \hat{z}. \tag{14.3}$$

We will assume that the operator G_z is *right invertible*. By this, we mean that there exists a right inverse G_z^{-1} such that $G_z G_z^{-1} = I_{\mathcal{R}(G_z)}$, where $I_{\mathcal{R}(G_z)}$ is the restriction of the identity operator to $\mathcal{R}(G_z)$. Also, we will use the set

$$\mathcal{D}_z \triangleq \{x \in \mathcal{R}(G_z) : G_z^{-1}x \in \mathcal{D}(G_y)\}. \tag{14.4}$$

Observe from Figure 14.2, that $\mathcal{D}_z$ is the set of signals z's such that $G_z^{-1}z$ gives a signal ν in the domain of the output operator G_y.

We define the *nonlinear filtering sensitivity*, $\mathbf{P}$, and the *nonlinear filtering complementary sensitivity*, $\mathbf{M}$, as

$$\begin{aligned} \mathbf{P} &\triangleq \mathbf{H}_{\tilde{z}\nu}|_{(w=0)} G_z^{-1}, \\ \mathbf{M} &\triangleq -\mathbf{H}_{\tilde{z}w}|_{(\nu=0)} G_y G_z^{-1}. \end{aligned} \tag{14.5}$$

We then have the following complementarity constraint for the filtering loop in Figure 14.1.

Theorem 14.1.1. Consider the plant (14.1) and the filter (14.2) with estimation error defined in (14.3). Assume that G_y is an unbiased operator. Then, the following complementarity constraint holds:

$$\mathbf{P} + \mathbf{M} = I_{\mathcal{D}_z}, \tag{14.6}$$

where $I_{\mathcal{D}_z}$ is the restriction of the identity operator to the set $\mathcal{D}_z$ given in (14.4).

Proof. From (14.3), (14.1) and (14.2), we have

$$\mathbf{H}_{\tilde{z}v}|_{(w=0)} = G_z - FG_y. \tag{14.7}$$

Next, we note that, $\hat{z} = Fw$ whenever $v = 0$, since G_y is unbiased. Then

$$\mathbf{H}_{\tilde{z}w}|_{(v=0)} = -\mathbf{H}_{\hat{z}w}|_{(v=0)} = F, \tag{14.8}$$

where the first equality follows using the fact that $\mathbf{H}_{zw} = 0$. From (14.7) and (14.8), we have

$$\mathbf{H}_{\tilde{z}v}|_{(w=0)} = G_z - \left(-\mathbf{H}_{\tilde{z}w}|_{(v=0)}\right) G_y.$$

Multiplying from the right by G_z^{-1} and using (14.5) yields

$$\mathbf{P} = I_{\mathcal{R}(G_z)} - \mathbf{M}.$$

The domain where the above relation holds, however, is not all of $\mathcal{R}(G_z)$, but the intersection of $\mathcal{R}(G_z)$ with the domain of $\mathbf{M}$. To compute $\mathcal{D}(\mathbf{M})$, note that

$$\mathbf{M} = F(G_y G_z^{-1}).$$

Since F is a stable operator, then $\mathcal{D}(\mathbf{M})$ is given by

$$\begin{aligned}\mathcal{D}(\mathbf{M}) &= \left(G_z^{-1}\right)^{-1} (\mathcal{D}(G_y)) \\ &= \{x \in \mathcal{D}(G_z^{-1}) : G_z^{-1}x \in \mathcal{D}(G_y)\} \\ &= \{x \in \mathcal{R}(G_z) : G_z^{-1}x \in \mathcal{D}(G_y)\} \\ &= \mathcal{D}_z.\end{aligned}$$

Hence (14.6) follows. □

Note that, if $\mathcal{D}(G_y) = \mathcal{D}(G_z)$, then $\mathcal{D}_z = \mathcal{R}(G_z)$. This is the case when the operators G_y and G_z share the same "instabilities". In the linear case, it would correspond to the situation where all the unstable modes of the plant, assumed observable from y, are also observable from the signal to be estimated, z.

Theorem 14.1.1 proves that the complementarity constraint shown to hold for linear filtering in Chapter 8, extends to the problem of nonlinear filtering under process noise and additive measurement noise.

14.2 Bounded Error Nonlinear Estimation

In this section, we will extend the concept of BEE, defined in Chapter 8, to nonlinear filters. First, we will refine the structure of the plant in Figure 14.2. Since z is an internal variable, it is reasonable to assume that the operators G_y and G_z have some common dynamics. We will model this as

$$\begin{aligned} G_y &= G_{yy} G_c, \\ G_z &= G_{zz} G_c. \end{aligned} \tag{14.9}$$

For convenience, we redraw the filtering loop as in Figure 14.3, indicating the special structure that we assume for the plant.

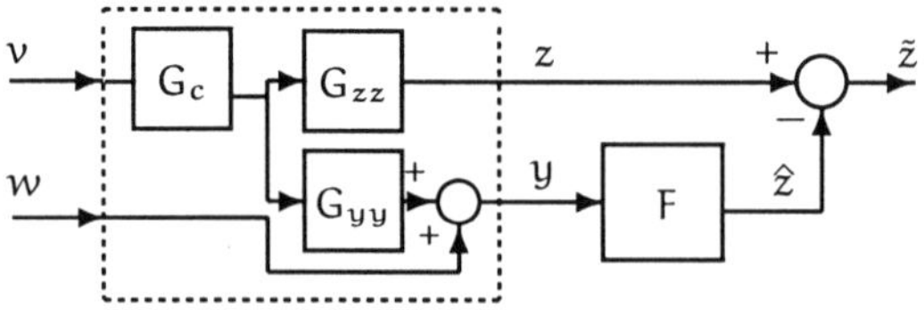

FIGURE 14.3. Filtering loop with special structure for the plant

From (14.7), (14.9) and (14.8), the filtering error operators have the form

$$\begin{aligned} \mathbf{H}_{\tilde{z}v}|_{(w=0)} &= \tilde{G} G_c, \\ \mathbf{H}_{\tilde{z}w}|_{(v=0)} &= F, \end{aligned} \tag{14.10}$$

where $\tilde{G} \triangleq G_{zz} - F G_{yy}$. We introduce the following definition.

Definition 14.2.1 (Bounded Error Nonlinear Estimator). A stable nonlinear operator F is said to be a bounded error nonlinear estimator (BENE) for the system (14.9), if the error operators (14.10) are stable nonlinear operators. ◦

Observe that, contrary to the linear case, the BENE concept applies only to the particular inputs v and w. Another interesting distinction is that there is no guarantee that the estimation error will be bounded when both inputs act simultaneously, since superposition is not valid for nonlinear systems.

In order to derive conditions for a nonlinear estimator to be a BENE, we need to work with the framework of §12.2 in Chapter 12. Using definition (12.5) to compute the domain of $\mathbf{H}_{\tilde{z}v}|_{(w=0)}$ given in (14.10), we have

$$\begin{aligned} \mathcal{D}(\mathbf{H}_{\tilde{z}v}|_{(w=0)}) &= \mathcal{D}(\tilde{G} G_c) \\ &= G_c^{-1}(\mathcal{D}(\tilde{G})) \cup \mathcal{A}(\tilde{G} G_c), \end{aligned} \tag{14.11}$$

where

$$\begin{aligned} G_c^{-1}(\mathcal{D}(\tilde{G})) &= \{x \in \mathcal{D}(G_c) : G_c x \in \mathcal{D}(\tilde{G})\}, \\ \mathcal{A}(\tilde{G}G_c) &= \{x \in \overline{\mathcal{D}(G_c)} : G_c x \in \mathcal{E}(\tilde{G})\}. \end{aligned} \tag{14.12}$$

We can now state a necessary and sufficient condition for a nonlinear operator to be a BENE.

Theorem 14.2.1. A stable nonlinear operator, F, is a BENE for the system (14.9) if and only if the following two conditions hold:

(i) $G_c^{-1}(\mathcal{D}(\tilde{G})) = \mathcal{D}(G_c)$,

(ii) $\mathcal{A}(\tilde{G}G_c) = \overline{\mathcal{D}(G_c)}$,

where $\tilde{G}$ is defined in (14.10), and where $G_c^{-1}(\mathcal{D}(\tilde{G}))$, $\mathcal{A}(\tilde{G}G_c)$ are given in (14.12).

Proof. Since F is stable, it follows from Definition 14.2.1 that F is a BENE for (14.9) if and only if $\mathbf{H}_{\tilde{z}\nu}|_{(w=0)}$ in (14.10) is a stable nonlinear operator. This, in turn, holds if and only if $\mathcal{D}(\tilde{G}G_c) = \mathcal{X}$ in (14.11). Since $G_c^{-1}(\mathcal{D}(\tilde{G}))$ and the added domain $\mathcal{A}(\tilde{G}G_c)$ in (14.12) have no intersection, the result then follows. □

Assuming that $\tilde{G}$ is stable and G_c is unstable, then the stability of $\mathbf{H}_{\tilde{z}\nu}|_{(w=0)}$ in (14.10) must be achieved by an unstable nonlinear zero-pole cancelation.

It is not difficult to see that condition (i) in Theorem 14.2.1 is equivalent to the following:

$$\mathcal{R}(G_c) \subset \mathcal{D}(\tilde{G}) .$$

Similarly, condition (ii) can be written as:

$$G_c(\overline{\mathcal{D}(G_c)}) \subset \mathcal{E}(\tilde{G}) ,$$

where $G_c(\overline{\mathcal{D}(G_c)})$ is the obvious notation for the image of the set $\overline{\mathcal{D}(G_c)}$ through G_c. Hence, we can alternatively say that F is a BENE if and only if: (i′) G_c maps every x in its domain to the domain of $\tilde{G}$, and (ii′) G_c maps every x outside its domain to the extended domain of $\tilde{G}$. Figure 14.4 illustrates this interpretation.

14.3 Sensitivity Limitations

Lemma 13.3.1 in Chapter 13 established that nonminimum phase and unstable characteristics of the open-loop system were inherited by the closed-loop operators **S** and **T**. These "nonlinear interpolation constraints"

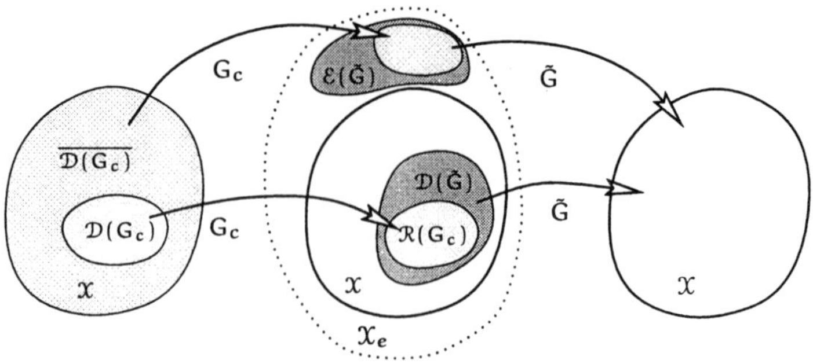

FIGURE 14.4. Set interpretation of a BENE.

extended the interpolation constraints for S and T in linear feedback control. In Chapters 8 and 9, we showed that similar interpolation constraints can be derived for the linear filtering sensitivities, P and M, achieved by BEEs.

By way of contrast, nonminimum phase and unstable characteristics of the system to be estimated, do not seem to necessarily constrain, at least within our framework, properties of the nonlinear filtering sensitivities achieved by a BENE. To see this, we will discuss the case where G_{zz} in (14.9) is the identity operator. This will simplify the analysis to the composition of two operators rather than three, and the conclusions are still similar.

Using $G_{zz} = I$ and (14.10) in the definitions (14.5), we obtain

$$\begin{aligned} \mathbf{P} &= \tilde{G} = I - FG_{yy}, \\ \mathbf{M} &= FG_{yy}. \end{aligned} \tag{14.13}$$

Next, assume that the output operator G_y in (14.9) is such that G_c is unstable and G_{yy} is D-NMP. If the filter, F, is a BENE, it follows from Theorem 14.1.1 that there exists an unstable nonlinear zero-pole cancelation in the composition $\tilde{G}G_c$. This means that $\tilde{G}$ is E-NMP. It does not follow, however, that **P** has a defect in the closure of its range unless $\tilde{G}$ is also D-NMP.

Turning to **M**, the fact that G_{yy} is D-NMP implies that the nonlinear operator F is acting on a set, $\mathcal{R}(G_{yy})$, which is a strict subset of $\mathcal{X}$. This yet does not necessarily lead to a defective range for the composition FG_{yy}.

We formalize the above discussion in the following result.

Theorem 14.3.1. Let the plant be given by (14.9) with $G_{zz} = I$. Suppose that **P** in (14.13) is a Lipschitz operator on $\mathcal{X}$. Then

(i) If G_{yy} is D-NMP, and the filter, F, maps sets whose closure are strict subsets of $\mathcal{X}$ into sets whose closure are also strict subsets of $\mathcal{X}$, then $\|\mathbf{P}\|_L \geq 1$.

(ii) If G_c is unstable, and $\tilde{G}$ is NMP, then $\|\mathbf{M}\|_L \geq 1$.

Proof. Note that the conditions in (i) imply that **M** is D-NMP, and the conditions in (ii) imply that **P** is NMP. The proof then follows similarly to the proof of Theorem 13.3.2 in Chapter 13. □

From Theorem 14.3.1 and the previous discussion, it seems plausible that the constraints on the filtering sensitivities, due to nonminimum phase and unstable plant dynamics, could be avoided by a proper selection of the filter operator, F.

14.4 Summary

For the problem of nonlinear filtering in the presence of process noise and additive measurement noise, we have defined in this chapter appropriate extensions of the filtering sensitivities considered in Chapters 8 and 9. These nonlinear filtering sensitivities are complementary operators when the external signals affecting the loop belong to a linear space. For bounded error nonlinear estimators, we established sensitivity limitations under conditions expressed in terms of nonminimum phase and unstable dynamics of the plant to be estimated. These limitations, however, do not seem to be unavoidable. Indeed, it appears possible to design nonlinear bounded error filters such that the sensitivities are not constrained by nonminimum phase and unstable dynamics of the plant.

Chapter 13 has dealt with the extension to nonlinear systems of limitations known to hold for LTI systems. It is possible that relaxation of linearity could circumvent some trade-offs that are unavoidable in the special case of LTI systems. Indeed, the results for nonlinear filtering of Chapter 14 hint at a positive answer to this question. This conjecture, however, still remains an open research question.

Notes and References

The results of this chapter are based on Seron (1995).

Appendix A
Review of Complex Variable Theory

A central result of Complex Variable Theory is the Cauchy Integral Theorem and its applications. In this appendix we will state and prove this fundamental result, which is the core of Cauchy's theory of complex integration. The theory of complex integration is one of the three avenues of approach to complex analysis, the other two being the theory of complex derivatives and the theory of power series. These three approaches are associated with the names of Cauchy, Riemann and Weierstrass respectively. The review given in this appendix summarizes the essential background on analytic functions and complex variable theory. The emphasis is on complex integration and complex derivatives, but a brief introduction to power series is also provided.

A.1 Functions, Domains and Regions

We will be primarily interested in complex-valued *functions* defined on *regions*. Before defining regions, we need some concepts related to sets. An open set is (pathwise) *connected* if each pair of points in it can be joined by a polygonal path (consisting of a finite number of line segments joined end to end) that lies entirely in the set (see Figure A.1). An open connected set of points is called a *domain*. A point s_0 is called a *boundary point* of a set U if every neighborhood of the point s_0 contains at least one point in the set U and at least one point not in the set U. A *region* is a domain together with none, some, or all of its boundary points. A region is said to be *closed* if it contains all its boundary points, and *bounded* if it can be enclosed in a circle of sufficiently large radius. If Ω is a region, the notation $\overline{\Omega}$ denotes the closed region formed by Ω together with *all* its boundary points. Domains and regions are subsets of the *extended complex plane*, i.e., the set of all finite complex numbers (the complex plane $\mathbb{C}$) and the *point at infinity*, ∞. We denote the extended complex plane by $\mathbb{C}_e = \mathbb{C} \cup \{\infty\}$.

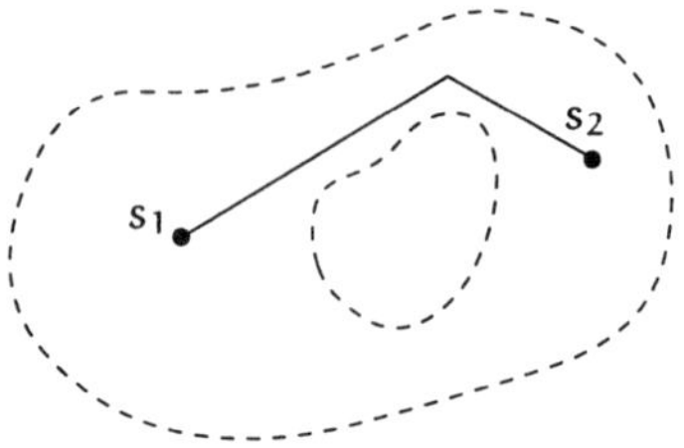

FIGURE A.1. Open connected set.

A function f is *continuous at a point* s_0 if the following condition is satisfied:

$$\lim_{s \to s_0} f(s) = f(s_0),$$

which implies that for each positive number ε there exists a positive number δ such that

$$|f(s) - f(s_0)| < \varepsilon \quad \text{whenever} \quad |s - s_0| < \delta. \tag{A.1}$$

A function of a complex variable is said to be *continuous in a region* Ω if it is continuous at each point in Ω.

A.2 Complex Differentiation

There is an important difference between real and complex derivatives. We recall that the existence of the derivative of a function of a real variable is essentially a mild smoothness condition. On the other hand, the existence of the derivative of a function of an independent variable having two "degrees of freedom" implies a great deal about the function. As we will see in this section, complex differentiability leads to a pair of differential equations — the Cauchy-Riemann equations — which must be satisfied by the function's real and imaginary parts.

Let $w = f(s)$ be a given complex function of the complex variable s, defined on a domain, D. Then w is said to have a *derivative* at $s_0 \in D$ if

$$\lim_{s \to s_0} \frac{f(s) - f(s_0)}{s - s_0} \tag{A.2}$$

exists, and is independent of the direction of the increment $s - s_0$. We denote this limit as $f'(s_0)$ or $df/ds|_{s_0}$. When the derivative of f at $s_0 \in D$ exists, f is said to be *differentiable at* s_0.

We then have the following necessary and sufficient condition for differentiability.

Theorem A.2.1. The function $f(s)$ is differentiable at the point $s_0 \in D$ if and only if the function increment $f(s) - f(s_0)$ can be written in the form

$$f(s) - f(s_0) = k(s - s_0) + \varepsilon(s, s_0)(s - s_0), \tag{A.3}$$

where $\varepsilon(s, s_0) \to 0$ as $s \to s_0$, and k is a constant independent of $s - s_0$ and ε.

Proof. If f has a derivative $f'(s_0)$ at s_0 then, by definition

$$\frac{f(s) - f(s_0)}{s - s_0} = f'(s_0) + \varepsilon(s, s_0), \tag{A.4}$$

where $\varepsilon(s, s_0) \to 0$ as $s \to s_0$. Multiplying (A.4) by $(s - s_0)$, we find that $f(s) - f(s_0)$ can be written in the form (A.3) with $k = f'(s_0)$. Conversely, if (A.3) holds, then dividing by $(s - s_0)$ and taking the limit as $s \to s_0$, we find that $f'(s_0)$ exists and equals k. □

The fact that the increment $s - s_0$ in (A.2) is complex, sets a very severe restriction on the class of functions that have a complex derivative, as we see from the following example.

Example A.2.1. Consider the function $f(s) = \bar{s} = \sigma - j\omega$, i.e., the complex conjugate of s. Then,

$$\frac{f(s) - f(s_0)}{s - s_0} = \frac{\overline{s - s_0}}{s - s_0}.$$

If $s - s_0$ is real, then $\overline{s - s_0} = s - s_0$ and the limit as $s \to s_0$ is 1. But if $s - s_0$ is purely imaginary, then $\overline{s - s_0} = -(s - s_0)$ and the limit is -1. Thus, this function *is not* differentiable anywhere, although its real and imaginary parts, σ and $-\omega$, are well behaved. ∘

In the above example, the nonexistence of $f'(s)$ was proven by taking limits first through real values and then through imaginary values. These two possibilities lead to the main condition that a complex function must satisfy in order that it have a derivative. These conditions are stated in the following result.

Theorem A.2.2 (Cauchy-Riemann equations). Let the function $f(s) = u(\sigma, \omega) + jv(\sigma, \omega)$, where $s = \sigma + j\omega$, be defined on a domain D. Then a necessary and sufficient condition for f to be differentiable at the point $s_0 = \sigma_0 + j\omega_0 \in D$ is that the functions u and v be differentiable (as functions of the two real variables σ and ω) at the point (σ_0, ω_0) and satisfy the Cauchy-Riemann equations,

$$\frac{\partial u}{\partial \sigma} = \frac{\partial v}{\partial \omega} \quad \text{and} \quad \frac{\partial u}{\partial \omega} = -\frac{\partial v}{\partial \sigma}, \tag{A.5}$$

at (σ_0, ω_0). Furthermore

$$f'(s_0) = \frac{\partial u}{\partial \sigma} + j\frac{\partial v}{\partial \sigma} = \frac{\partial v}{\partial \omega} + j\frac{\partial v}{\partial \sigma} = \frac{\partial u}{\partial \sigma} - j\frac{\partial u}{\partial \omega} = \frac{\partial v}{\partial \omega} - j\frac{\partial u}{\partial \omega}, \tag{A.6}$$

where the partial derivatives are evaluated at (σ_0, ω_0).

Proof. We prove that the conditions are necessary. Let $\Delta_w = f(s_0 + \Delta_s) - f(s_0) = \Delta_u + j\Delta_v$. Since f is differentiable at s_0, we have from Theorem A.2.1

$$\Delta_w = k\Delta_s + \varepsilon\Delta_s, \qquad k = f'(s_0), \tag{A.7}$$

where ε goes to zero as Δ_s goes to zero. Then, writing $\Delta_s = \Delta_\sigma + j\Delta_\omega$, $k = a + jb$ and $\varepsilon = \varepsilon_1 + j\varepsilon_2$, we have

$$\Delta_u + j\Delta_v = (a + jb)(\Delta_\sigma + j\Delta_\omega) + (\varepsilon_1 + j\varepsilon_2)(\Delta_\sigma + j\Delta_\omega).$$

So

$$\Delta_u = a\Delta_\sigma - b\Delta_\omega + \varepsilon_1\Delta_\sigma - \varepsilon_2\Delta_\omega,$$
$$\Delta_v = b\Delta_\sigma + a\Delta_\omega + \varepsilon_2\Delta_\sigma + \varepsilon_1\Delta_\omega,$$

where ε_1, $\varepsilon_2 \to 0$ as Δ_σ, $\Delta_\omega \to 0$, since

$$|\Delta_s| = \sqrt{(\Delta_\sigma)^2 + (\Delta_\omega)^2}, \; |\varepsilon_1| \le |\varepsilon|, \; |\varepsilon_2| \le |\varepsilon|.$$

It follows that the functions u and v are differentiable at (σ_0, ω_0) and

$$\frac{\partial u}{\partial \sigma} = a = \frac{\partial v}{\partial \omega}, \qquad \frac{\partial u}{\partial \omega} = -b = -\frac{\partial v}{\partial \sigma}, \tag{A.8}$$

which immediately imply (A.5) and (A.6).

The proof of sufficiency follows by reversing the preceding argument. □

A.3 Analytic functions

The Cauchy-Riemann equations represent a necessary and sufficient condition for pointwise differentiability of a complex function. We will be generally interested, however, in differentiability throughout a region. This motivates the following definition.

Definition A.3.1 (Analytic Function). A function f is said to be *analytic*[1] at a point s_0 if f is differentiable throughout some neighborhood of s_0. A function is analytic in a region, if it is analytic at every point of the region. ○

In the case of a domain (open, connected, nonempty set), differentiability and analyticity are equivalent, but in other cases, differentiability is required in a larger open set (for example, f is analytic for $|s| \leq 1$ if f is differentiable for $|s| < 1 + \delta$, where $\delta > 0$).

A remarkable result from the theory of functions of a complex variable is that every analytic function is infinitely differentiable and, furthermore, has a power series expansion about each point of its domain. We will show this in §A.7.1 and §A.7.2.

As we know from calculus, a sufficient condition for the differentiability of the functions $u(\sigma, \omega)$ and $v(\sigma, \omega)$ on a domain D is that the partial derivatives

$$\frac{\partial u}{\partial \sigma}, \frac{\partial u}{\partial \omega}, \frac{\partial v}{\partial \sigma}, \frac{\partial v}{\partial \omega} \tag{A.9}$$

exist and are continuous on D. Therefore, a sufficient condition for the function $f = u + jv$ to be analytic on D is that the partial derivatives exist, are continuous and satisfy the Cauchy-Riemann equations on D. Conversely, it follows from Theorem A.2.2 that, if $f = u + jv$ is analytic on a domain D, then u and v have partial derivatives (A.9) satisfying the Cauchy-Riemann conditions (A.5).[2]

We illustrate by some examples.

Example A.3.1. Consider $f(s) = s^2 \triangleq u(\sigma, \omega) + jv(\sigma, \omega)$. Then $u(\sigma, \omega) = \sigma^2 - \omega^2$, $v(\sigma, \omega) = 2\sigma\omega$, and thus

$$\begin{aligned} \frac{\partial u}{\partial \sigma} &= 2\sigma , & \frac{\partial v}{\partial \sigma} &= 2\omega , \\ \frac{\partial u}{\partial \omega} &= -2\omega , & \frac{\partial v}{\partial \omega} &= 2\sigma . \end{aligned}$$

Hence the Cauchy-Riemann equations (A.5) hold for all σ and ω, and the function is clearly analytic in the complex plane. ○

Example A.3.2. Consider $f(s) = |s|^2$. We then have that $u(\sigma, \omega) = \sigma^2 + \omega^2$ and $v(\sigma, \omega) = 0$. Here the Cauchy-Riemann equations require $2\sigma = 0$, $2\omega = 0$, and hence they require $s = 0$. Since $f'(s)$ does not exist throughout a neighborhood of $s = 0$, this function is not analytic. Nevertheless

[1] Also known as *holomorphic* or *regular*.

[2] In fact, these partial derivatives are continuous, but this does not follow from Theorem A.2.2, unless analyticity is defined as continuous-differentiability. We will use this restricted definition of analyticity to give a compact proof of Cauchy's Integral Theorem in §A.5.2.

$f'(0)$ exists, as shown by the equation

$$\lim_{s \to 0} \frac{f(s) - f(0)}{s} = \lim_{s \to 0} \frac{|s|^2}{s} = 0.$$

○

Example A.3.3. Consider a rational function of the form:

$$H(s) = \frac{K(s - q_1) \cdots (s - q_m)}{(s - p_1) \cdots (s - p_n)} \triangleq \frac{N(s)}{D(s)}. \tag{A.10}$$

Then

$$\frac{dH}{ds} = \frac{D\frac{dN}{ds} - N\frac{dD}{ds}}{D^2}.$$

These derivatives clearly exist save when $D = 0$. Hence H is analytic save at the points where D has zeros. ○

Example A.3.4. Consider H as in (A.10). Then

$$\begin{aligned} \frac{d \log H}{ds} &= \left[\frac{D}{N}\right] \left\{ \frac{D\frac{dN}{ds} - N\frac{dD}{ds}}{D^2} \right\} \\ &= \frac{1}{N}\frac{dN}{ds} - \frac{1}{D}\frac{dD}{ds}. \end{aligned} \tag{A.11}$$

Hence $\log H$ is analytic save at the points where N and D have zeros. ○

The two previous examples were concerned with functions that are analytic on a region except for some points. Those points where the function is not analytic are called *singular points* or *singularities*, and are analyzed in more detail in §A.8.

A.3.1 Harmonic Functions

Let $f = u + jv$ be analytic on a domain D. Then it satisfies the Cauchy-Riemann equations (A.5) on D. Suppose further that u and v have continuous second partial derivatives (we will show in §A.7.1 that they are in fact infinitely differentiable). Differentiating the Cauchy-Riemann equations again we get

$$\frac{\partial^2 u}{\partial \sigma^2} = \frac{\partial^2 v}{\partial \sigma \, \partial \omega}, \qquad \frac{\partial^2 u}{\partial \omega^2} = -\frac{\partial^2 v}{\partial \omega \, \partial \sigma}.$$

Hence,

$$\frac{\partial^2 u}{\partial \sigma^2} + \frac{\partial^2 u}{\partial \omega^2} = 0. \tag{A.12}$$

A similar relation holds for v. Equation (A.12) is known as *Laplace's equation* and the LHS the *Laplacian* of u, usually denoted by $\nabla^2 u$. We then have the following definition.

Definition A.3.2 (Harmonic Function). Let u be a real-valued function defined on a domain D. Then u is said to be *harmonic* if u has continuous second partial derivatives that satisfy Laplace's equation (A.12). ∘

It follows that the real and imaginary parts of an analytic function are harmonic functions. A harmonic function v related to u by the Cauchy-Riemann equations is said to be the *harmonic conjugate* of u.

The following example of a harmonic function is of particular interest in the sequel.

Example A.3.5. Consider again $\log H$, where H is the rational function of Example A.3.3. We have seen that $\log H$ is analytic on any complex domain that does not contain zeros of either the numerator or denominator of H. It follows that, on such domains, the function $\log |H| = \operatorname{Re} \log H$ is harmonic. ∘

Finally, a continuous real-valued function $u(\sigma, \omega)$, defined in some region, is said to be *subharmonic* if it satisfies

$$\nabla^2 u \triangleq \frac{\partial^2 u}{\partial \sigma^2} + \frac{\partial^2 u}{\partial \omega^2} \geq 0.$$

A.4 Complex Integration

In contrast to the real case, complex integration requires the specification of not only the limits of integration but also the particular curve that connects these points. We thus start this section with a review of curves.

A.4.1 Curves

A *curve*[3] in the extended complex plane is a set of points

$$\{\sigma(\zeta) + j\omega(\zeta), \quad \zeta \in [a, b]\}$$

where σ and ω are continuous real-valued functions of the parameter ζ in some interval $[a, b]$. A concise way of denoting a curve is by means of a complex-valued function $s = s(\zeta)$, where

$$s(\zeta) = \sigma(\zeta) + j\omega(\zeta). \tag{A.13}$$

[3]Note that there are slight differences in the literature on various definitions used in this section.

We will then say that the curve is *defined by* the function $s(\zeta)$.

A curve defined by $s = s(\zeta)$, with ζ in $[a, b]$, is a *closed* curve if its extreme points coincide, i.e., $s(a) = s(b)$. A curve is *simple* if it does not intersect itself, with the possible exception of the extreme points if the curve is also closed. A closed simple curve is sometimes called a Jordan curve. According to the *Jordan curve theorem* (e.g., Churchill & Brown 1984, p. 83), a closed simple curve divides $\mathbb{C}_e$ into two domains: an interior domain, the *interior*, which is bounded, and an exterior domain, the *exterior*. Points of the interior of a Jordan curve are called *interior points* or points *within* the curve.

The derivative $s'(\zeta)$, or $d[s(\zeta)]/d\zeta$, of s with respect to the parameter ζ is defined as

$$s'(\zeta) = \sigma'(\zeta) + j\,\omega'(\zeta) = d\sigma/d\zeta + j\,d\omega/d\zeta,$$

provided the derivatives $\sigma'(\zeta)$ and $\omega'(\zeta)$ both exist.

A curve is said to be *differentiable* if σ and ω are differentiable functions of the parameter ζ. A continuously differentiable curve is called an *arc*.

Let C, given by $s = s(\zeta)$, $\zeta \in [a, b]$, be an arc. Then the real valued function

$$|s'(\zeta)| = \sqrt{[\sigma'(\zeta)]^2 + [\omega'(\zeta)]^2}$$

is integrable over the interval $[a, b]$, and the arc C is said to have *length*[4]

$$\ell = \int_a^b |s'(\zeta)|\,d\zeta. \tag{A.14}$$

An arc is *smooth* if among its various parametric representations, there is at least one representation (A.13) such that the continuous derivative $s'(\zeta)$ is never zero on the interval $[a, b]$. The geometric meaning of smoothness is clear from the fact that if $s'(\zeta) \neq 0$ for all ζ in $[a, b]$, then the curve has a tangent at every point, whose angle of inclination is $\arg s'(\zeta)$.

A *contour* is a curve consisting of a finite number of arcs joined end to end. The length of a contour is the sum of the lengths of the arcs that constitute the contour. If each of the arcs that form the contour is smooth, then the contour is *piecewise smooth*. A piecewise smooth contour may not have a tangent at the points where the arcs are joined to each other, in which case it has a corner. For example, a rectangle on the plane is a piecewise smooth contour. Figure A.2 illustrates the curves just defined.

For the purposes of this book, it suffices to assume that all the curves considered are piecewise smooth contours. Most of the contours that we use are in fact piecewise smooth and closed.

---

[4]This expression arises from the definition of arc length in calculus.

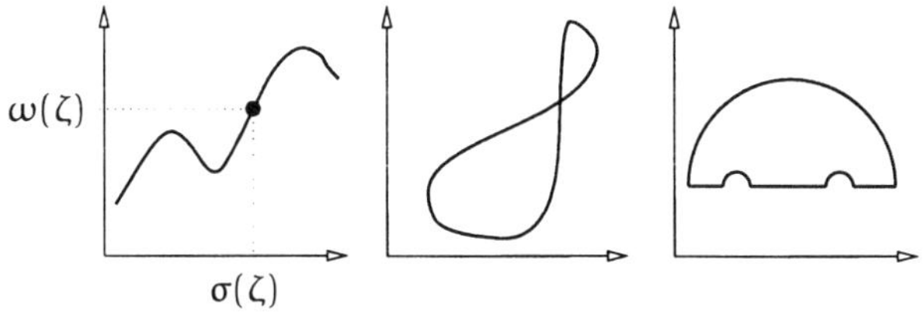

FIGURE A.2. Simple curve, closed but not simple curve, and a contour.

A.4.2 Integrals

In this subsection we define complex integration and relate it to line and area integrals of functions of two real variables.

As a preliminary step, we define the integral of a complex-valued function of a *real* variable ζ over a given interval $a \leq \zeta \leq b$. Let $s(\zeta) = \sigma(\zeta) + j\omega(\zeta)$, $\zeta \in [a, b]$, be piecewise continuous on $[a, b]$. The integral of s from a to b is defined as

$$\int_a^b s(\zeta)\, d\zeta = \int_a^b \sigma(\zeta)\, d\zeta + j \int_a^b \omega(\zeta)\, d\zeta. \tag{A.15}$$

It is easy to show (e.g., Churchill & Brown 1984, p. 80) that the integral (A.15) satisfies the inequality

$$\left| \int_a^b s(\zeta)\, d\zeta \right| \leq \int_a^b |s(\zeta)|\, d\zeta, \tag{A.16}$$

which also holds in cases when the integrals are improper, provided both sides of (A.16) exist.

By *complex integration* we refer to the integration of complex-valued functions of the complex variable s. Recall that a complex-valued function $f(s)$ is said to be continuous on the arc C, defined by $s = s(\zeta)$, if the function $f(s(\zeta))$ is continuous for ζ in $[a, b]$[5]. Then for a function f continuous[6] on C, the *integral of* f *on* C is defined by

$$\int_C f(s)\, ds = \int_a^b f(s(\zeta))\, s'(\zeta)\, d\zeta. \tag{A.17}$$

When f is continuous on a contour C, i.e., continuous on each arc C_i, $i = 1, \ldots n$, that forms the contour, then we define the integral by

$$\int_C f(s)\, ds = \int_{C_1} f(s)\, ds + \int_{C_2} f(s)\, ds + \cdots + \int_{C_n} f(s)\, ds.$$

[5]Note that f is continuous on C if it is continuous in a region of the extended complex plane containing C.

[6]Piecewise continuity is, in fact, sufficient.

Complex integration is a linear operation, that is, the following property holds

$$\int_C [\alpha f(s) + \beta g(s)]\, ds = \alpha \int_C f(s)\, ds + \beta \int_C g(s)\, ds,$$

whenever α and β are complex constants and f and g are (piecewise) continuous on the arc or contour C. Also, if F is a complex function with continuous derivative[7] f in a domain containing the arc C, then the fundamental theorem of calculus holds in the form

$$\int_C f(s)\, ds = F(A) - F(B), \tag{A.18}$$

where A and B are the initial and terminal points of C respectively.

Complex integrals can also be represented as integrals of real-valued functions of two real variables. Indeed, by separating into real and imaginary parts, any function f can be expressed in terms of two real functions, i.e., $f = u + jv$. Thus, if $s = \sigma + j\omega$, we can write

$$f(s) = u(\sigma, \omega) + jv(\sigma, \omega).$$

We can then formally replace $f = u + jv$ and $ds = d\sigma + j\, d\omega$ in the expression $\int_C f(s)\, ds$, to obtain

$$\begin{aligned}\int_C f(s)\, ds &= \int_C \big[u(\sigma,\omega) + jv(\sigma,\omega)\big](d\sigma + j\, d\omega) \\ &= \int_C \big[u(\sigma,\omega)\, d\sigma - v(\sigma,\omega)\, d\omega\big] + j \int_C \big[u(\sigma,\omega)\, d\omega + v(\sigma,\omega)\, d\sigma\big].\end{aligned} \tag{A.19}$$

Hence, we can represent a complex integral using two *line integrals* of the kind

$$\int_C p(\sigma,\omega)\, d\sigma + q(\sigma,\omega)\, d\omega, \tag{A.20}$$

where p and q are real-valued functions of the real variables σ and ω. That (A.19) is equivalent to (A.17) follows immediately from definition (A.15). Now, assuming that p and q are continuous in some domain D containing the arc C given by $s = s(\zeta)$, $\zeta \in [a, b]$, then the line integrals along C from point $A = s(a)$ to point $B = s(b)$ are defined as

$$\begin{aligned}\int_A^B p(\sigma,\omega)\, d\sigma &= \int_a^b p\,(\sigma(\zeta), \omega(\zeta))\; \sigma'(\zeta)\, d\zeta, \\ \int_A^B q(\sigma,\omega)\, d\omega &= \int_a^b q\,(\sigma(\zeta), \omega\,(\zeta))\, \omega'(\zeta)\, d\zeta.\end{aligned} \tag{A.21}$$

[7]See definition of complex derivative in §A.2.

The line integral (A.20) can be interpreted as a vector integral (e.g., Widder 1961). Indeed, consider the unit tangent vector, $\vec{t}$, and the unit outer normal, $\vec{n}$, to the curve C, as shown in Figure A.3. Then $\vec{t}$ has

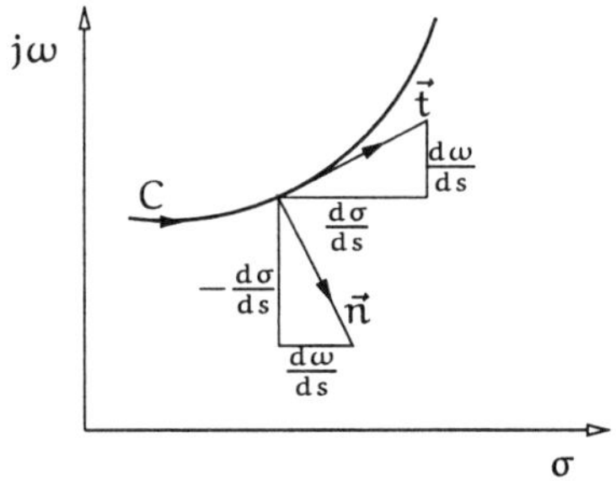

FIGURE A.3. Tangent and outer normal vectors.

components $d\sigma/ds$, $d\omega/ds$, and $\vec{n}$ has components $d\omega/ds$, $-d\sigma/ds$ (Kaplan 1973, p. 237). Consider next the vector $(q, -p)$, and note that

$$(q, -p) \cdot \vec{n} = q\,d\omega/ds + p\,d\sigma/ds,$$

where "·" denotes scalar product of vectors. It is then easy to see that

$$\int_C p\,d\sigma + q\,d\omega = \int_C (q, -p) \cdot \vec{n}\,ds. \tag{A.22}$$

We next turn to the definition of area (or double) integral of a function of two real variables. Let $p(\sigma, \omega)$ be a function defined over a closed and bounded region Ω of the extended complex plane. Suppose we subdivide Ω by drawing a grid parallel to the real and imaginary axes as shown in Figure A.4. We thus form a finite number of closed square

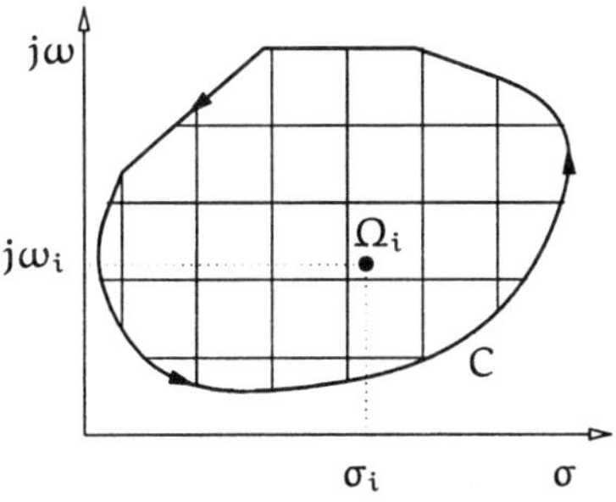

FIGURE A.4. Subdivision of the region Ω.

subregions where each point of Ω lies in at least one such subregion and each subregion contains points of Ω. Consider those square regions that contain only points inside Ω (i.e., only "full" squares). Number them from 1 to m and denote by Ω_i the area of the i-th square. Choosing an arbitrary point (σ_i, ω_i) in the i-th square, then the *double integral* of $p(\sigma, \omega)$ over the region Ω is defined as

$$\iint_{\Omega} p(\sigma, \omega)\, d\sigma\, d\omega = \lim \sum_{i=1}^{m} p(\sigma_i, \omega_i)\Omega_i, \tag{A.23}$$

where the limit is taken as the grid becomes infinitely fine, i.e., m goes to infinite and the maximum diagonal of all squares approaches 0.

The existence of the limit in (A.23) can be demonstrated (see e.g., Widder 1961) when p is continuous and Ω satisfies simple conditions, e.g., when Ω can be split into a finite number of square subregions. In fact, the result also holds when the subregions are not necessarily square, but each of them can be described by inequalities of any of the following forms

$$a \le \sigma \le b, \quad g_1(\sigma) \le \omega \le g_2(\sigma), \tag{A.24}$$
$$c \le \omega \le d, \quad g_3(\omega) \le \sigma \le g_4(\omega), \tag{A.25}$$

where g_1, g_2, g_3 and g_4 are continuous functions. The first of these two situations is represented in Figure A.5.

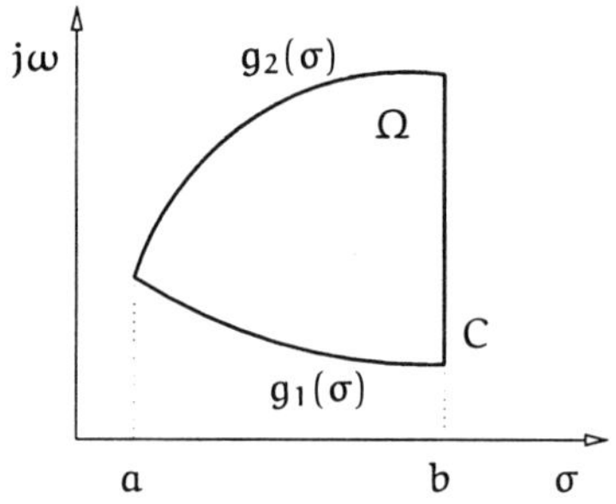

FIGURE A.5. Simple region.

In both cases, the double integral (A.23) can be computed as an *iterated integral*, i.e., if Ω is given by (A.24), we have

$$\iint_{\Omega} p(\sigma, \omega)\, d\sigma\, d\omega = \int_a^b \int_{g_1(\sigma)}^{g_2(\sigma)} p(\sigma, \omega)\, d\omega\, d\sigma,$$

or alternatively, if Ω is given by (A.25),

$$\iint_{\Omega} p(\sigma, \omega)\, d\sigma\, d\omega = \int_c^d \int_{g_3(\omega)}^{g_4(\omega)} p(\sigma, \omega)\, d\sigma\, d\omega.$$

Integrals in Limiting Cases

In many of the applications to control and filtering theory discussed in the main body of the book, we consider integrals over the semicircular contour shown in Figure A.6, where the radius R can be made either indefinitely large or indefinitely small. This requires evaluation of integrals in limiting

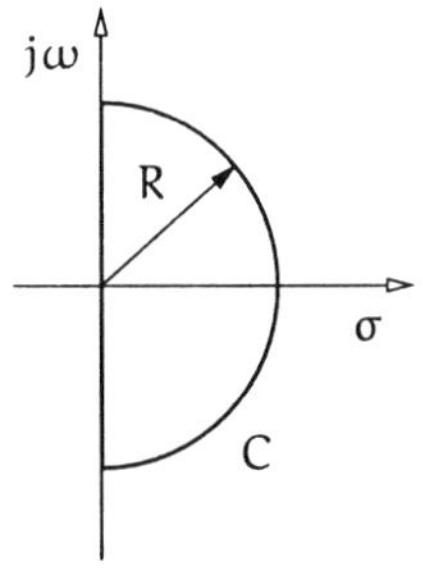

FIGURE A.6. Typical path of integration.

cases, a task that is circumvented by appropriate estimates or bounds on the integrals in question, as we show next.

Let f be any function defined on a domain D, and consider a contour, C, in D, defined by $s = s(\zeta)$, $\zeta \in [a, b]$. Suppose that f is piecewise continuous on C and let f_m be the largest value of $|f(s)|$ over C[8]. It is easy to see[9] that

$$\left|\int_C f(s)\, ds\right| \leq \int_a^b f_m |s'(\zeta)|\, d\zeta = f_m\, \ell, \tag{A.26}$$

where ℓ is the length of the contour C.

The bound for the integral given in (A.26) can be used to obtain estimates of the integral in various limiting situations, as the following examples shows.

Example A.4.1. Consider the integral of powers of s over the contour C shown in Figure A.6. The path length of this contour is πR. Hence if $f(s)$ varies as s^{-2}, then the magnitude of $f(s)$ on C, and hence f_m in (A.26), must vary as R^{-2}. Thus the integral on C vanishes as R goes to ∞. The same result holds, of course, if $f(s)$ varies as any larger negative power of s. ∘

---

[8]Such a value f_m will always exist. Indeed, if f is continuous on the arc C, then the real-valued function $|f(s(\zeta))|$ is continuous on the closed bounded interval $[a, b]$, and hence it always reaches a maximum value f_m on that interval. Hence $|f(s)|$ has a maximum value on C when f is continuous. It is immediate that the same is true when f is piecewise continuous on C.

[9]Using definition (A.17), property (A.16) and the definition of the length of a contour given in §A.4.1.

Example A.4.2. If the semicircle of Figure A.6 has very small, rather than very large radius then the path length vanishes in the limit. It is then clear from (A.26) that the integral also vanishes if $f(s)$ either approaches a constant value or behaves in the vicinity of the origin as any positive power of s. On the other hand, if $f(s)$ behaves as s^{-2} or any larger negative power of s, then f_m will increase rapidly as R diminishes, and thus we can say nothing about the integral on the basis of (A.26). ∘

When not only powers of s are involved but also exponential functions, a handy result is found in Jordan's Lemma, which is reviewed next.

Lemma A.4.1 (Jordan's Lemma). Let C be the contour of Figure A.6 and let τ be a positive real number. Then

$$\int_C \left|e^{-s\tau}\right| |ds| < \frac{\pi}{\tau}. \tag{A.27}$$

Proof. On C we have

$$\int_C \left|e^{-s\tau}\right| |ds| = \int_{-\pi/2}^{\pi/2} e^{-\tau R\cos\theta} R\, d\theta = 2\int_0^{\pi/2} e^{-\tau R\cos\theta} R\, d\theta. \tag{A.28}$$

Now, it is easy to see that $-\cos\theta \leq \frac{2}{\pi}\theta - 1$ for $\theta \in [0, \pi/2]$, which, when used in (A.28), gives

$$\int_C \left|e^{-s\tau}\right| |ds| \leq 2\int_0^{\pi/2} e^{\tau R(2\theta/\pi - 1)} R\, d\theta = \frac{\pi}{\tau}\left(1 - e^{-R\tau}\right) < \frac{\pi}{\tau},$$

and the proof is completed. □

The following example shows the use of Jordan's Lemma to evaluate an integral in a limiting case.

Example A.4.3. Consider the function

$$f(s) = \frac{e^{-s\tau}}{s}, \qquad \tau > 0.$$

and let C be the semicircle of the Jordan's lemma. Then

$$\left|\int_C \frac{e^{-s\tau}}{s} ds\right| \leq \int_C \frac{|e^{-s\tau}|}{|s|} |ds| < \frac{\pi}{R\tau},$$

and hence

$$\lim_{R\to\infty} \int_C \frac{e^{-s\tau}}{s} ds = 0.$$

Obviously the same result is true for $f(s) = e^{-s\tau}/s^k$, $k \geq 1$. ∘

A.5 Main Integral Theorems

In complex integration, a natural question arises whether the value of the integral is affected by the choice of the path between integration limits, or whether the integral is invariant with respect to the path. We will see in §A.5.2 that Cauchy's Integral Theorem connects this question to important properties of the function to be integrated. As a prelude, we next give Green's Theorem, which is part of the necessary background to prove the Cauchy Integral Theorem.

A.5.1 Green's Theorem

We prove in this section Green's Theorem, which applies to real-valued functions of two real variables defined on a simply connected domain.

Informally, a domain D is said to be *simply connected* if it has no holes. More precisely, D is simply connected if for every simple closed curve $C \subset D$, the region Ω formed from C plus its interior lies wholly in D.

We then have the following theorem for functions defined over a simply connected domain.

Theorem A.5.1 (Green's Theorem). Let D be a simply connected domain and let C be a piecewise smooth simple closed contour in D. Let $p(\sigma, \omega)$, $q(\sigma, \omega)$ be functions that are continuous and have continuous first partial derivatives in D. Then

$$\oint_C p \, d\sigma + q \, d\omega = \iint_\Omega \left(\frac{\partial q}{\partial \sigma} - \frac{\partial p}{\partial \omega} \right) d\sigma \, d\omega, \tag{A.29}$$

where Ω is the closed region bounded by C.

Proof. We first consider a simple case in which Ω has the form shown in Figure A.5, i.e., it is representable in both of the forms (A.24) and (A.25). Then the double integral

$$\iint_\Omega \frac{\partial p}{\partial \omega} \, d\omega \, d\sigma$$

can be written as the iterated integral

$$\iint_\Omega \frac{\partial p}{\partial \omega} \, d\sigma \, d\omega = \int_a^b \int_{g_1(\sigma)}^{g_2(\sigma)} \frac{\partial p}{\partial \omega} \, d\omega \, d\sigma.$$

One can now integrate to achieve

$$\begin{aligned}\iint_{\Omega} \frac{\partial p}{\partial \omega}\, d\sigma\, d\omega &= \int_a^b [p(\sigma, g_2(\sigma)) - p(\sigma, g_1(\sigma))]\, d\sigma \\ &= \int_a^b p(\sigma, g_2(\sigma))\, d\sigma + \int_b^a p(\sigma, g_1(\sigma))\, d\sigma \\ &= -\oint_C p(\sigma, \omega)\, d\sigma.\end{aligned}$$

By a similar argument we obtain

$$\iint_{\Omega} \frac{\partial q}{\partial \sigma}\, d\sigma\, d\omega = \oint_C q\, d\omega.$$

The result follows on adding the two double integrals for the simple type of region Ω represented as in (A.24) and (A.25).

It is easy to prove the result for more complex regions that can be decomposed into a finite number of simple regions. For the most general case, it is necessary to approximate the region by the latter and then perform a limiting process. □

Green's formula (A.29) connects a double integral over a region with a line integral over its boundary, and it is at the basis of the proof of Cauchy's Integral Theorem. We next give a version of this formula that is used in Chapter 4 to obtain integral constraints for multivariable systems.

For a real-valued function $f(\sigma, \omega)$ with continuous second partial derivatives in some domain, let ∇f and $\nabla^2 f$ be the *gradient* and *Laplacian* of f, respectively, given by

$$\nabla f = \left(\frac{\partial f}{\partial \sigma}, \frac{\partial f}{\partial \omega}\right), \qquad \text{and} \qquad \nabla^2 f = \frac{\partial^2 f}{\partial \sigma^2} + \frac{\partial^2 f}{\partial \omega^2}.$$

We then have the following corollary.

Corollary A.5.2. Let C be a piecewise smooth, simple closed contour in a simply connected domain, D, and let Ω be the closed region bounded by C. Let $f(\sigma, \omega)$, $g(\sigma, \omega)$ be functions that are continuous and have continuous second partial derivatives in D. Then

$$\oint_C \left(f \frac{\partial g}{\partial \vec{n}} - g \frac{\partial f}{\partial \vec{n}}\right) ds = \iint_{\Omega} (f\, \nabla^2 g - g\, \nabla^2 f)\, d\sigma\, d\omega, \tag{A.30}$$

where $\partial f / \partial \vec{n} = \nabla f \cdot \vec{n}$ is the directional derivative of f along $\vec{n}$, the exterior normal of C.

Proof. We use the identity

$$\operatorname{div}(f \nabla g) = f \nabla^2 g + \nabla f \cdot \nabla g,$$

where the symbol div denotes the divergence of a vector field (see e.g., Kaplan 1973). Integrating the above equation over Ω, we have

$$\iint_{\Omega} \operatorname{div}(f \nabla g)\, d\sigma\, d\omega = \iint_{\Omega} f \nabla^2 g\, d\sigma\, d\omega + \iint_{\Omega} \nabla f \cdot \nabla g\, d\sigma\, d\omega. \tag{A.31}$$

Using the definitions of divergence and gradient, and equations (A.29) and (A.22), the LHS above can be written as

$$\begin{aligned}\iint_{\Omega} \operatorname{div}(f \nabla g)\, d\sigma\, d\omega &= \iint_{\Omega} \left[\frac{\partial}{\partial\sigma}\left(f\frac{\partial g}{\partial\sigma}\right) + \frac{\partial}{\partial\omega}\left(f\frac{\partial g}{\partial\omega}\right)\right] d\sigma\, d\omega \\ &= \oint_C f \nabla g \cdot \vec{n}\, ds \\ &= \oint_C f \frac{\partial f}{\partial \vec{n}}\, ds.\end{aligned}$$

Hence, (A.31) can alternatively be expressed as

$$\oint_C f \frac{\partial f}{\partial \vec{n}}\, ds = \iint_{\Omega} f \nabla^2 g\, d\sigma\, d\omega + \iint_{\Omega} \nabla f \cdot \nabla g\, d\sigma\, d\omega.$$

Using the above identity with f and g interchanged, and subtracting, leads to (A.30). □

A.5.2 The Cauchy Integral Theorem

In §A.3, we introduced the class of analytic functions, which are at the core of the results presented in the book. For these functions, a fundamental result — Cauchy's Integral Theorem — can be stated which allows one to evaluate the integral of such functions on closed contours.

Before addressing this result, let us motivate it with a simple example of integration of powers of s over a circle around the origin.

Example A.5.1. Let $f(s) = s^{-1}$ and consider the contour C shown in Figure A.7. The integral over C is computed in the clockwise direction, i.e., from s_1 to s_2. On C, we have that $s = Re^{j\theta}$, and so $ds = jRe^{j\theta}\, d\theta$. Hence

$$\begin{aligned}\int_C \frac{ds}{s} &= \int_{\theta_1}^{\theta_2} \frac{jRe^{j\theta}\, d\theta}{Re^{j\theta}} \\ &= -j\,(\theta_1 - \theta_2)\,.\end{aligned} \tag{A.32}$$

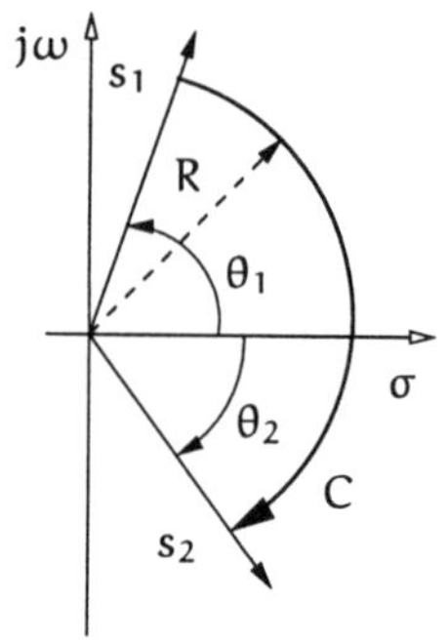

FIGURE A.7. Path of integration used in Example A.5.1.

Note that this result is independent of the value of R. We will next use (A.32) to evaluate the integral of powers of s around a full circle of radius R centered at the origin of the complex plane. In this case,

$$\begin{aligned}\oint s^m ds &= \int_{-\pi}^{\pi} \left(R^m e^{jm\theta}\right) jRe^{j\theta}\, d\theta \\ &= jR^{m+1}\int_{-\pi}^{\pi} [\cos(m+1)\theta + j\sin(m+1)\theta]\, d\theta \\ &= \begin{cases} 0 & \text{for } m \neq -1, \\ -j2\pi & \text{for } m = -1, \text{ (integration clockwise).} \end{cases}\end{aligned} \tag{A.33}$$

In particular, note that the integral of positive power of s over the circle vanishes. This is not casual, as we see next. ○

Theorem A.5.3 (Cauchy Integral Theorem). If f is analytic on some simply connected domain D, then

$$\oint_C f(s)\, ds = 0,$$

where C is any piecewise smooth simple closed contour in D.

Proof. We will prove this theorem assuming the restricted definition of analyticity referred to in the footnote on page 279, i.e., a function is analytic if it is continuously differentiable.[10]

Let $f = u + jv$. From (A.19), we have

$$\oint_C f\, ds = \oint_C (u\, d\sigma - v\, d\omega) + j\oint_C (v\, d\sigma + u\, d\omega).$$

[10]This restricted definition allows us to obtain a simple proof of the theorem. Moreover, there is no loss of generality in this assumption since the original definition not requiring continuity ultimately leads to the conclusion that the derivative of the function is, in fact, continuous.

Since f is analytic (and hence f' is continuous) on a simply connected domain D, it follows from Theorem A.2.2 that u and v have continuous partial derivatives (A.9). Since D is simply connected, we can apply Green's theorem (Theorem A.5.1) to the real and imaginary parts of the integral in the above equation to obtain

$$\oint_C f\,ds = \iint_\Omega \left(-\frac{\partial v}{\partial \sigma} - \frac{\partial u}{\partial \omega}\right) d\sigma\, d\omega + j \iint_\Omega \left(\frac{\partial u}{\partial \sigma} - \frac{\partial v}{\partial \omega}\right) d\sigma\, d\omega,$$

where Ω is the region bounded by C. Since the partial derivatives of u and v satisfy the Cauchy-Riemann conditions (A.5) (Theorem A.2.2), the integrands of these two double integrals are zero throughout D and hence the result follows. □

The above proof follows the original setting used by Cauchy in his proof in the early part of the last century. Several decades later, it was discovered by Edouard Goursat that the hypothesis of continuity of the derivative in Cauchy's theorem can be dispensed with, which led to the modern, more general version of this important result[11]. The relaxation of this assumption allows us to use Cauchy's Integral Theorem to show that, in fact, continuity of the derivative follows from analyticity of the function.

Theorem A.5.3 is readily illustrated using Example A.5.1, as seen below.

Example A.5.2. Suppose that f(s) is the polynomial

$$f(s) = a_0 + a_1 s + \cdots + a_n s^n.$$

Then f(s) is analytic on and within a circle about the origin, and, according to the Cauchy integral theorem, the integral on the complete circle must vanish. But this was indeed the result of Example A.5.1, which showed that the integral of each term of the polynomial vanishes. ∘

The simple piecewise smooth closed contour in Theorem A.5.3 can be replaced by less trivial contours without compromising the result. These extensions are discussed in the following subsection.

A.5.3 Extensions of Cauchy's Integral Theorem

As a first extension, we note that the contour in Theorem A.5.3 can be replaced by a piecewise smooth closed contour C that is *not necessarily simple*. For if C intersects itself a finite number of times, it consists of a finite number of simple closed contours, as illustrated in Figure A.2 in §A.4.1. By applying the Cauchy Integral Theorem to each of those simple closed

[11] Some authors even refer to this theorem as the Cauchy-Goursat Theorem (e.g., Levinson & Redheffer 1970, Churchill & Brown 1984).

contours, the desired result for C is obtained. Also, a portion of C may be traversed twice in opposite directions since the integrals along these portions in the two directions cancel each other.

Another useful extension of Cauchy Integral Theorem covers the case where the contour C is the *oriented boundary of a multiply connected domain.* To see how this extension can be established, let C_0 be a simple closed contour and let C_i, $i = 1, 2, \cdots, n$, be a finite number of simple closed contours inside C_0 such that the interiors of each C_i have no points in common. Let Ω be the closed region consisting of all points within and on C_0 except for points interior to each C_i (Figure A.8). Let C denote the

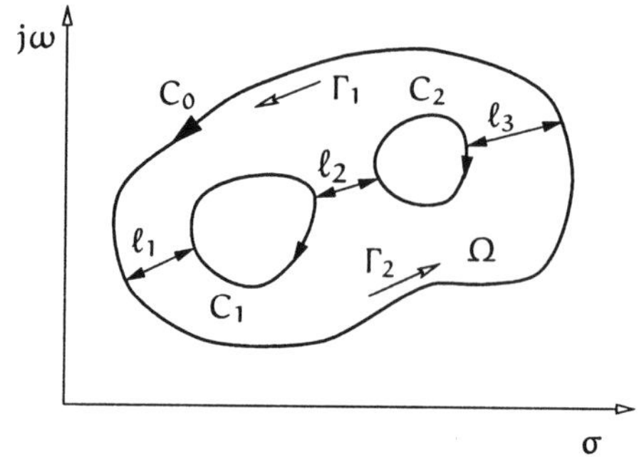

FIGURE A.8. Multiply connected domain.

entire connected boundary of Ω consisting of C_0 and all the contours C_i, described in a direction such that the interior points of Ω lie to the left of C. Next, we introduce a polygonal path ℓ_1, consisting of a finite number of line segments joined end to end, to connect the outer contour C_0 to the inner contour C_1. We introduce another polygonal path ℓ_2 that connects C_1 to C_2; and we continue in this manner, with ℓ_{n+1} connecting C_n to C_0. As indicated in Figure A.8, two simple closed contours Γ_1 and Γ_2 can be formed, each consisting of polygonal paths ℓ_i or $-\ell_i$ and pieces of C_0 and C_i. Then, if f is analytic throughout Ω, the Cauchy Integral Theorem can be applied to f on Γ_1 and Γ_2, and the sum of the integrals over those contours is found to be zero. Since the integrals in opposite directions along each path ℓ_i cancel, only the integral along C remains. Thus,

$$\int_C f(s)\, ds = 0.$$

The following examples show how the extension of Theorem A.5.3 to the boundary of a multiply connected domain can be used.

Example A.5.3. Let C_0 be a simple closed piecewise smooth contour lying on the interior of a simple closed piecewise smooth contour C, where both C and C_0 are equally oriented, e.g., in the counter-clockwise direction (Figure A.9). Let f be analytic in the closed region bounded by C and C_0. Then, the extension of the Cauchy Integral Theorem to its boundary gives that the integral of f around the outer contour, C, minus the integral around the interior contour, C_0, must equal zero, i.e.,

$$\oint_C f(s)\,ds = \oint_{C_0} f(s)\,ds.$$

○

Example A.5.4. When integration on closed contours is extended to functions having isolated singularities, the value of the integral is not zero, in general, but each singularity contributes a term called the *residue*.[12]

Say that $f(s)$ can be expanded as

$$f(s) = \frac{c_{-1}}{s - s_0} + c_0 + c_1(s - s_0) + c_2(s - s_0)^2 + \cdots . \tag{A.34}$$

The number c_{-1} is called the *residue* of f at s_0.

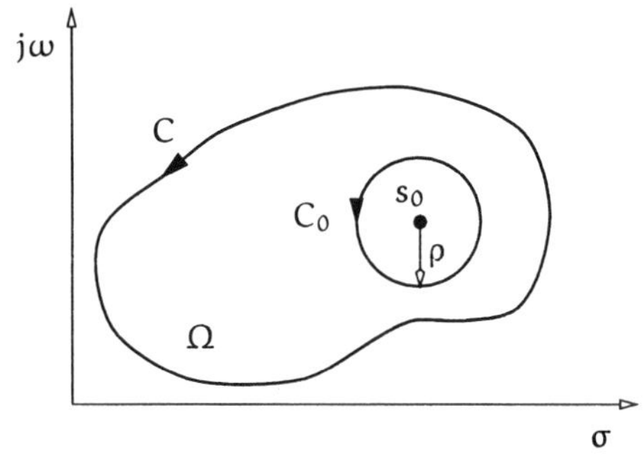

FIGURE A.9. Region for Example A.5.4.

Now consider the region Ω with boundary consisting of the piecewise smooth contour C and the smaller circle C_0 centered at s_0, as shown in Figure A.9. Since $f(s)$ is analytic on Ω, Example A.5.3 shows that the integral around the outer curve, C in Figure A.9, equals the integral around the inner circle, C_0. The counter-clockwise circular integral around s_0 is $j2\pi c_{-1}$ using (A.33). Thus, we may conclude that

$$\oint_C f(s)\,ds = j2\pi c_{-1}. \tag{A.35}$$

○

[12]The residue is formally defined in §A.9.1; here we give only a preview that will motivate Cauchy's Integral Formula.

A.5.4 The Cauchy Integral Formula

A general problem of central importance in this book is that of relating the values assumed by an analytic function within a given region, to its values on the boundary of the region. A remarkable tool for dealing with this problem is found in Cauchy's Integral Formula.

Theorem A.5.4 (Cauchy's Integral Formula). Let f be analytic within and on a simple closed piecewise smooth contour, C. If s_0 is any point interior to C, then

$$\oint_C \frac{f(s)}{s - s_0}\, ds = j2\pi f(s_0), \tag{A.36}$$

where the integral is computed in the counter-clockwise direction.

Proof. Since f is continuous at s_0, given $\varepsilon > 0$ there is a $\delta > 0$ such that

$$|f(s) - f(s_0)| < \varepsilon \quad \text{whenever} \quad |s - s_0| < \delta.$$

Choose $0 < \rho < \delta$ such that the counter-clockwise oriented circle $|s - s_0| = \rho$, denoted by C_0 in Figure A.9, is interior to C. Then

$$|f(s) - f(s_0)| < \varepsilon \quad \text{whenever} \quad |s - s_0| = \rho. \tag{A.37}$$

Next, observe that the function $f(s)/(s - s_0)$ is analytic at all points within and on C except at s_0. Hence, by the Cauchy Integral Theorem for multiply connected domains, its integral around the oriented boundary of the region between C and C_0 has value zero, i.e.,

$$\oint_C \frac{f(s)}{s - s_0}\, ds = \oint_{C_0} \frac{f(s)}{s - s_0}\, ds.$$

Subtracting the constant term $f(s_0) \oint_{C_0} ds/(s - s_0)$, which equals $j2\pi f(s_0)$ by (A.33), from both sides of the above equation yields

$$\oint_C \frac{f(s)}{s - s_0}\, ds - j2\pi f(s_0) = \oint_{C_0} \frac{f(s) - f(s_0)}{s - s_0}\, ds. \tag{A.38}$$

Referring to (A.37) and noting that the length of C_0 is $2\pi\rho$, we may apply property (A.16) to the RHS of (A.38), to obtain

$$\left| \oint_{C_0} \frac{f(s) - f(s_0)}{s - s_0}\, ds \right| < \frac{\varepsilon}{\rho} 2\pi\rho = 2\pi\varepsilon.$$

From (A.38), then

$$\left| \oint_C \frac{f(s)}{s - s_0}\, ds - j2\pi f(s_0) \right| < 2\pi\varepsilon.$$

Since the LHS of this inequality is a nonnegative number that is less than an arbitrarily small positive number, it must be equal to zero, giving the desired result. □

We note from Theorem A.5.4 that the value $f(s_0)$ can be obtained by integrating $f(s)/(s-s_0)$ on a contour encircling s_0. Hence we can determine the value of an analytic function inside a region by its behavior on the boundary. We extensively exploit this result in the book to examine the characteristics that a function must have on the boundary (typically the imaginary axis) when it is known to achieve certain values in the interior.

The Cauchy integral formula can be extended to cases in which the simple closed contour C is replaced by the oriented boundary of a multiply connected domain, as described in §A.5.3. The following example shows how this can be accomplished.

Example A.5.5. Let C_0 and C_1 be two counter-clockwise oriented, concentric circles, where C_1 is smaller than C_0 (Figure A.10). Assume that f is analytic on both circles and throughout the annular domain between them. Let s_0 be a point inside the annulus and construct a counter-clockwise oriented circle γ about s_0, small enough to be completely contained in the annular domain, as shown in Figure A.10. It then follows from the adaptation of the Cauchy integral theorem to the boundary of a multiply connected domain that

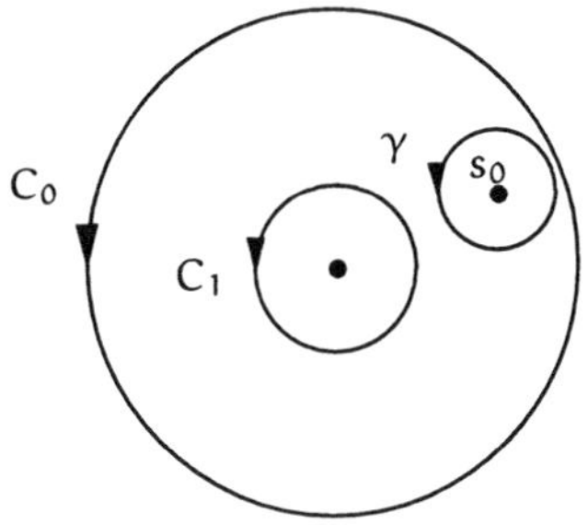

FIGURE A.10. Annular domain for Example A.5.5.

$$\oint_{C_0} \frac{f(s)}{s-s_0}\,ds - \oint_{C_1} \frac{f(s)}{s-s_0}\,ds - \oint_{\gamma} \frac{f(s)}{s-s_0}\,ds = 0.$$

But, according to Cauchy's integral formula, the value of the third integral above is $2\pi j\, f(s_0)$. Hence,

$$f(s_0) = \frac{1}{2\pi j}\oint_{C_0} \frac{f(s)}{s-s_0}\,ds - \frac{1}{2\pi j}\oint_{C_1} \frac{f(s)}{s-s_0}\,ds. \qquad \text{(A.39)}$$

○

In the following section we discuss an important application of the Cauchy's integral theorem and formula.

A.6 The Poisson Integral Formula

An application of the integral theorems of the previous section, central to the purposes of this book, is found in the Poisson integral formula for both the half plane and the unit disk.

A.6.1 Formula for the Half Plane

In the case of the right half plane, we are faced with integration over a contour that becomes arbitrarily long. To deal with this, we will consider a class of functions with restricted behavior at infinity, and then use the bounding technique of (A.26) to estimate the integral over the infinite contour.

In particular, given a function f, define

$$m(R) = \sup_{\theta} |f(Re^{j\theta})|, \qquad \theta \in [-\pi/2, \pi/2]. \tag{A.40}$$

Then $f(s)$ is said to be of class $\mathcal{R}$ if

$$\lim_{R\to\infty} \frac{m(R)}{R} = 0. \tag{A.41}$$

For example, the functions considered in Examples A.4.1 and A.4.3 are in this class. More generally, if f is analytic and of bounded magnitude in the CRHP, then f is of class $\mathcal{R}$.

The Poisson integral formula for the half plane is given in the following result.

Theorem A.6.1 (Poisson Integral Formula for the Half Plane). Let f be analytic in the CRHP and suppose that f is of class $\mathcal{R}$. Let $s_0 = \sigma_0 + j\omega_0$ be a point in the complex plane with $\sigma_0 > 0$. Then

$$f(s_0) = \frac{1}{\pi}\int_{-\infty}^{\infty} f(j\omega)\frac{\sigma_0}{\sigma_0^2 + (\omega_0 - \omega)^2}\, d\omega. \tag{A.42}$$

Proof. Let f be as in the statement of the theorem and let $s_0 = \sigma_0 + j\omega_0$ be any point such that $\sigma_0 > 0$. Consider the clockwise oriented semicircular contour C shown in Figure A.11, where R is large enough so that s_0 is interior to the contour.[13] Thus C consists of the segment $s = j\omega$, $\omega \in [-R, R]$, together with the arc C_R given by $s = Re^{j\theta}$, $\theta \in [-\pi/2, \pi/2]$.

Since f is analytic on and inside C, then Cauchy's integral formula (A.36) gives

$$f(s_0) = -\frac{1}{2\pi j}\oint_C \frac{f(s)}{s - s_0}\, ds.$$

---

[13] Recall that the interior is the domain *bounded* by the curve, and it is defined independent of the orientation.

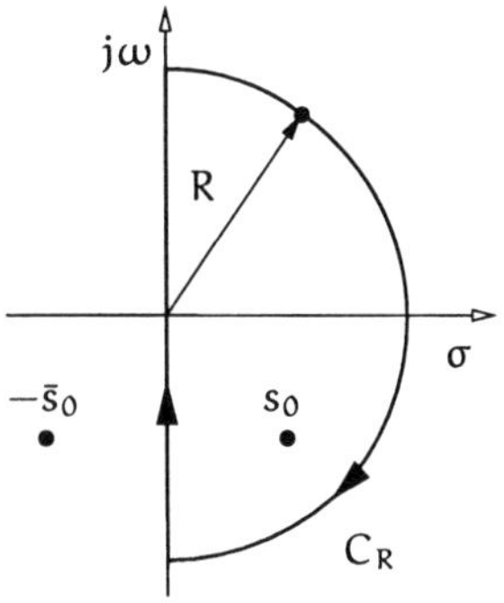

FIGURE A.11. Contour for the Poisson integral formula

Now consider the point $-\bar{s}_0$, which is outside C. Thus, the Cauchy integral theorem gives

$$0 = \frac{1}{2\pi j}\oint_C \frac{f(s)}{s+\bar{s}_0}\,ds.$$

Adding the above two equations, we obtain

$$f(s_0) = -\frac{1}{2\pi j}\oint_C f(s)\frac{s_0+\bar{s}_0}{(s-s_0)(s+\bar{s}_0)}\,ds, \tag{A.43}$$

which can be decomposed into the sum of two integrals as

$$\begin{aligned} f(s_0) = &-\frac{1}{\pi}\int_{-R}^{R} f(j\omega)\frac{\sigma_0}{(j\omega-s_0)(j\omega+\bar{s}_0)}\,d\omega \\ &-\frac{1}{\pi j}\int_{C_R} f(s)\frac{\sigma_0}{(s-s_0)(s+\bar{s}_0)}\,ds. \end{aligned} \tag{A.44}$$

As $R \to \infty$, the first integral in (A.44) becomes

$$\frac{1}{\pi}\int_{-\infty}^{\infty} f(j\omega)\frac{\sigma_0}{\sigma_0^2+(\omega_0-\omega)^2}\,d\omega. \tag{A.45}$$

Comparing to (A.42), it thus remains to show that the second integral in (A.44) vanishes as $R \to \infty$. Using (A.40) and the fact that, for R sufficiently large, the denominator in the second integral has magnitude R^2, we have

$$\left|\frac{1}{\pi j}\int_{C_R} f(s)\frac{\sigma_0}{(s-s_0)(s+\bar{s}_0)}\,ds\right| \le \frac{1}{\pi}\frac{m(R)\,\sigma_0\,\pi R}{R^2},$$

which tends to zero as $R \to \infty$ since f is of class $\mathcal{R}$ and hence satisfies (A.41). The result then follows. □

If f is analytic in the ORHP and on the imaginary axis except for singularities of a particular type, then the Poisson integral formula is still valid, as shown next.

Lemma A.6.2. Let f be analytic in the CRHP, except for singular points s_k on the imaginary axis that satisfy

$$\lim_{\substack{s \to s_k \\ \mathrm{Re}\, s \geq 0}} (s - s_k) f(s) = 0. \tag{A.46}$$

Suppose further that f is of class $\mathcal{R}$. Then the Poisson integral formula (A.42) holds at each complex point $s_0 = \sigma_0 + j\omega_0$ with $\sigma_0 > 0$.

Proof. Consider the contour shown in Figure A.12, i.e., a semicircle of radius R encircling the point s_0 and such that the portion of curve on the imaginary axis has semicircular indentations of radius δ into the ORHP at each singularity s_k of f.

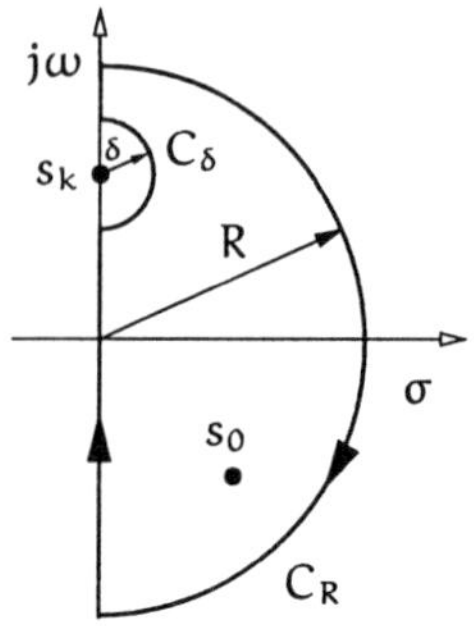

FIGURE A.12. Contour for f with singularities on the $j\omega$-axis.

We next proceed as in the proof of Theorem A.6.1, but in this case the integral (A.43) will be decomposed into the sum of the integral over the semicircular contour C_R, the integrals over the small semicircular contours of radii δ, and the integrals over the remaining portions of the imaginary axis. This last sum of integrals over the portions of imaginary axis between the small semicircles will tend to (A.45) as $R \to \infty$ and $\delta \to 0$. The integral over C_R vanishes as in the proof of Theorem A.6.1. It thus remains to show that the integrals over each semicircle of radius δ also vanish as $\delta \to 0$.

Consider then one of the semicircles C_δ in Figure A.12, centered at s_k, say. On this contour, $s = s_k + \delta e^{j\theta}$, $\theta \in [-\pi/2, \pi/2]$. Then

$$\int_{C_\delta} \frac{f(s)\,\sigma_0}{(s - s_0)(s + \bar{s}_0)}\, ds = \int_{-\pi/2}^{\pi/2} \frac{f(s_k + \delta e^{j\theta})\,\delta e^{j\theta}\,\sigma_0}{(s_k + \delta e^{j\theta} - s_0)(s_k + \delta e^{j\theta} + \bar{s}_0)}\, d\theta.$$

Note that (A.46) implies that $\lim_{\delta\to 0} f(s_k + \delta e^{j\theta})\, \delta e^{j\theta} = 0$. Hence the integrand on the RHS above vanishes and the result follows. □

Other forms of the Poisson formula are obtained by separating into real and imaginary parts. For example, if $f(s) = u(\sigma, \omega) + jv(\sigma, \omega)$, then the formula (A.42) gives two real equations of the same structure, namely

$$
\begin{aligned}
u(\sigma, \omega) &= \frac{1}{\pi} \int_{-\infty}^{\infty} u(0, \omega)\, \frac{\sigma_0}{\sigma_0^2 + (\omega_0 - \omega)^2}\, d\omega, \\
v(\sigma, \omega) &= \frac{1}{\pi} \int_{-\infty}^{\infty} v(0, \omega)\, \frac{\sigma_0}{\sigma_0^2 + (\omega_0 - \omega)^2}\, d\omega.
\end{aligned}
\tag{A.47}
$$

Since u and v are harmonic when f is analytic, each of the formulae (A.47) is the Poisson integral formula for harmonic functions.

Note that the integrals in (A.47) are improper integrals of the form

$$
I = \int_{-\infty}^{\infty} w(\omega)\, d\omega.
$$

Every such integral will be evaluated based on its *Cauchy principal value*, i.e.,

$$
I = \lim_{R\to\infty} \int_{-R}^{R} w(\omega)\, d\omega.
$$

Existence of the Cauchy principal value of an integral does not, in general, guarantee the existence of the two limits $\lim_{R\to\infty} \int_{-R}^{0} w(\omega)\, d\omega$ and $\lim_{R\to\infty} \int_{0}^{R} w(\omega)\, d\omega$. However, if $w(\omega)$ is even, i.e., $w(-\omega) = w(\omega)$, then existence of the Cauchy principal value implies existence of these two limits.

A useful result for a particular harmonic function is given next.[14]

Corollary A.6.3. Let f be analytic and nonzero in the CRHP except for possible zeros on the imaginary axis and/or zeros at infinity. Assume that $\log f$ is in class $\mathcal{R}$. Then, at each complex point $s_0 = \sigma_0 + j\omega_0$, $\sigma_0 > 0$,

$$
\log |f(s_0)| = \frac{1}{\pi} \int_{-\infty}^{\infty} \log |f(j\omega)| \frac{\sigma_0}{\sigma_0^2 + (\omega_0 - \omega)^2}\, d\omega. \tag{A.48}
$$

Proof. If f is as in the statement of the theorem, then $\log f$ is analytic in the CRHP except for singularities at the imaginary zeros of f and/or at zeros of f at infinity. If f has imaginary zeros s_k, it is not difficult to prove that $\log f$ satisfies (A.46), and hence Lemma A.6.2 shows that these singularities do not affect the Poisson integral. If f has zeros at infinity, then $\log f$ has a

[14]This result, to the best of our knowledge, first appeared in Freudenberg & Looze (1985).

singularity at infinity, but the contour of integration in Figure A.12 has an indentation around the point at infinity (i.e., the large semicircle into the ORHP). The fact that $\log f$ is in class $\mathcal{R}$ shows that the integral on this semicircle vanishes as the radius tends to infinity. Then (A.42) holds for $\log f(s)$ and (A.47) holds for $\log |f(s)| = \operatorname{Re} \log f(s)$. □

Note that zeros of f at infinity can be treated as zeros of f on the imaginary axis, since both types of zeros are in fact singularities of $\log f$ on the contour of integration that encircles the ORHP. The procedure that we take to deal with these singularities consists of two steps: first, indentations around these singularities have to be made on the contour of integration; second, precautions should be taken to show that the integrals on those indentations converge (go to zero in our case) as the indentations vanish, i.e., condition (A.46) is assumed for imaginary zeros, and the property of $\log f$ being in class $\mathcal{R}$ is assumed for zeros at infinity.

A.6.2 *Formula for the Disk*

Let $s = re^{j\theta}$, and consider the unit circle $|s| = 1$ described in a counterclockwise sense by $s = e^{j\theta}$, with $-\pi \le \theta \le \pi$. We then have the following result.

Theorem A.6.4 (Poisson Integral Formula for the Disk). If f is a function analytic on $\overline{\mathbb{D}}$, then for any interior point $s_0 = r_0 e^{j\theta_0}$, $r_0 < 1$,

$$f(r_0 e^{j\theta_0}) = \frac{1}{2\pi} \int_{-\pi}^{\pi} f(e^{j\theta}) \frac{1 - r_0^2}{1 - 2r_0 \cos(\theta - \theta_0) + r_0^2} \, d\theta. \tag{A.49}$$

Proof. Let f be a function analytic on $\overline{\mathbb{D}}$. For any point $s_0 = r_0 e^{j\theta}$ interior to $\overline{\mathbb{D}}$, the Cauchy integral formula expresses $f(s_0)$ as

$$f(s_0) = \frac{1}{2\pi} \int_{-\pi}^{\pi} \frac{f(e^{j\theta})}{e^{j\theta} - s_0} e^{j\theta} \, d\theta. \tag{A.50}$$

Introduce the inverse of the point s_0 with respect to the unit circle, say s_1, which lies on the same ray from the origin as does s_0, but is exterior to $\overline{\mathbb{D}}$, i.e., $s_1 = 1/\bar{s}_0$ (see Figure A.13). It then follows from the Cauchy integral theorem that the integral in (A.50) equals 0 when s_0 is replaced by s_1 in the integrand, i.e.,

$$0 = \frac{1}{2\pi} \int_{-\pi}^{\pi} \frac{f(e^{j\theta})}{e^{j\theta} - s_1} e^{j\theta} \, d\theta. \tag{A.51}$$

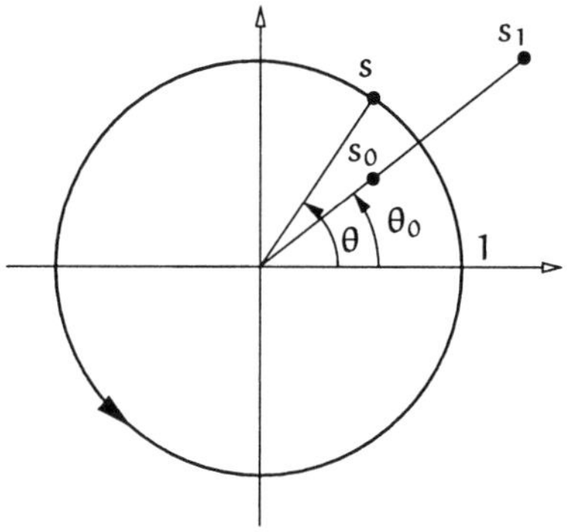

FIGURE A.13. The unit disk.

Subtracting (A.51) from (A.50), and replacing $s_1 = 1/\bar{s}_0$, yields

$$\begin{aligned} f(s_0) &= \frac{1}{2\pi}\int_{-\pi}^{\pi} f(e^{j\theta}) \left(\frac{e^{j\theta}}{e^{j\theta} - s_0} - \frac{e^{j\theta}}{e^{j\theta} - s_1} \right) d\theta \\ &= \frac{1}{2\pi}\int_{-\pi}^{\pi} f(e^{j\theta}) \left(\frac{e^{j\theta}}{e^{j\theta} - s_0} + \frac{\bar{s}_0}{e^{-j\theta} - \bar{s}_0} \right) d\theta. \end{aligned}$$

The term within parenthesis in the integrand above can be written as

$$\begin{aligned} \frac{e^{j\theta}}{e^{j\theta} - s_0} + \frac{\bar{s}_0}{e^{-j\theta} - \bar{s}_0} &= \frac{1 - r_0^2}{|e^{j\theta} - r_0 e^{j\theta_0}|^2} \\ &= \frac{1 - r_0^2}{1 - 2\cos(\theta - \theta_0) + r_0^2}, \end{aligned}$$

where $1 - 2\cos(\theta - \theta_0) + r_0^2 > 0$ represents the distance between the points $s = e^{j\theta}$ and s_0. Hence, the formula (A.49) follows. □

Taking the real or imaginary part of (A.49) we obtain the Poisson integral formula for harmonic functions on the unit circle.

For our purposes, we will require a version of Theorem A.6.4 that reconstructs the values of a harmonic function at any point *outside* the unit disk from the values on the border. This follows as an easy corollary.

Corollary A.6.5. If u is a function harmonic outside $\mathbb{D}$, then for any point exterior to the unit disk $s_0 = r_0 e^{j\theta_0}$, $r_0 > 1$,

$$u(r_0 e^{j\theta_0}) = \frac{1}{2\pi}\int_{-\pi}^{\pi} u(e^{j\theta}) \frac{r_0^2 - 1}{1 - 2r_0\cos(\theta - \theta_0) + r_0^2} d\theta. \tag{A.52}$$

Proof. Straightforward from Theorem A.6.4 on taking the real part of (A.49) and considering $u(1/s)$. □

Both versions of the Poisson integral formulae, for the half plane and the disk, were obtained following the same procedure: the Cauchy integral formula was applied to a point inside the contour of interest and then added to (subtracted from) the Cauchy integral theorem applied to a particular point outside the contour of interest. The reader may wonder if the same procedure applied to other points outside the contour would lead to other interesting relationships. The answer is indeed yes (see §2.3.2 in Chapter 2 for other applications).

A.7 Power Series

This section gives three applications of the Cauchy integral formula that are related with derivatives and series expansions of analytic functions.

A.7.1 Derivatives of Analytic Functions

Theorem A.5.4 has an immediate application in showing that an analytic function possesses derivatives of all orders, and these derivatives are themselves analytic, as we see next.

Theorem A.7.1. Let f be analytic in an arbitrary domain D. Then its derivatives of all orders exist in D and are analytic functions. Moreover, if C is any closed contour contained in D, then the n-th derivative of f at any point s_0 inside C is computed as

$$f^{(n)}(s_0) = \frac{n!}{2\pi j} \oint_C \frac{f(s)}{(s - s_0)^{n+1}}\, ds. \tag{A.53}$$

Proof. Let f be analytic in a domain D and let α be any point of D. We will show that f has all derivatives at α. Since α is arbitrary, this will establish the existence of all derivatives in D.

If C is a sufficiently small circle with α as center (see Figure A.14), then it follows from (A.36) that for any point s_0 inside C

$$f(s_0) = \frac{1}{2\pi j} \oint_C \frac{f(s)}{s - s_0}\, ds, \tag{A.54}$$

where C is now traversed in the counter-clockwise direction. If (A.54) is differentiated formally n times with respect to s_0, we have

$$f^{(n)}(s_0) = \frac{n!}{2\pi j} \oint_C \frac{f(s)}{(s - s_0)^{n+1}}\, ds. \tag{A.55}$$

We will next show the validity of (A.55).

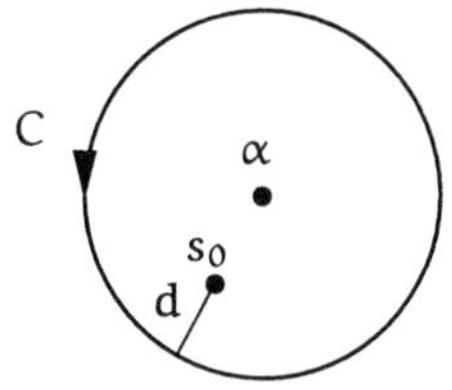

FIGURE A.14. Contour for Theorem A.7.1.

Let Δ_s be such that $0 < |\Delta_s| < d$, where d is the shortest distance from s_0 to points on C, as shown in Figure A.14. Using (A.54) for s_0 and $s_0 + \Delta_s$, we have

$$\begin{aligned} \frac{f(s_0+\Delta_s)-f(s_0)}{\Delta_s} &= \frac{1}{2\pi j}\oint_C \left(\frac{1}{s-s_0-\Delta_s} - \frac{1}{s-s_0}\right)\frac{f(s)}{\Delta_s}\,ds \\ &= \frac{1}{2\pi j}\oint_C \frac{f(s)}{(s-s_0)^2}\,ds + J, \end{aligned} \tag{A.56}$$

where

$$J = \frac{\Delta_s}{2\pi j}\oint_C \frac{f(s)}{(s-s_0-\Delta_s)(s-s_0)^2}\,ds.$$

Let f_m be the maximum value of $|f(s)|$ on C and let ℓ be the length of C. Since $|s - s_0| \geq d$ by construction, and

$$|s-s_0-\Delta_s| \geq ||s-s_0| - |\Delta_s|| \geq d - |\Delta_s|,$$

we readily obtain the following bound for J in (A.56):

$$|J| \leq \frac{|\Delta_s| f_m \ell}{2\pi j (d - |\Delta_s|) d^2},$$

where the last fraction approaches zero as Δ_s approaches zero. Taking limits in (A.56), it then follows that

$$\lim_{\Delta_s \to 0} \frac{f(s_0+\Delta_s)-f(s_0)}{\Delta_s} = \frac{1}{2\pi j}\oint_C \frac{f(s)}{(s-s_0)^2}\,ds,$$

and the expression (A.55) is then seen to hold for $n = 1$.

Repeating the above procedure starting with (A.55) for $n = 1$ establishes the existence of f'' and proves (A.55) for $n = 2$. This shows that if f is analytic so is f'. The result for $f^{(n)}$ then follows by induction. □

Note that, if we agree that $f^{(0)}(s) = f(s)$ and that $0! = 1$, then the formula (A.53) for $n = 0$ is the Cauchy integral formula.

In particular, when a function

$$f(s) = u(\sigma, \omega) + jv(\sigma, \omega)$$

is analytic in a domain D, the analyticity of f' ensures the continuity of f' there. Then, since

$$f' = \frac{\partial u}{\partial \sigma} + j\frac{\partial v}{\partial \sigma} = \frac{\partial v}{\partial \omega} - j\frac{\partial u}{\partial \omega},$$

Theorem A.2.2 shows that the first order partial derivatives of u and v are continuous in D. Similarly, Theorem A.7.1 shows that the partial derivatives of u and v of all orders are continuous in D. This result was anticipated in §A.3.1 in the discussion of harmonic functions.

As was the case with the Cauchy integral formula, (A.53) can be extended to the case in which the circle C is replaced by the oriented boundary of a multiply connected domain.

A.7.2 Taylor Series

We will next give a result that adapts the familiar *Taylor series* expansion from calculus, to functions of a complex variable.

We briefly review some terminology. The partial sums of a *power series*

$$a_0 + a_1(s - s_0) + a_2(s - s_0)^2 + \cdots,$$

where the a_i are complex numbers, are defined by

$$\Sigma_n(s) = a_0 + a_1(s - s_0) + a_2(s - s_0)^2 + \cdots + a_n(s - s_0)^n.$$

The partial sums form a sequence of polynomials

$$\{\Sigma_n(s)\}, \qquad n = 0, 1, 2, \cdots.$$

Let $f(s)$ and a sequence of functions $\{\Sigma_n(s)\}$ be given in a region Ω of the complex plane. Then the sequence is said to *converge uniformly* to the function f in Ω if, given any $\varepsilon > 0$, there exists an integer N, which can depend on ε (but not on $s \in \Omega$), such that

$$|f(s) - \Sigma_n(s)| < \varepsilon, \quad \text{for } n \geq N \text{ and } s \text{ in } \Omega. \tag{A.57}$$

The uniform convergence of $\{\Sigma_n(s)\}$ to f in a region Ω is sometimes stated in the form

$$f(s) = \lim_{n\to\infty} \Sigma_n(s), \quad \text{uniformly in the region } \Omega. \tag{A.58}$$

In particular, if $\Sigma_n(s)$ are the partial sums of a power series, (A.58) is sometimes written as

$$f(s) = \sum_{k=0}^{\infty} a_k(s - s_0)^k, \quad \text{uniformly in the region } \Omega.$$

Theorem A.7.2 (Taylor Series). Let $f(s)$ be analytic in a domain D. Let s_0 be in D and let R be the radius of the largest circle with center at s_0 and having its interior in D. Then the power series

$$f(s) = \sum_{k=0}^{\infty} a_k (s - s_0)^k, \tag{A.59}$$

converges uniformly to $f(s)$ for all s such that $|s - s_0| \leq r < R$. The coefficients in (A.59) are given by

$$a_k = \frac{f^{(k)}(s_0)}{k!} = \frac{1}{2\pi j} \oint_{C_1} \frac{f(s)}{(s - s_0)^{k+1}}\, ds, \quad k = 0, 1, \cdots, \tag{A.60}$$

where C_1 is any circle with center at s_0 and radius r_1 such that $r < r_1 < R$.

Proof. Let R be as in the statement of the theorem and let s be a point such that $|s - s_0| \leq r < R$. Pick r_1 such that $r < r_1 < R$, defining a circle C_1 (Figure A.15).

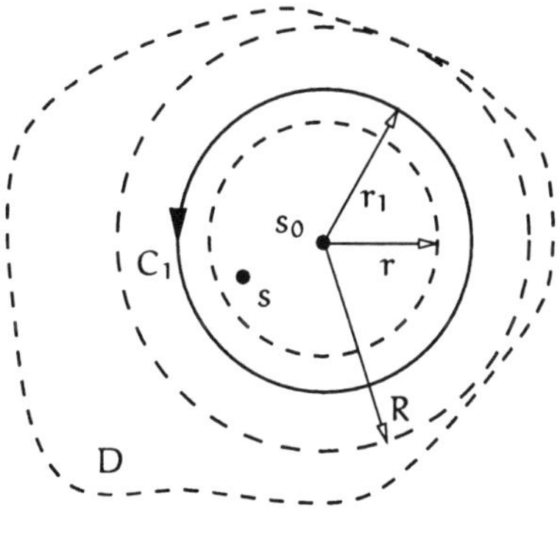

FIGURE A.15. Contours for Theorem A.7.2.

Since s is interior to C_1 and f is analytic on and within C_1, Cauchy's integral formula (A.36) holds, i.e.,

$$f(s) = \frac{1}{2\pi j} \oint_{C_1} \frac{f(\xi)}{\xi - s}\, d\xi\,, \tag{A.61}$$

where the contour is traversed counter-clockwise.

Next notice that we can write

$$\frac{1}{\xi - s} = \frac{1}{\xi - s_0} \frac{1}{1 - \dfrac{s - s_0}{\xi - s_0}}\,,$$

which, for all s such that $\left|\frac{s-s_0}{\xi-s_0}\right| \le \frac{r}{r_1} < 1$, converges uniformly to

$$\frac{1}{\xi - s} = \sum_{k=0}^{\infty} \frac{(s-s_0)^k}{(\xi - s_0)^{k+1}} \,. \tag{A.62}$$

Substituting the RHS of (A.62) into (A.61) yields

$$f(s) = \frac{1}{2\pi j}\oint_{C_1} f(\xi) \sum_{k=0}^{\infty} \frac{(s-s_0)^k}{(\xi - s_0)^{k+1}}\, d\xi \,.$$

Due to the uniform convergence of (A.62) in $|s-s_0| \le r$, we can interchange integration and summation to obtain

$$f(s) = \sum_{k=0}^{\infty} \left(\frac{1}{2\pi j}\oint_{C_1} \frac{f(\xi)}{(\xi - s_0)^{k+1}}\, d\xi \right)(s-s_0)^k .$$

Note that, using (A.55), the coefficient between parenthesis corresponds to $f^{(k)}(s_0)/k!$. We then have

$$f(s) = \sum_{k=0}^{\infty} \frac{f^{(k)}(s_0)}{k!}(s-s_0)^k, \quad \text{uniformly in } |s-s_0| \le r < R,$$

thus proving the theorem. □

Example A.7.1. Let $f(s) = \cos s$. The derivatives are $-\sin s$, $-\cos s$, $\sin s$, $\cos s$, etc., and hence $f(0) = 1$, $f'(0) = 0$, $f''(0) = -1$, $f'''(0) = 0$. Since $f^{(iv)}(s) = f(s)$, the sequence repeats, so that

$$\cos s = 1 - \frac{s^2}{2!} + \frac{s^4}{4!} - \frac{s^6}{6!} + \cdots , \tag{A.63}$$

is the Taylor series of $f(s) = \cos s$ about the point $s_0 = 0$. Since $\cos s$ is analytic in $|s| < R$ for every R, Theorem A.7.2 shows that the convergence is uniform in $|s| \le r$ for each fixed $r < \infty$.

Similarly,

$$e^s = 1 + s + \frac{s^2}{2!} + \frac{s^3}{3!} + \cdots , \tag{A.64}$$

$$\sin s = s - \frac{s^3}{3!} + \frac{s^5}{5!} - \cdots , \tag{A.65}$$

uniformly for $|s| \le r < \infty$. ∘

The Taylor series shows that the values $f^{(k)}(s_0)$, $k = 0, 1, \cdots$, at a point s_0 of a domain D in which f is analytic, determine $f(s)$ in a disk $|s - s_0| < R$ centered at s_0. Thus, if $f(s)$ is known on some infinitely differentiable short arc in $|s-s_0| < R$, then f is uniquely determined in $|s-s_0| < R$, since by differentiation of f on the arc, its derivatives at a point are also known. This is not implied by the Cauchy integral formula, since in that case, knowledge of f on the entire boundary of the region was required to determine f inside.

A.7.3 Laurent Series

If a function f has an isolated singular point at s_0, we cannot apply Theorem A.7.2 at that point. It is often possible, however, to find a series representation for $f(s)$ involving both positive and negative powers of $s - s_0$, as shown in the following theorem.

Theorem A.7.3 (Laurent Series). Let C_0 and C_1 be two circles centered at a point s_0, counter-clockwise oriented, and such that C_1 is smaller than C_0 (Figure A.16). Let f be analytic on both circles and throughout the annular domain between them. Then the Laurent series

$$f(s) = f_1(s) + f_2(s) \triangleq \sum_{k=0}^{\infty} a_k (s - s_0)^k + \sum_{k=1}^{\infty} \frac{b_k}{(s - s_0)^k}, \qquad \text{(A.66)}$$

where

$$a_k = \frac{1}{2\pi j} \oint_{C_0} \frac{f(s)}{(s - s_0)^{k+1}} \, ds, \quad k = 0, 1, \cdots, \qquad \text{(A.67)}$$

and

$$b_k = \frac{1}{2\pi j} \oint_{C_1} \frac{f(s)}{(s - s_0)^{-k+1}} \, ds, \quad k = 1, 2, \cdots, \qquad \text{(A.68)}$$

converges uniformly to $f(s)$ in any closed annulus contained in the domain enclosed between C_0 and C_1.

Proof. Let s be any point interior to the annular domain between C_0 and C_1, as shown in Figure A.16. It then follows, as in Example A.5.5, that

$$f(s) = \frac{1}{2\pi j} \oint_{C_0} \frac{f(\xi)}{\xi - s} \, d\xi + \frac{1}{2\pi j} \oint_{C_1} \frac{-f(\xi)}{\xi - s} \, d\xi. \qquad \text{(A.69)}$$

Select a closed annulus with internal radius r_1 and external radius r, contained in the domain enclosed between C_0 and C_1, and such that $r_1 \leq |s - s_0| \leq r$, as depicted in Figure A.16.

We proved in Theorem A.7.2 that the first integral in (A.69) converges to $f_1(s)$ in (A.66) uniformly for all s in $|s - s_0| \leq r$.

As for the second integral, we note that $-1/(\xi - s) = 1/(s - \xi)$. Thus, interchanging s and ξ in (A.62), and following the proof of Theorem A.7.2,

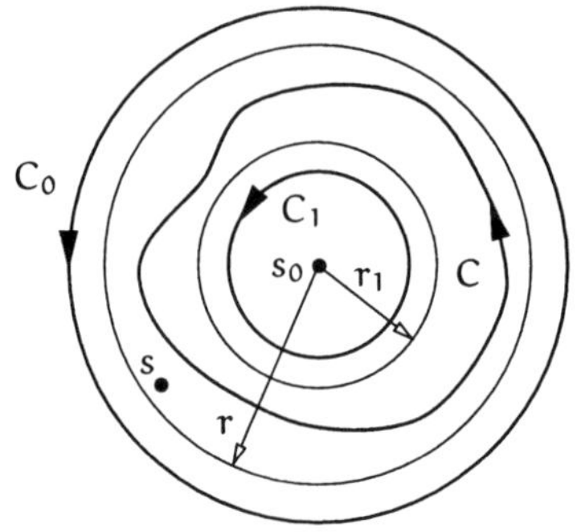

FIGURE A.16. Contours for Theorem A.7.3.

we show that the second integral in (A.69) converges to $f_2(s)$ in (A.66) uniformly for all s in $|s - s_0| \geq r_1$.

It follows that (A.69) converges uniformly to $f(s) = f_1(s)+f_2(s)$ in (A.66) for all s in the annulus $r_1 \leq |s - s_0| \leq r$, thus proving the theorem. □

Note that the two integrands $f(s)/(s - s_0)^{k+1}$ and $f(s)/(s - s_0)^{-k+1}$ in the coefficient expressions (A.67) and (A.68) are analytic throughout the annular domain $R_1 < |s - s_0| < R_0$ and on its boundary. Then, as in Example A.5.3, any simple closed piecewise smooth contour C around that domain in the counter-clockwise direction, as shown in Figure A.16, can be used as a path of integration in place of the circular paths C_0 and C_1. Thus, the Laurent series (A.66) can be written as

$$f(s) = \sum_{k=-\infty}^{\infty} c_k (s - s_0)^k, \tag{A.70}$$

where

$$c_k = \frac{1}{2\pi j} \oint_C \frac{f(s)}{(s - s_0)^{k+1}}\, ds, \quad k = 0, \pm 1, \pm 2, \cdots. \tag{A.71}$$

Corresponding to the decomposition $f = f_1 + f_2$ in (A.66), are the two parts of the Laurent series, namely,

$$f_1(s) = \sum_{k=0}^{\infty} a_k (s - s_0)^k, \qquad f_2(s) = \sum_{k=1}^{\infty} \frac{b_k}{(s - s_0)^k}.$$

The function $f_1(s)$ involves nonnegative powers of $s - s_0$ and is called the *regular part* of $f(s)$ at s_0. The function $f_2(s)$ involves negative powers of $s - s_0$ and is called the *singular* or *principal part* of $f(s)$ at s_0. In the following section, we show that there is an intimate connection between the principal part and the nature of the singularity at s_0.

A.8 Singularities

A.8.1 Isolated Singularities

In this subsection we examine functions that are analytic in a punctured disk (an open disk with the center removed). We will then consider that the center of the disk is an isolated singularity and classify it according to the behavior of the function near that point.

Denote by $B(s_0, R)$ the disk of center s_0 and radius R, i.e., $B(s_0, R) = \{s : |s - s_0| < R\}$. We then have the following definition.

Definition A.8.1 (Isolated Singularities). A function f has an *isolated singularity* at $s = s_0$ if there is an $R > 0$ such that f is defined and analytic in $B(s_0, R) - \{s_0\}$ but not in $B(s_0, R)$.

Let $s = s_0$ be an isolated singularity of f. Then:

(i) $s = s_0$ is called a *removable singularity* if there is an analytic function $g : B(s_0, R) \to \mathbb{C}$ such that $g(s) = f(s)$ for $0 < |s - s_0| < R$;

(ii) $s = s_0$ is called a *pole* if $\lim_{s \to s_0} |f(s)| = \infty$, that is, for any $K > 0$ there is a number $\varepsilon > 0$ such that $|f(s)| \geq K$ whenever $0 < |s - s_0| < \varepsilon$;

(iii) $s = s_0$ is called an *essential singularity* if s_0 is neither a pole nor a removable singularity.

○

An alternative definition of a pole is as follows: an isolated singularity $s = s_0$ is called a *pole* of f if

$$f(s) = \frac{g(s)}{(s - s_0)^m}, \tag{A.72}$$

where $m \geq 1$ is an integer, $g(s)$ is analytic in a neighborhood of s_0 and $g(s_0) \neq 0$. This definition and (ii) are equivalent and can be derived from each other. Also, if f has a pole at $s = s_0$ and m is the smallest positive integer such that $(s - s_0)^m f(s)$ has a removable singularity at $s = s_0$, then f has a *pole of order m* at $s = s_0$.

In the following theorem, we show how the Laurent series expansion is used to classify isolated singularities.

Theorem A.8.1. Let $s = s_0$ be an isolated singularity of f and let $f(s) = \sum_{k=-\infty}^{\infty} c_k (s - s_0)^k$ be its Laurent series expansion in $0 < |s - s_0| < R$. Then:

(i) $s = s_0$ is a removable singularity if and only if $c_k = 0$ for $k \leq -1$;

(ii) $s = s_0$ is a pole if and only if $c_{-m} \neq 0$ and $c_k = 0$ for $k \leq -(m + 1)$;

(iii) $s = s_0$ is an essential singularity if and only if $c_k \neq 0$ for infinitely many negative integers k.

Proof. (i) If $c_k = 0$ for $k \leq -1$ then let $g(s)$ be defined in $B(s_0, R)$ by $g(s) = \sum_{k=0}^{\infty} c_k(s-s_0)^k$. Thus, g is analytic and agrees with f in the punctured disk $0 < |s - s_0| < R$. Thus s_0 is a removable singularity according to Definition A.8.1. The converse is equally as easy.

(ii) Suppose that $c_k = 0$ for $k \leq -(m+1)$. Then $(s-s_0)^m f(s)$ has a Laurent expansion that has no negative powers of $s - s_0$. By part (i), $(s-s_0)^m f(s)$ has a removable singularity at $s = s_0$. The converse is established by an equally straightforward argument.

(iii) Since f has an essential singularity at $s = s_0$ when it has neither a removable singularity nor a pole, part (iii) follows from parts (i) and (ii).

□

We next give some examples of the different types of singularities.

Example A.8.1. The function $f(s) = \sin s/s$ has a removable singularity at $s = 0$. Indeed, using L'Hospital's rule (e.g., Levinson & Redheffer 1970, p. 152) we have that

$$\lim_{s \to 0} \frac{\sin s}{s} = \lim_{s \to 0} \cos s = 1,$$

and hence $f(s) = g(s)$ for $|s| > 0$, where $g : \mathbb{C} \to \mathbb{C}$ is the analytic function defined as

$$g(s) = \begin{cases} \dfrac{\sin s}{s} & \text{if } s \neq 0, \\ 1 & \text{if } s = 0. \end{cases}$$

○

Example A.8.2. As an example of poles, consider the rational function (A.10) in Example A.3.3. Assume that the numbers $q_1, \cdots, q_m$, $p_1, \cdots, p_n$ are all different. Then H has *simple* (i.e., of order one) poles at the n zeros of its denominator. ○

Example A.8.3. The function $e^{1/s}$ has an essential singularity at $s = 0$. Indeed, since for real positive σ, with $\nu = 1/\sigma$, and for all $m > 0$,

$$\lim_{\sigma \to 0} \sigma^m e^{1/\sigma} = \lim_{\nu \to 0} \frac{e^{\nu}}{\nu^m} = \infty,$$

it follows that $e^{1/s}$ is not bounded at $s = 0$ nor can it have a pole of order m for any m at $s = 0$. Hence it has an essential singularity. ○

If f has a removable singularity at $s = s_0$, it follows from the definition that f can be redefined to be analytic in the disk $B(s_0, R)$ by assigning a value to f at $s = s_0$. Because an analytic function is continuous, it is clear that we need only define $f(s_0)$ so as to make f continuous at $s = s_0$. This implies that $\lim_{s \to s_0} f(s)$ must exist (see Example A.8.1).

If f has a pole at $s = s_0$, then $\lim_{s \to s_0} |f(s)|$ exists and equals infinity.

On the other hand, f has an essential singularity at $s = s_0$ when the limit $\lim_{s \to s_0} |f(s)|$ fails to exist. Moreover, a remarkable feature of essential singularities is that the function f in any neighborhood of an essential singularity assumes all values except possibly one. This fact is known as Picard's theorem (e.g., Conway 1973). For example, given any complex number $\gamma \neq 0$ and any small $\delta > 0$, then it is easy to show that $e^{1/s}$ takes on the value γ an infinite number of times in $0 < |s| < \delta$.

We end this section with a discussion of singularities at infinity. The behavior of a function $f(s)$ at $s = \infty$ is defined by considering the behavior of $f(1/\xi)$ at the point $\xi = 0$. For example, $f(s)$ is continuous at $s = \infty$ if $f(1/\xi)$ is continuous at $\xi = 0$. Let $f(s)$ be analytic for $R < |s| < \infty$ but not analytic in $R < |s| \leq \infty$. Then, by using $s = 1/\xi$ and considering $\xi = 0$, it follows that the point $s = \infty$ is an isolated singular point. This may be a removable singularity, a pole, or an essential singularity of f.

As a final remark, notice that the Laurent expansion of a function $f(s)$ at $s = \infty$ is computed by first obtaining the Laurent expansion of the function $g(\xi) = f(1/\xi)$ at $\xi = 0$, and then evaluating this series at $\xi = 1/s$.

Example A.8.4. Consider a function f analytic at $s = \infty$. Then, the function $g(\xi) = f(1/\xi)$ is analytic at $\xi = 0$, and so admits a Taylor series $g(\xi) = \sum_{k=0}^{\infty} a_k \xi^k$ that converges uniformly in $|\xi| \leq r$ for some $r > 0$. Therefore, f is represented, uniformly in $|s| \geq 1/r$, by the (Laurent) series expansion at $s = \infty$

$$f(s) = \sum_{k=-\infty}^{0} c_k s^k,$$

where $c_{-k} = a_k$. Notice that, as opposed to the case of a function analytic at a *finite* point, this series has only nonpositive powers of s. This is precisely because, in defining regular and principal parts at infinity, the roles of positive and negative powers of s are interchanged. ∘

A.8.2 Branch Points

A special kind of singularity arises when *multiple-valued functions* are involved. A particular function of this type, which is of central interest in this book, is the logarithm

$$\log s = \log |s| + j \arg s.$$

The function $\log s$ is multiple-valued due to the presence of $\arg s$. Indeed, at every point s in the complex plane $\arg s$ has infinitely many values differing by $2\pi k$, with $k = 0, \pm 1, \pm 2, \ldots$.

The function $\log s$ can be defined to be *single-valued* by restricting its domain; for example, by letting $-\pi < \arg s < \pi$. The function so obtained is called the *principal branch* of $\log s$. Other branches are obtained by considering different restrictions on $\arg s$, i.e., for $k = \pm 1, \pm 2, \ldots$. Any of these branches can be shown to be analytic in their domain of definition, i.e., the extended complex plane with the negative real axis deleted (including infinity). Such a domain is called a *cut plane* and the deleted portion is a *branch cut*. Every point of a branch cut is a singularity of the branch function (since the function cannot be defined to be continuous there) and, moreover, it is *nonisolated*.

A singular point common to *any* branch cut is called a *branch point*. For example, the branch points of $\log s$ are $s = 0$ and $s = \infty$. Special care must be taken when performing contour integration of functions with branch points, as we will see in §A.9.2.

Example A.8.5. The function $\log H$ considered in Example A.3.4 has branch points at the zeros and poles of the transfer function H, i.e., at the zeros of N and D. ∘

Example A.8.6. Since the principal branch of $\log s$ is analytic in its cut plane, it admits a Taylor series expansion around any interior point s_0. Then, since

$$\frac{d^k}{ds^k} \log s = (-1)^{k-1} \frac{(k-1)!}{s^k}, \quad k = 1, 2, \cdots,$$

it follows from (A.59) and (A.60), with $s_0 = 1$, that

$$\log s = \sum_{k=0}^{\infty} (-1)^{k-1} \frac{(s-1)^k}{k}. \tag{A.73}$$

The distance between the point $s_0 = 1$ and the boundary of the cut plane equals 1, and thus the expansion (A.73) is valid for $|s-1| < 1$. Replacing $s-1$ by s we obtained the following series for $\log(1+s)$, called the *logarithmic series*:

$$\log(1+s) = \sum_{k=0}^{\infty} (-1)^{k-1} \frac{(s-1)^k}{k}, \quad \text{in } |s| < 1. \tag{A.74}$$

A useful observation that is used in Chapter 3 is obtained by noting that, if $|s| < 1$ then

$$\begin{aligned} \left| 1 - \frac{\log(1+s)}{s} \right| &= \left| \frac{1}{2}s - \frac{1}{3}s^2 + \cdots \right| \\ &\le \frac{1}{2} \left(|s| + |s|^2 + \cdots \right) \\ &= \frac{1}{2} \frac{|s|}{1-|s|}. \end{aligned}$$

If we further require that $|s| < 1/2$ then $|1 - \log(1+s)/s| \leq 1/2$. This gives that for $|s| < 1/2$

$$\frac{1}{2}|s| \leq |\log(1+s)| \leq \frac{3}{2}|s|.$$

$\circ$

Example A.8.7. As an application of the logarithmic series, we will show that the function $f : \mathbb{C} \times \mathbb{R} \to \mathbb{C}$, given by

$$f(s,\eta) = \eta \log\left(1 + \frac{s}{\eta}\right),$$

satisfies

$$\lim_{\eta \to \infty} f(s,\eta) = s, \tag{A.75}$$

uniformly on any compact set in $\mathbb{C}$.[15]

Indeed, let s belong to a compact set in $\mathbb{C}$. Then there exists a finite constant $K > 0$ such that $|s| \leq K$ for all s in the set. For $\eta > K$, the following expansion, of the form (A.74), holds

$$\log\left(1 + \frac{s}{\eta}\right) = \frac{s}{\eta} - \frac{s^2}{2\eta^2} + \cdots .$$

(A.75) then follows on multiplying both sides of the above series by η and taking the limit as $\eta \to \infty$. $\circ$

A.9 Integration of Functions with Singularities

A.9.1 Functions with Isolated Singularities

The Cauchy integral theorem tells us that if a function is analytic at all points interior to and on a simple closed contour C, the value of the integral of the function around that contour is zero. As we anticipated in Example A.5.4, if the function fails to be analytic at a finite number of points interior to C, there is a specific number, called the residue, which each of those points contributes to the value of the integral.

Let then f have an isolated singularity at $s = s_0$ and let

$$f(s) = \sum_{k=-\infty}^{\infty} c_k (s - s_0)^k$$

[15] A compact set in $\mathbb{C}$ is bounded.

be its Laurent series expansion. The *residue* of f at $s = s_0$, denoted by $\operatorname{Res}_{s=s_0} f(s)$, is defined as the coefficient c_{-1}, i.e.,

$$\operatorname*{Res}_{s=s_0} f(s) \triangleq c_{-1}. \tag{A.76}$$

According to (A.71), we have that

$$\operatorname*{Res}_{s=s_0} f(s) = \frac{1}{2\pi j} \oint_C f(s)\, ds, \tag{A.77}$$

where the curve C encircles the point $s = s_0$ in the counter-clockwise direction.

The generalization of (A.77) to the case where f has a finite number of singular points interior to C is given in the following theorem.

Theorem A.9.1 (Residue Theorem). Let C be a simple closed piecewise smooth contour, counter-clockwise oriented. Let f be analytic on and within C except for a finite number of singular points $s_1, s_2, \cdots, s_n$ interior to C. If $\operatorname{Res}_{s=s_1} f(s), \operatorname{Res}_{s=s_2} f(s), \cdots, \operatorname{Res}_{s=s_n} f(s)$ denote the residues of f at those respective points, then

$$\oint_C f(s)\, ds = 2\pi j \sum_{k=1}^{n} \operatorname*{Res}_{s=s_k} f(s). \tag{A.78}$$

Proof. Let the singular points $s_1, s_2, \cdots, s_n$ be centers of counter-clockwise oriented circles C_k, which are interior to C, and are so small that no two of the circles have points in common, as shown in Figure A.17.

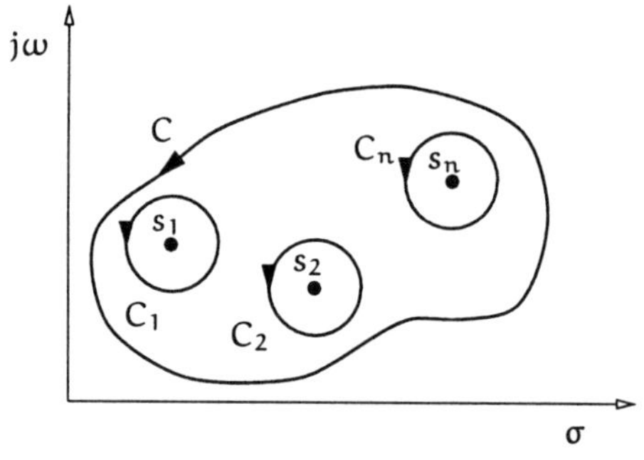

FIGURE A.17. Contours for Theorem A.9.1.

The circles C_k together with the contour C form the boundary of a closed region throughout which f is analytic and whose interior is a multiply connected domain. Hence, according to the extension of the Cauchy integral theorem to such regions

$$\oint_C f(s)\, ds - \oint_{C_1} f(s)\, ds - \cdots - \oint_{C_n} f(s)\, ds = 0.$$

but this reduces to (A.78) by using (A.77) for each of the integrals around the circles. □

Residue at Infinity

We have just introduced the definition of residue of a function at an isolated singular point. This definition could in fact be extended to *any* point that is at the center of a punctured disk where the function is analytic — whether the function is analytic or not at the point itself. Evidently, the residue of a function at a finite regular point is zero.

Analogously, we can define the residue of a function at $s = \infty$ as

$$\operatorname*{Res}_{s=\infty} f(s) = \frac{1}{2\pi j} \oint f(s) ds, \tag{A.79}$$

where the integral is computed *clockwise* along a circle of radius R so large that the only possible singularity of f in $|s| \geq R$ is the point $s = \infty$. The clockwise direction is used because the point $s = \infty$ is thus "enclosed" by the contour of integration.[16]

Let the Laurent expansion at infinity of f be

$$f(s) = \cdots + \frac{c_{-k}}{s^k} + \cdots + \frac{c_{-1}}{s} + c_0 + c_1 s + \cdots ,$$

uniformly convergent in $R \leq |s| < \infty$. Then, in virtue of Example A.5.1,

$$\operatorname*{Res}_{s=\infty} f(s) = -c_{-1}, \tag{A.80}$$

which is consistent with the definition of the residue at a finite point given by (A.76).

It is interesting to note that, for a function f having a finite number, n say, of isolated singularities, the sum of all the residues in the extended complex plane is zero. Indeed, integrating f around a circle large enough to contain all finite singularities, the Residue Theorem together with (A.79) give

$$\sum_{k=1}^{n} \operatorname*{Res}_{s=s_k} f(s) = -\operatorname*{Res}_{s=\infty} f(s). \tag{A.81}$$

Example A.9.1. Consider the function

$$f(s) = \frac{1}{s-1} .$$

[16] This assumes the convention that a counter-clockwise oriented curve "encloses" its interior whereas a clockwise oriented curve "encloses" its exterior.

This function has a single pole at $s = 1$ and it is analytic at $s = \infty$. The residue of $f(s)$ at $s = 1$ is $\mathrm{Res}_{s=1} f(s) = 1$. To obtain the residue at infinity, let

$$g(\xi) = f(1/\xi) = \frac{\xi}{1-\xi}\,.$$

The Laurent expansion of $g(\xi)$ at $\xi = 0$ is

$$g(\xi) = \sum_{k=1}^{\infty} \xi^k\,, \quad \text{in } 0 < |\xi| < 1,$$

and so the Laurent expansion of $f(s)$ at $s = \infty$ is

$$f(s) = \sum_{1}^{\infty} \frac{1}{s^k}\,, \quad \text{in } 1 < |s| < \infty.$$

The residue of f at $s = \infty$ is then $\mathrm{Res}_{s=\infty} f(s) = -1$. ∘

Notice that the residue at infinity can be nonzero even if the function is analytic at $s = \infty$.

We end this subsection with an example of application to SISO control systems.

Example A.9.2. Let L be a proper rational function and consider the sensitivity function $S = 1/(1 + L)$ (see e.g., (1.5) in Chapter 1). Assume further that the closed-loop system is stable, i.e., the numerator of $1 + L$ is Hurwitz and $L(\infty) \neq -1$. We are interested in evaluating the residue at infinity of the function $\log S$, which is analytic at infinity by the assumptions of closed-loop stability and the properness of L.

Since L is a proper rational function, it has a Laurent expansion at infinity of the form

$$L(s) = L(\infty) + \frac{c_{-1}}{s} + \frac{c_{-2}}{s^2} + \cdots\,, \tag{A.82}$$

in $|s| > r$, for some $r > 0$. Then $\log S = -\log(1 + L)$ can be written as

$$\begin{aligned}\log S(s) &= -\log\left[1 + L(\infty) + \frac{c_{-1}}{s} + \frac{c_{-2}}{s^2} + \cdots\right] \\ &= -\log[1 + L(\infty)] - \log\left[1 + \frac{c_{-1}}{1+L(\infty)}\frac{1}{s} + \frac{c_{-2}}{1+L(\infty)}\frac{1}{s^2} + \cdots\right] \\ &= \log S(\infty) - \log\left[1 + \frac{S(\infty)c_{-1}}{s} + \frac{S(\infty)c_{-2}}{s^2} + \cdots\right].\end{aligned}$$

Then

$$\log\frac{S(s)}{S(\infty)} = -\log[1 + \tilde{L}(s)], \tag{A.83}$$

where

$$\tilde{L}(s) = \frac{S(\infty)c_{-1}}{s} + \frac{S(\infty)c_{-2}}{s^2} + \cdots .$$

The power series expansion of $\log(1+s)$ for $|s| < 1$ is (see Example A.8.6)

$$\log(1+s) = s - \frac{s^2}{2} + \cdots, \qquad \text{in } |s| < 1.$$

Using this expansion in (A.83) yields

$$\begin{aligned}
\log \frac{S(s)}{S(\infty)} &= -\tilde{L}(s) + \frac{\tilde{L}^2(s)}{2} + \cdots, \qquad \text{in } |\tilde{L}(s)| < 1 \\
&= -\frac{S(\infty)c_{-1}}{s} - \frac{S(\infty)c_{-2}}{s^2} + \cdots, \qquad \text{in } |s| > \tilde{r},
\end{aligned}$$

for some $\tilde{r} > r$. We have thus obtained the Laurent expansion at infinity of the function $\log S$, whose residue at infinity can be computed from (A.80) as

$$\begin{aligned}
\operatorname*{Res}_{s=\infty} \log S(s) &= \operatorname*{Res}_{s=\infty} \log \frac{S(s)}{S(\infty)} \\
&= S(\infty)c_{-1} \\
&= \frac{1}{S(\infty)} \lim_{s\to\infty} s[S(\infty) - S(s)],
\end{aligned} \tag{A.84}$$

where the last line follows from (A.82) on noting that

$$c_{-1} = \lim_{s\to\infty} s[L(s) - L(\infty)] = \lim_{s\to\infty} s\left[\frac{1}{S(s)} - \frac{1}{S(\infty)}\right].$$

If L in (A.82) has relative degree one, then

$$\operatorname*{Res}_{s=\infty} \log S(s) = \lim_{s\to\infty} s[1 - S(s)].$$

Alternatively, if L has relative degree two or more, then

$$\operatorname*{Res}_{s=\infty} \log S(s) = 0.$$

∘

A.9.2 *Functions with Branch Points*

When integration is to be performed around a region where the integrand has branch points, the residue theorem cannot be used in the form of Theorem A.9.1 but an extended version that handles this situation can be derived (Levinson & Redheffer 1970, Theorem 9.1). Since our requirements of integration of functions with branch points are limited to the logarithm, we will concentrate on an example of contour integration of this function around a branch cut.

Consider the semicircular contour of Figure A.6, and the integral on this contour of function $\log S$, where S is a stable system transfer function. If $S(s)$ has a zero at $s = p$ in the ORHP then $\log S(s)$ has a branch point at $s = p$. It is then necessary to indent the contour to avoid the branch cut, as shown in Figure A.18 for p real.

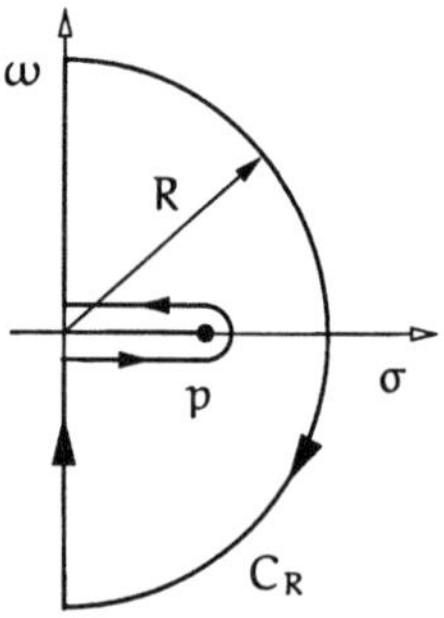

FIGURE A.18. Contour for $\log(s-p)$.

For simplicity, we will next assume that $S(s)$ has a simple zero at $s = p$ on the positive real axis. Consider the integral of $\log S(s)$ on the indentation, C, shown in detail in Figure A.19.

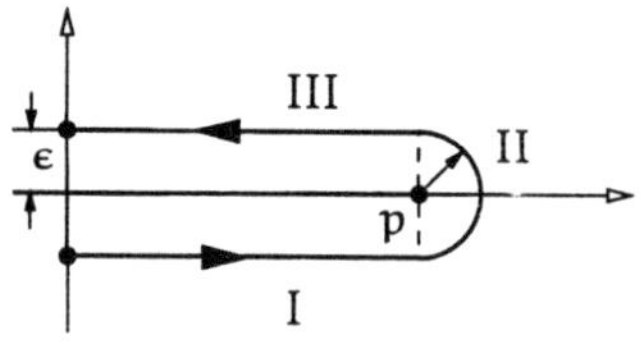

FIGURE A.19. Indentation around branch cut.

Since $\log S(s) = \log(s-p) + \log \bar{S}(s)$, where $\bar{S}(s)$ is analytic on and inside the indentation, then

$$\int_C \log S(s)\, ds = \int_C \log(s-p)\, ds + \int_{-j\varepsilon}^{j\varepsilon} \log \bar{S}(j\omega)\, d\omega.$$

We will thus concentrate on $\int_C \log(s-p)\, ds$.

Since C is in the domain of analyticity of $\log(s-p)$, the fundamental theorem of calculus, given by (A.18), yields

$$\begin{aligned}\int_C \log(s-p)\,ds &= [(s-p)\log(s-p)-(s-p)]\Big|_{-j\varepsilon}^{j\varepsilon}\\ &= jp[\arg(-p-j\varepsilon)-\arg(-p+j\varepsilon)].\end{aligned}$$

The limit when $\varepsilon \to 0$ is then

$$\lim_{\varepsilon\to 0}\int_C \log(s-p)\,ds = -j2\pi p.$$

As an aside, note that the integral around the semicircle II in Figure A.19 goes to zero when $\varepsilon \to 0$. Indeed,

$$\begin{aligned}\left|\int_{II}\log(s-p)\,ds\right| &= \left|\int_{-\pi/2}^{\pi/2}(\log\varepsilon + j\theta)\varepsilon\, e^{j\theta}\,d\theta\right|\\ &\le \varepsilon\log\varepsilon\int_{-\pi/2}^{\pi/2} + \varepsilon\int_{-\pi/2}^{\pi/2}|\theta d\theta|,\end{aligned}$$

which goes to 0 as $\varepsilon \to 0$. Since the total integral on C is non zero, this means that the integrals on I and III, which are in different sides of the branch cut, *do not* cancel. This is in contrast to what happens with functions with at most isolated singularities.

It is easy to see that if the function S has zeros $p_1, p_2, \ldots, p_n$ with the same imaginary part (possibly different than zero and possibly with the same real part), the same branch cut can be used for all of them to obtain

$$\int_C \log[(s-p_1)(s-p_2)\cdots(s-p_n)]\,ds = -j2\pi\sum_{i=1}^{n}\operatorname{Re} p_i. \tag{A.85}$$

A.10 The Maximum Modulus Principle

Another result that relates characteristics of a function, analytic in a region, with its behavior on the boundary of the region is the *maximum modulus principle* of analytic functions. This is interesting but less informative than Poisson integrals. We will give two versions of this principle.

Theorem A.10.1 (Maximum Principle). Let f be analytic in a domain D. Then $|f|$ cannot have a maximum anywhere in D unless f is a constant.

Proof. Assume that f is not a constant in D and suppose that s_0 is a point in D such that $|f(s_0)|$ is maximum. Let n be the smallest integer such that

$f^{(n)}(s_0) \neq 0$, which exists since f is not constant. Then the Taylor series of f around s_0 has the form[17]

$$f(s) = f(s_0) + \frac{f^{(n)}(s_0)}{n!}(s - s_0)^n + o\left((s - s_0)^n\right). \tag{A.86}$$

Let h be a complex number such that

$$h^n = n!\frac{f(s_0)}{f^{(n)}(s_0)}\varepsilon^n, \quad \varepsilon > 0,$$

where ε is small enough such that $s = s_0 + h$ is inside the disk of convergence of the Taylor series (A.86). Then, evaluating this series at $s = s_0 + h$, we have

$$f(s_0 + h) = (1 + \varepsilon^n)f(s_0) + o(\varepsilon^n).$$

This implies

$$|f(s_0 + h)| > |f(s_0)|,$$

which contradicts the assumption that $|f(s_0)|$ was a maximum. Since s_0 is generic, it follows that $|f|$ cannot achieve a maximum anywhere in D, and the result follows. □

Theorem A.10.2 (Maximum Principle, second version). Let $f(s)$ be analytic in a bounded region R and let $|f(s)|$ be continuous in the closed region $\overline{R}$. Then $|f(s)|$ assumes its maximum on the boundary of the region.

Proof. The theorem trivially holds if $f(s)$ is constant. Suppose then that $f(s)$ is not a constant. Since $|f(s)|$ is continuous, by a well-known theorem in real variables $|f(s)|$ assumes a maximum somewhere in the closed bounded region $\overline{R}$. By Theorem A.10.1, this maximum cannot be assumed at any interior point and hence must be assumed on the boundary. □

The maximum modulus principle is used in Chapter 3 to show the "push-pop" or "water-bed" phenomenon in linear control theory.

A.11 Entire Functions

An entire function, f, is a function defined and analytic for all finite values of the complex variable s. An entire function that is not a polynomial is called an entire *transcendental* function.

In Chapters 10 and 11, we use the following result on zeros of transcendental functions of a particular form.

[17] Here we use the notation $g(s) = o(s^\mu)$ to mean that $g(s)/s^\mu \to 0$ when s is near to some given limit.

Lemma A.11.1. Consider an entire function having the particular form

$$f(s) = g_1(s)e^{-s\tau} + g_2(s), \tag{A.87}$$

where $\tau > 0$ and $g_1(s)$ and $g_2(s)$ are polynomials. Let

$$\delta = \deg(g_2) - \deg(g_1). \tag{A.88}$$

We then have the following implications.

(i) If $\delta > 0$, then the high frequency zeros ($s \to \infty$) of $f(s)$ have negative real part.

(ii) If $\delta < 0$, then the high frequency zeros of $f(s)$ have positive real part.

(iii) If $\delta = 0$, then the high frequency zeros of $f(s)$ converge to the sequence

$$s_k = -\frac{1}{\tau}\log\eta + jk\frac{2\pi}{\tau}, \qquad k = 0, \pm 1, \pm 2, \cdots . \tag{A.89}$$

And η is the ratio between the highest order coefficients of $g_2(s)$ and $g_1(s)$, i.e.,

$$\eta = \lim_{s\to\infty} \frac{g_2(s)}{g_1(s)} . \tag{A.90}$$

Proof. The zeros of f in (A.87) satisfy the identity

$$e^{-s\tau} = -\frac{g_2(s)}{g_1(s)} .$$

Taking logarithms yields

$$s = -\frac{1}{\tau}\log\left|\frac{g_2(s)}{g_1(s)}\right| - j\frac{1}{\tau}\arg\left[\frac{g_2(s)}{g_1(s)}\right] + jk\frac{2\pi}{\tau} . \tag{A.91}$$

Assume $\delta > 0$. Then clearly, for s a high frequency zero of f, the argument of the log function in (A.91) will be greater than one and thus the zero will have negative real part. This shows that case (i) is true. The proof of case (ii) follows similarly.

Finally, for $\delta = 0$, we note that, for large s, the right hand side of (A.91) converges to $(-\log|\eta| - j\arg(\eta) + jk2\pi)/\tau$, with η given in (A.90). Hence (A.89) follows. □

Example A.11.1. Even in the case of $\delta > 0$, the function f may have zeros with positive real part. For example, let f be given by

$$f(s) = e_1 s + e_2 - be^{a\tau}e^{-s\tau}, \tag{A.92}$$

where e_1, e_2, b, a are positive real constants. Assume further that $b > e_2$. Consider the functions

$$\begin{aligned} p(s) &= e_2 - b e^{a\tau} e^{-s\tau}, \\ q(s) &= e_1 s, \end{aligned} \tag{A.93}$$

and the closed curve $\gamma = \gamma_1 \cup \gamma_2$ defined as

$$\begin{aligned} \gamma_1 &= \{s = j\omega : \ \omega \in [-R, R]\}, \\ \gamma_2 &= \{s = Re^{j\theta} : \ \theta \in [-\pi/2, \pi/2]\}. \end{aligned} \tag{A.94}$$

On γ_1 we have

$$\begin{aligned} &|p(s)| = |e_2 - b e^{a\tau} e^{j\omega\tau}| \geq ||e_2| - |b e^{a\tau}|| = b e^{a\tau} - e_2, \\ &|q(s)| = e_1 \omega \leq e_1 R. \end{aligned} \tag{A.95}$$

Then $|q(s)| < |p(s)|$ on γ_1 if

$$R < R_1 \triangleq \frac{b e^{a\tau} - e_2}{e_1}. \tag{A.96}$$

On the other hand, on γ_2, we have

$$\begin{aligned} &|p(s)| \geq |e_2 - |b e^{a\tau} e^{-R\tau \cos\theta}||, \\ &|q(s)| = e_1 R. \end{aligned} \tag{A.97}$$

Some simple calculations show that $|q(s)| < |p(s)|$ on γ_2 if

$$R < R_2 \triangleq \frac{b e^{a\tau} - e_2}{e_1 + \tau}. \tag{A.98}$$

Since $\tau > 0$, we have that $R_2 < R_1$. It follows that, if $R < R_2$, $|q(s)| < |p(s)|$ on the semicircular contour γ. Note also that both p and q are analytic on and inside γ. By Rouche's Theorem (Conway 1973) we then know that p and $p + q = f$ have the same number of zeros inside γ.

The zeros of $p(s)$ are given by

$$s_k = -\frac{1}{\tau} \log \frac{e_2}{b e^{a\tau}} + jk\frac{2\pi}{\tau}. \tag{A.99}$$

Since $e_2 < b e^{a\tau}$, the real parts of s_k are positive. this establishes the claim that f can have low frequency zeros having positive real part. (Actually, the imaginary parts are spaced by $2\pi/\tau$. Hence, as τ increases, then R_2 given in (A.98) increases and the spacing between zeros decreases. It follows that the number of roots of f in the ORHP increases with τ). ∘

Notes and References

The material of this chapter is based mainly on Levinson & Redheffer (1970) and Churchill & Brown (1984). These are standard textbooks on Complex Variable Theory. Some specific results and definitions were obtained from Widder (1961), Kaplan (1973), Markushevich (1965) and Conway (1973).

In particular, §A.4.2 contains important input from Widder (1961); §A.5.1 and §A.8 are largely based on Widder (1961) and Conway (1973), respectively.

Equations (A.5), which are usually called the Cauchy-Riemann equations, are of central importance in the theory of analytic functions. However, it should be noted that this universally encountered attribution is not historically justified. In fact, equations (A.5) had already been studied in the eighteenth century by D'Alembert (1717-1783) and Euler (1707-1783), in research devoted to the application of functions of a complex variable to hydrodynamics (D'Alembert and Euler), and to cartography and integral calculus (Euler) (Markushevich 1965, p. 111).

Appendix B
Proofs of Some Results in the Chapters

B.1 Proofs for Chapter 4

In this section, we prove the Bode integral constraints for the logarithm of the singular values of the sensitivity function, given in Theorem 4.2.2 of Chapter 4. We follow Chen (1995). A few preliminary technical results are needed in order to prove this theorem.

Fact B.1.1. Let $f : \mathbb{R} \times \mathbb{R} \to \mathbb{R}$ given by

$$f(\eta, \omega) = \frac{\eta^2}{\eta^2 + \omega^2}.$$

Then $f(\eta, \omega) \to 1$ as $\eta \to \infty$, uniformly on any compact interval. ∘

Fact B.1.2 (Levinson & Redheffer (1970), p.337). Consider $f : \mathbb{C} \times \mathbb{R} \to \mathbb{C}$. Suppose that $f(s, \eta)$ is analytic in a domain $D \subset \mathbb{C}$, and that $f(s, \eta) \to g(s)$ uniformly in D as $\eta \to \infty$. Write $s = re^{j\theta}$. Then, $\partial f(re^{j\theta}, \eta)/\partial r \to \partial g(re^{j\theta})/\partial r$ uniformly in D as $\eta \to \infty$. ∘

Lemma B.1.3. The following limit converges uniformly on any compact set

$$\lim_{\eta \to \infty} \eta \log \left| \frac{\eta + s}{\eta - s} \right| = 2 \operatorname{Re} s. \tag{B.1}$$

Proof. The result follows from Example A.8.7 in Appendix A, and the observation that

$$\eta \log\left|\frac{\eta+s}{\eta-s}\right| = \mathrm{Re}\,\eta\log\left(1+\frac{s}{\eta}\right) - \mathrm{Re}\,\eta\log\left(1-\frac{s}{\eta}\right).$$

□

Lemma B.1.4. The following limit converges uniformly on any compact set

$$\lim_{\eta\to\infty}\eta\frac{\partial}{\partial R}\left(\log\left|\frac{\eta+Re^{j\theta}}{\eta-Re^{j\theta}}\right|\right) = 2\cos\theta. \tag{B.2}$$

Proof. Immediate from Lemma B.1.3 and Fact B.1.2. □

Consider next the sensitivity function S corresponding to the feedback system of Figure 4.1. Let p_i, $i = 1,\ldots,n_p$, be the poles of the open-loop system L in the ORHP, repeated according to their geometric multiplicities. Recall that, if the open-loop system is unstable, then S can be factored as

$$S = S_m \prod_{i=1}^{n_p} B_i, \tag{B.3}$$

where S_m is minimum phase and B_i is the all-pass factor corresponding to the pole p_i.

We make the following assumption.

Assumption B.1.

(i) The closed-loop system of Figure 4.1 is stable.

(ii) $\lim_{R\to\infty}\sup_{\substack{s\in\overline{\mathbb{C}^+}\\|s|\geq R}} R\,\overline{\sigma}(L(s)) = 0.$

(iii) The singular values of S_m in (B.3), i.e., $\sigma_i(S_m(s))$, $i = 1,\cdots,n$, have continuous second order derivatives for all $s\in\overline{\mathbb{C}^+}$.

∘

We first prove the Bode integral for open-loop stable system, i.e., $S_m = S$ in (B.3).

Theorem B.1.5 (Bode Integral for S – Stable systems). Let S be the sensitivity function of the feedback loop of Figure 4.1. Assume that the open-loop system, L, is stable. Then, under Assumption B.1 (with $S_m = S$),

$$\int_0^\infty \log\sigma_j(S(j\omega))\,d\omega = \frac{1}{2}\iint_{\overline{\mathbb{C}^+}} \sigma\nabla^2\log\sigma_j(S(\sigma+j\omega))\,d\sigma\,d\omega. \tag{B.4}$$

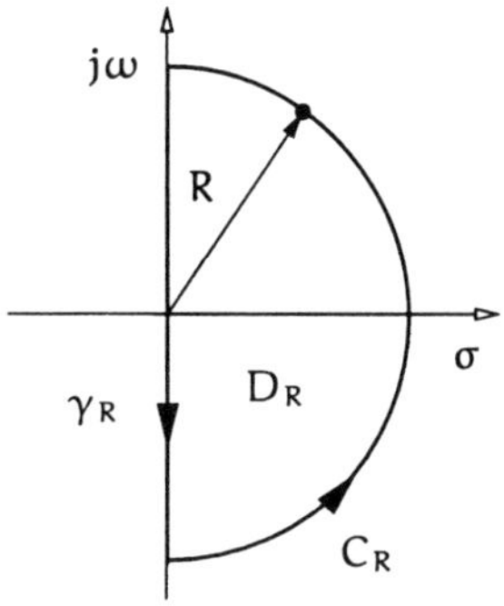

FIGURE B.1. Contour for the Poisson integral formula

Proof. Consider the half disc D_R depicted in Figure B.1. Denote by $\gamma_R = \{-j\omega : -R \le \omega \le R\}$ and $C_R = \{Re^{j\theta} : -\pi/2 \le \theta \le \pi/2\}$. Then the boundary of D_R is given by $\gamma_R \cup C_R$. Let $\eta > R$ and consider the real function $g : \mathbb{C} \to \mathbb{R}$ defined by

$$g(s) = \log \left| \frac{\eta + s}{\eta - s} \right| .$$

It is easy to see that g is harmonic in D_R, and hence for any $s \in D_R$, $\nabla^2 g(s) = 0$. Also, $g(j\omega) = 0$, $\forall \omega \in \mathbb{R}$. Let

$$f(s) = \log \sigma_j(S(s)).$$

Since Assumption B.1 holds, then each $\sigma_j(S)$ has continuous second order derivative in D_R, and so does $\log \sigma_j(S)$ since $\sigma_j(S(s)) > 0$. We can now apply Green's formula (A.30) of Appendix A to f and g, with the contour $C = \gamma_R \cup C_R$ and the domain $\Omega = D_R$. We have

$$\int_{\gamma_R} f \frac{\partial g}{\partial \vec{n}}\, ds + \int_{C_R} \left(f \frac{\partial g}{\partial \vec{n}} - g \frac{\partial f}{\partial \vec{n}} \right) ds = - \iint_{D_R} g \nabla^2 f \, d\sigma \, d\omega. \tag{B.5}$$

We next compute each of the integrals in (B.5). Notice that, on γ_R, the outer normal $\vec{n}$ is in the direction of the negative real axis. Thus, we have that $\partial g / \partial \vec{n} = -\partial g / \partial \sigma$, and

$$\begin{aligned} \frac{\partial g}{\partial \sigma} &= \frac{\partial}{\partial \sigma} \left(\operatorname{Re} \log \frac{\eta + s}{\eta - s} \right) \\ &= \operatorname{Re} \frac{\partial}{\partial \sigma} \left(\log \frac{\eta + s}{\eta - s} \right) \\ &= \operatorname{Re} \left(\frac{1}{\eta + s} + \frac{1}{\eta - s} \right). \end{aligned}$$

Setting $\sigma = 0$ in the last equality gives, on γ_R

$$\frac{\partial g}{\partial \vec{n}} = -\frac{2\eta}{\eta^2 + \omega^2},$$

which leads to

$$\int_{\gamma_R} f\frac{\partial g}{\partial \vec{n}}\, ds = -\int_{-R}^{R} \log \sigma_j(S(j\omega))\frac{2\eta}{\eta^2+\omega^2}\, d\omega.$$

Multiplying both sides of the above equality by η, taking limits as $\eta \to \infty$, and using the uniform convergence theorem (Levinson & Redheffer 1970, p. 335) and Fact B.1.1, it follows that

$$\begin{aligned}\lim_{\eta\to\infty} \eta \int_{\gamma_R} f\frac{\partial g}{\partial \vec{n}}\, ds &= -\lim_{\eta\to\infty}\int_{-R}^{R} \log \sigma_j(S(j\omega))\frac{2\eta^2}{\eta^2+\omega^2}\, d\omega \\ &= \int_{-R}^{R} \log \sigma_j(S(j\omega)) \lim_{\eta\to\infty}\frac{2\eta^2}{\eta^2+\omega^2}\, d\omega \\ &= -2\int_{-R}^{R} \log \sigma_j(S(j\omega))\, d\omega. \qquad \text{(B.6)}\end{aligned}$$

Turning to the second integral in (B.5), we have

$$\int_{C_R} \left(f\frac{\partial g}{\partial \vec{n}} - g\frac{\partial f}{\partial \vec{n}}\right) ds = I_1 - I_2,$$

where

$$\begin{aligned}I_1 &\triangleq \int_{-\pi/2}^{\pi/2} R\frac{\partial}{\partial R}\left(\log\left|\frac{\eta + Re^{j\theta}}{\eta - Re^{j\theta}}\right|\right) \log \sigma_j(S(Re^{j\theta}))\, d\theta, \\ I_2 &\triangleq \int_{-\pi/2}^{\pi/2} R\log\left|\frac{\eta + Re^{j\theta}}{\eta - Re^{j\theta}}\right| \frac{\partial}{\partial R}[\log \sigma_j(S(Re^{j\theta}))]\, d\theta.\end{aligned}$$

As before, we have, using the uniform convergence theorem and Lemmas B.1.3 and B.1.4,

$$\begin{aligned}\lim_{\eta\to\infty} \eta I_1 &= \lim_{\eta\to\infty} \eta \int_{-\pi/2}^{\pi/2} R\frac{\partial}{\partial R}\left(\log\left|\frac{\eta + Re^{j\theta}}{\eta - Re^{j\theta}}\right|\right) \log \sigma_j(S(Re^{j\theta}))\, d\theta \\ &= 2\int_{-\pi/2}^{\pi/2} R \log \sigma_j(S(Re^{j\theta})) \cos\theta\, d\theta,\end{aligned}$$

and

$$\begin{aligned}\lim_{\eta\to\infty} \eta I_2 &= \lim_{\eta\to\infty} \eta \int_{-\pi/2}^{\pi/2} R\log\left|\frac{\eta + Re^{j\theta}}{\eta - Re^{j\theta}}\right| \frac{\partial}{\partial R}[\log \sigma_j(S(Re^{j\theta}))]\, d\theta \\ &= 2\int_{-\pi/2}^{\pi/2} R^2 \frac{\partial}{\partial R}[\log \sigma_j(S(Re^{j\theta}))] \cos\theta\, d\theta.\end{aligned}$$

These inequalities lead to

$$\lim_{\eta\to\infty} \eta \int_{C_R} \left(f\frac{\partial g}{\partial \vec{n}} - g\frac{\partial f}{\partial \vec{n}} \right) ds = 2I_3 - 2I_4, \tag{B.7}$$

where

$$I_3 \triangleq \int_{-\pi/2}^{\pi/2} R \log \sigma_j(S(Re^{j\theta})) \cos\theta \, d\theta,$$

$$I_4 \triangleq \int_{-\pi/2}^{\pi/2} R^2 \frac{\partial}{\partial R}[\log \sigma_j(S(Re^{j\theta}))] \cos\theta \, d\theta.$$

As for the integral on the RHS of (B.5), we use the same limiting technique to obtain

$$\lim_{\eta\to\infty} \eta \iint_{D_R} g\nabla^2 f \, d\sigma \, d\omega = 2 \iint_{D_R} \sigma\nabla^2 \log \sigma_j(S(\sigma + j\omega)) \, d\sigma \, d\omega. \tag{B.8}$$

Combining (B.6)-(B.8) yields

$$\iint_{D_R} \sigma\nabla^2 \log \sigma_j(S(\sigma + j\omega)) \, d\sigma \, d\omega = \int_{-R}^{R} \log \sigma_j(S(j\omega)) \, d\omega - I_3 + I_4.$$

The next step is to take limits of both sides of the above equation as $R \to \infty$. This gives

$$\begin{aligned} \iint_{\overline{\mathbb{C}^+}} \sigma\nabla^2 \log \sigma_j(S(\sigma + j\omega)) \, d\sigma \, d\omega = & \int_{-\infty}^{\infty} \log \sigma_j(S(j\omega)) \, d\omega \\ & - \lim_{R\to\infty} I_3 + \lim_{R\to\infty} I_4. \end{aligned} \tag{B.9}$$

We claim that $\lim_{R\to\infty} I_3 = 0$. To see this, we first note that for $j = 1, 2, \ldots, n$, we have that $\underline{\sigma}(S(s)) \le \sigma_j(S(s)) \le \overline{\sigma}(S(s))$ for any $s \in \overline{\mathbb{C}^+}$. Furthermore, under Assumption B.1 (ii), $\overline{\sigma}(L(s)) < 1$ for any $|s| > R$ if $R > 0$ is sufficiently large. As a result, I_3 can be bounded as follows

$$\begin{aligned} |I_3| &\le \int_{-\pi/2}^{\pi/2} R|\log \sigma_j(S(Re^{j\theta}))| \cos\theta \, d\theta \\ &\le 2 \max_{\theta\in[-\pi/2,\pi/2]} R|\log \sigma_j(S(Re^{j\theta}))| \\ &\le 2 \max_{\theta\in[-\pi/2,\pi/2]} \max \{R|\log \overline{\sigma}(S(Re^{j\theta}))|, R|\log \underline{\sigma}(S(Re^{j\theta}))|\} \\ &\le 2 \max_{\theta\in[-\pi/2,\pi/2]} \max \{R|\log[1 - \overline{\sigma}(L(Re^{j\theta}))]|, R|\log[1 + \overline{\sigma}(L(Re^{j\theta}))]|\}. \end{aligned}$$

From Assumption B.1 (ii) and the logarithm series expansions (see Example A.8.6 in Appendix A)

$$\log[1-\overline{\sigma}(L(Re^{j\theta}))] = -\overline{\sigma}(L(Re^{j\theta})) - \frac{1}{2}\overline{\sigma}^2(L(Re^{j\theta})) + \cdots,$$
$$\log[1+\overline{\sigma}(L(Re^{j\theta}))] = \overline{\sigma}(L(Re^{j\theta})) - \frac{1}{2}\overline{\sigma}^2(L(Re^{j\theta})) + \cdots,$$

which hold for $|s| > R$ and $R > 0$ sufficiently large, we conclude that

$$\lim_{R\to\infty} R|\log[1-\overline{\sigma}(L(Re^{j\theta}))]| = 0,$$
$$\lim_{R\to\infty} R|\log[1+\overline{\sigma}(L(Re^{j\theta}))]| = 0,$$

and thus

$$\lim_{R\to\infty} I_3 = 0, \tag{B.10}$$

proving our claim. We next show that

$$\lim_{R\to\infty} I_4 = 0. \tag{B.11}$$

We do this by showing that $\lim_{R\to\infty} I_4 = -\lim_{R\to\infty} I_3$. First we write, using the Leibniz rule (see e.g., Kaplan 1973, p. 219)

$$\int_{-\pi/2}^{\pi/2} \frac{\partial}{\partial R}[\log \sigma_j(S(Re^{j\theta}))] \cos\theta \, d\theta = \frac{d}{dR}\int_{-\pi/2}^{\pi/2} \log \sigma_j(S(Re^{j\theta})) \cos\theta \, d\theta.$$

Then we have

$$\begin{aligned}
\lim_{R\to\infty} I_4 &= \lim_{R\to\infty} R^2 \frac{d}{dR}\int_{-\pi/2}^{\pi/2} \log \sigma_j(S(Re^{j\theta})) \cos\theta \, d\theta \\
&= -\lim_{R\to\infty} \frac{\frac{d}{dR}\int_{-\pi/2}^{\pi/2} \log \sigma_j(S(Re^{j\theta})) \cos\theta \, d\theta}{d(1/R)/dR} \\
&= -\lim_{R\to\infty} R\int_{-\pi/2}^{\pi/2} \log \sigma_j(S(Re^{j\theta})) \cos\theta \, d\theta \\
&= -\lim_{R\to\infty} I_3,
\end{aligned} \tag{B.12}$$

where the third equality holds from L'Hospital's rule (e.g., Widder 1961, p. 260). Thus, (B.11) follows on using (B.12) and (B.10).

Finally, combining (B.9)-(B.11), and observing that

$$\int_{-\infty}^{\infty} \log \sigma_j(S(j\omega))\, d\omega = 2\int_0^{\infty} \log \sigma_j(S(j\omega))\, d\omega,$$

due to conjugate symmetry of $S(j\omega)$, (B.4) is then obtained. □

It remains to establish Theorem 4.2.2 for open-loop unstable systems. This is done in the following theorem by using the factorization of S given in (B.3).

Theorem B.1.6 (Bode Integral for S – Unstable systems). Let S be factorized as in (B.3). Then, under Assumption B.1,

$$\int_0^{\infty} \log \sigma_j(S(j\omega))\, d\omega = F_j + K_j,$$

where

$$F_j = \frac{1}{2} \iint_{\overline{\mathbb{C}^+}} \sigma \nabla^2 \log \sigma_j(S_m(\sigma + j\omega))\, d\sigma\, d\omega, \qquad \text{and}$$

$$K_j = \lim_{R\to\infty} \int_{-\pi/2}^{\pi/2} R \log \sigma_j \left(\prod_{i=1}^{n_p} B_i^{-1}(Re^{j\theta}) \right) \cos\theta\, d\theta.$$

Proof. We use (the proof of) Theorem B.1.5 for the minimum-phase factor S_m in (B.3). If then follows from (B.9) and (B.12) that

$$\int_{-\infty}^{\infty} \log \sigma_j(S_m(j\omega))\, d\omega = 2F_j + 2 \lim_{R\to\infty} I_3, \tag{B.13}$$

where

$$I_3 = \int_{-\pi/2}^{\pi/2} R \log \sigma_j(S_m(Re^{j\theta})) \cos\theta\, d\theta.$$

Since each B_i in (B.3) is all-pass for $i = 1, \ldots, n_p$, we have that $S(j\omega)S^*(j\omega) = S_m(j\omega)S_m^*(j\omega)$, and hence

$$\sigma_j(S_m(j\omega)) = \sigma_j(S(j\omega)), \qquad j = 1, \ldots, n. \tag{B.14}$$

Also, for $|s| \geq R \geq \max_{1\leq i\leq n_p} |p_i|$, both $B_i^{-1}(s)$ and $S(s)$ are well-defined. By using norm inequalities (see e.g., Golub & Van Loan 1983), we have

$$\underline{\sigma}(S(Re^{j\theta}))\sigma_j(\Pi) \leq \sigma_j(S_m(j\omega)) \leq \overline{\sigma}(S(Re^{j\theta}))\sigma_j(\Pi),$$

where

$$\Pi \triangleq \prod_{i=1}^{n_p} B_i^{-1}(Re^{j\theta}).$$

This leads to

$$I_3 \geq \int_{-\pi/2}^{\pi/2} R \log \underline{\sigma}(S(Re^{j\theta})) \cos\theta\, d\theta + \int_{-\pi/2}^{\pi/2} R \log \sigma_j(\Pi) \cos\theta\, d\theta,$$

and

$$I_3 \le \int_{-\pi/2}^{\pi/2} R \log \overline{\sigma}(S(Re^{j\theta})) \cos\theta \, d\theta + \int_{-\pi/2}^{\pi/2} R \log \sigma_j (\Pi) \cos\theta \, d\theta.$$

As shown in the proof of Theorem B.1.5, however, the first terms on the RHSs of the two inequalities above tend to zero as $R \to \infty$, and thus

$$\lim_{R\to\infty} I_3 = \lim_{R\to\infty} \int_{-\pi/2}^{\pi/2} R \log \sigma_j \left(\prod_{i=1}^{n_p} B_i^{-1}(Re^{j\theta}) \right) \cos\theta \, d\theta. \tag{B.15}$$

The proof is completed by substituting (B.14) and (B.15) into (B.13), and by using the conjugate symmetry of $S(j\omega)$. □

B.2 Proofs for Chapter 6

B.2.1 Proof of Lemma 6.2.2

In this subsection, we provide the proof of Lemma 6.2.2 on steady-state frequency response of sampled-data systems. First, we need two preliminary lemmas.

Lemma B.2.1. Suppose that h is the pulse response of a hold device as described on page 138 of §6.1.2, i.e., h is a function of bounded variation with finite support on the interval $[0, \tau]$; let $H = \mathcal{L}h$ be its frequency response function. Then, there exist finite constants c_1 and c_2 such that

$$|s\,H(s)| \le c_1 + c_2 e^{-\operatorname{Re} s\tau} \quad \text{if } s \text{ is in } \mathbb{C}^- , \tag{B.16}$$

$$|s\,H(s)| \le c_2 + c_1 e^{-\operatorname{Re} s\tau} \quad \text{if } s \text{ is in } \overline{\mathbb{C}^+} . \tag{B.17}$$

Proof. Using the definition of H and integration by parts, we can write

$$\begin{aligned} |s\,H(s)| &= \left| \int_0^\tau s\, e^{-s\zeta} h(\zeta)\, d\zeta \right| \\ &= \left| h(0^+) - e^{-s\tau} h(\tau^-) + \int_0^\tau e^{-s\zeta} \dot{h}(\zeta)\, d\zeta \right| \\ &\le \|h\|_\infty \left(1 + e^{-\operatorname{Re} s\tau}\right) + \int_0^\tau e^{-\operatorname{Re} s\zeta} \left|\dot{h}(\zeta)\right| d\zeta , \end{aligned} \tag{B.18}$$

where $\|h\|_\infty = \sup_{[0,\tau]} |h(t)|$, and $\dot{h}$ denotes dh/dt, which exists almost everywhere on $[0, \tau]$ because h is of bounded variation. If s is in $\mathbb{C}^-$, then $e^{-\operatorname{Re} s\zeta} \le e^{-\operatorname{Re} s\tau}$ for $\zeta \in [0, \tau]$, and we have from (B.18) that

$$|s\,H(s)| \le \|h\|_\infty + \left(\|h\|_\infty + \|\dot{h}\|_1 \right) e^{-\operatorname{Re} s\tau} ,$$

where $\|\dot{h}\|_1 = \int_0^\tau |\dot{h}(\zeta)| d\zeta$, which is finite also because h is assumed of bounded variation (e.g., see Rudin (1987, p. 157)). The bound (B.16) then follows. If s is in $\overline{\mathbb{C}^+}$, then (B.17) follows from (B.18) on using the bound $e^{-\operatorname{Re} s\zeta} \leq 1$ for $\zeta \in [0, \tau]$. □

Lemma B.2.2. Suppose that G is a proper transfer function, which may include a time delay, and let H be the frequency response of a hold, as in Lemma B.2.1. Let ρ be a finite number in $\mathbb{C}^-$ that is not a pole of G, and define $\rho_k \triangleq \rho + jk\omega_s$, with $k = \pm 1, \pm 2, \ldots$ Assume further that ρ_k is not a pole of G for any k. Then,

$$\lim_{K \to \infty} \sum_{k=-K}^{K} |G(\rho_k)H(\rho_k)|^2 < \infty . \tag{B.19}$$

Proof. Since G is assumed to be proper, then $|G(\rho_k)|$ converges to a finite constant as $k \to \infty$. Denote this constant by M_G. Using this bound, and B.16 from Lemma B.2.1 yields

$$\lim_{K \to \infty} \sum_{k=-K}^{K} |G(\rho_k)H(\rho_k)|^2 \leq M_G^2 (c_1 + c_2 e^{-\rho\tau})^2 \sum_{k=-\infty}^{\infty} \frac{1}{|\rho_k|^2} < \infty ,$$

and the result follows. □

Proof of Lemma 6.2.2. We consider only the disturbance response, calculations for the noise response follow similarly. For convenience, we recall the expression for Y^d from (6.9) in Chapter 6, which we rewrite as

$$Y^d(s) = D(s) - G(s)H(s)K_d(e^{s\tau})S_d(e^{s\tau})V_d(e^{s\tau}), \tag{B.20}$$

where $V_d(e^{s\tau})$ is given by (cf. the proof of Lemma 6.2.1)

$$V_d(e^{s\tau}) = \frac{1}{\tau} \sum_{k=-\infty}^{\infty} F_k(s) D_k(s).$$

To evaluate the steady-state response to $d(t) = e^{j\omega t}$, we must first evaluate the inverse Laplace transform of Y^d, and then discard all terms due to those poles lying in $\mathbb{C}^-$. Inverting the Laplace transform requires that we evaluate the Bromwich integral (Levinson & Redheffer 1970)

$$y^d(t) = \frac{1}{2\pi j} \int_{\gamma - j\infty}^{\gamma + j\infty} e^{st} Y^d(s) \, ds , \tag{B.21}$$

where $\gamma > 0$. This integral may be evaluated using the residue theorem (Theorem A.9.1 in Appendix A).

It follows from (B.20) that Y^d has poles due to the disturbance located along the imaginary axis at $s = j(\omega + k\omega_s)$, $k = 0, \pm 1, \pm 2, \ldots$. By the assumption of closed loop stability all other poles of Y^d lie in $\mathbb{C}^-$. Using (B.20), it may be shown that these poles have the following properties:

(i) they all lie to the right of some vertical line $\operatorname{Re} s = c < 0$,

(ii) there are finitely many poles due to G and no poles due to H,

(iii) there are finitely many sequences of poles due to $K_d(e^{s\tau})$, $S_d(e^{s\tau})$, and $F(s + jk\omega_s)$, $k = 0, \pm 1, \pm 2, \ldots$ lying on vertical lines and spaced at intervals equal to ω_s.

Next, it is straightforward to verify from (B.20) and Definition 6.2.1 that the residues of $e^{st}Y^d$ at the $j\omega$-axis poles are given by

$$\lim_{s \to j(\omega + k\omega_s)} \big(s - j(\omega + k\omega_s)\big) e^{st} Y^d(s) = \begin{cases} S^0(j\omega)e^{j\omega t} & \text{if } k = 0\,, \\ -T_k(j\omega)e^{j(\omega + k\omega_s)t} & \text{if } k \neq 0\,. \end{cases} \tag{B.22}$$

We need not calculate explicitly the residues at the other poles; as we will show, they do not contribute to the steady-state response.

Consider the contours of integration C_n, $n = 1, 2, 3, \ldots$ depicted in Figure B.2, and chosen so that (i) C_1 encloses only that $j\omega$-axis pole lying in Ω_N, (ii) the horizontal line $\operatorname{Im} s = R_1$ does not contain any OLHP poles of Y^d, and (iii) $R_{n+1} = R_n + \omega_s$.

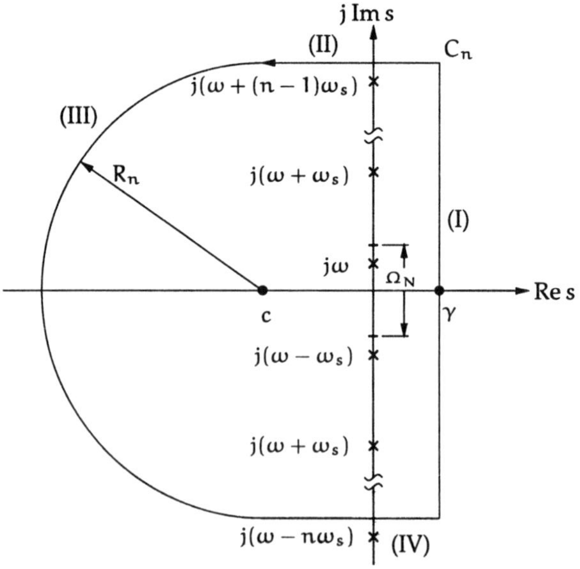

FIGURE B.2. Contours of integration.

Figure B.2 and subsequent calculations are appropriate for the case that ω is in Ω_N (modifications to the general case are straightforward). Our construction of the contour of integration guarantees that for n sufficiently large no poles of Y^d will lie on C_N. Hence the residue theorem may be applied to yield

$$\frac{1}{2\pi j}\left\{\int_I e^{st}Y^d(s)\,ds + \int_{II} e^{st}Y^d(s)\,ds + \int_{III} e^{st}Y^d(s)\,ds + \int_{IV} e^{st}Y^d(s)\,ds\right\}$$
$$= S^0(j\omega)e^{j\omega t} - \sum_{\substack{k=-n \\ k\neq 0}}^{n} T_k(j\omega)e^{j(\omega+k\omega_s)t} + \Psi(t)\,, \tag{B.23}$$

where $\Psi(t)$ denotes the contribution of the poles in $\mathbb{C}^-$.

We now sketch a proof that as $t \to \infty$, $\Psi(t) \to 0$. First, it is clear that the contribution to Ψ from each pole of G converges to zero. Consider next the contribution of one of the finitely many sequences of poles described in (iii) above. Let this sequence be denoted $\rho_k \triangleq \rho + jk\omega_s$, $k = 0, \pm 1, \pm 2, \ldots$, and $\operatorname{Re}\rho < 0$. We will assume that ρ is real for notational simplicity, and will also assume for simplicity that each ρ_k is a simple pole. Then, for any fixed value of t, the contribution to Ψ from this sequence of poles is given by

$$y_\rho(t) \triangleq e^{\rho t} \lim_{K\to\infty} \sum_{k=-K}^{K} \operatorname*{Res}_{s=\rho_k} Y^d(s)e^{jk\omega_s t}\,, \tag{B.24}$$

where $\operatorname{Res}_{s=\rho_k} Y^d(s) = \lim_{s\to\rho_k}(s-\rho_k)Y^d(s)$ (see §A.9 in Appendix A). From (B.20) we have

$$\operatorname*{Res}_{s=\rho_k} Y^d(s) = -G(\rho_k)H(\rho_k) \lim_{s\to\rho_k}(s-\rho_k)K_d(e^{sT})S_d(e^{sT})V_d(e^{sT})\,. \tag{B.25}$$

Because $K_d(e^{sT})$, $S_d(e^{sT})$, and $V_d(e^{sT})$ are each periodic in s along vertical lines, it may be shown that the limit on the RHS of (B.25) is independent of k. Denote the common value of this limit by $-L_\rho$. Then (B.24) becomes

$$y_\rho(t) = e^{\rho t}L_\rho \lim_{K\to\infty} \sum_{k=-K}^{K} G(\rho_k)H(\rho_k)e^{jk\omega_s t}\,. \tag{B.26}$$

By Lemma B.2.2, the sequence $\{G(\rho_k)H(\rho_k)\}$ is square-summable. Therefore, by the Riesz-Fischer Theorem (Riesz & Sz.-Nagy 1990, p.70), the series in (B.26) converges to a bounded periodic function of t. Since $\rho < 0$, it thus follows that $y_\rho(t) \to 0$ as $t \to \infty$. Since there are only finitely many sequences of the form (B.26), we then have that $\Psi(t) \to 0$.

The desired result (6.17) will hold if it may be shown that the last three integrals on the LHS of (B.23) converge to zero as $n \to \infty$. We now show that the integral (II) converges to zero; similar calculations apply to (IV). Consider values of s such that $s = \sigma + jR_n$, $c \le \sigma \le \gamma$, and R_n is sufficiently large that $R_n > \omega$ and that C_n encloses all poles of G. It may be shown that there exist constants M and M_G, independent of n, such that $|K_d(e^{s\tau})S_d(e^{s\tau})V_d(e^{s\tau})| \leq M$ and $|G(s)| \leq M_G$ for all such s. Also, it follows from Lemma B.2.1 that, for $t \geq \tau$, there exists some constant c_3 such that

$$|se^{st}H(s)| \leq c_3 e^{\gamma t}, \quad \text{for all } s \text{ with } \operatorname{Re} s \leq \gamma. \tag{B.27}$$

Using these bounds in conjunction with (B.20) and the fact that $D(s) = 1/(s - j\omega)$, yields

$$|e^{st}Y^d(s)| \leq \frac{e^{\gamma t}(1 + MM_G c_3)}{R_n - \omega}. \tag{B.28}$$

Using (B.28) in integral (II) yields

$$\left|\int_{II} e^{st}Y^d(s)\, ds\right| \leq \frac{(\gamma - c)e^{\gamma t}(1 + MM_G c_3)}{R_n - \omega},$$

which converges to zero as $R_n \to \infty$.

It remains to show that the integral (III) converges to zero as $R_n \to \infty$. Parametrize (III) by $s = c + R_n e^{j\theta}$, with $\pi/2 \leq \theta \leq 3\pi/2$, and define $\xi = s - c$; contour (III) is then a semicircle φ_n centered at the origin of the ξ-plane and extended into the left half plane. We then have that

$$\begin{aligned}\left|\int_{III} e^{st}\, Y^d(s)\, ds\right| &= \left|\int_{\varphi_n} e^{ct}\, e^{\xi(t-\tau)}\, e^{\xi\tau}\, Y^d(\xi + c)\, d\xi\right| \\ &\leq e^{ct} \int_{\varphi_n} \left|e^{\xi\tau}\, Y^d(\xi + c)\right| \left|e^{\xi(t-\tau)}\, d\xi\right|. \end{aligned} \tag{B.29}$$

Now, following similar steps to those used to obtain (B.28), $\left|e^{\xi\tau}\, Y^d(\xi + c)\right|$ can be bounded on φ_n as

$$\left|e^{\xi\tau}\, Y^d(\xi + c)\right| \leq \frac{e^{(\gamma - c)\tau}(1 + MM_G c_3)}{R_n - \omega},$$

This bound and the application of Jordan's Lemma (Lemma A.4.1 from Appendix A) to obtain a bound on the integral $\int_{\varphi_n} |e^{\xi(t-\tau)}\, d\xi|$, for $t > \tau$, gives, from (B.29),

$$\left|\int_{III} e^{st}\, Y^d(s)\, ds\right| \leq \frac{e^{\gamma t}(1 + MM_G c_3)\pi}{(R_n - \omega)(t - \tau)},$$

which converges to zero as $R_n \to \infty$, concluding the proof. □

The following subsections give the proofs of the results in §6.2.2 on robust stability of sampled-data systems.

B.2.2 *Proof of Lemma 6.2.4*

It is necessary for closed loop stability that

$$\tilde{S}_d(z) = [I + K_d(z)(F\tilde{G}H)_d(z)]^{-1} \tag{B.30}$$

have no poles in $\mathbb{D}^c$. Rearranging yields

$$\tilde{S}_d(z) = [I + S_d(z)K_d(z)(FW\Delta GH)_d(z)]^{-1}S_d(z). \tag{B.31}$$

Since the nominal system is stable, $\tilde{S}_d$ will have no poles in $\mathbb{D}^c$ if and only if

$$\det[I + S_d(e^{j\omega\tau})K_d(e^{j\omega\tau})(FW\Delta GH)_d(e^{j\omega\tau})] \neq 0 \quad \text{for all } \omega. \tag{B.32}$$

The proof proceeds by contradiction. We follow the argument used in Chen & Desoer (1982, Theorem 2). Denote $Q(j\omega) \triangleq T^0(j\omega)W(j\omega)$, and suppose that (6.23) is violated. Then there exists a frequency ω_1 such that $\sigma_1 \triangleq \overline{\sigma}(Q(j\omega_1)) > 1$, where $\overline{\sigma}(\cdot)$, recall, denotes the maximum singular value. Performing a singular value decomposition of $Q(j\omega_1)$ yields

$$Q(j\omega_1) = U\,\mathrm{diag}[\sigma_1 \ldots]\,V^*,$$

where $U \triangleq \{u_{ij}\}$ and $V \triangleq \{v_{ij}\}$ are unitary matrices. Now assume for the moment that there exists an admissible $\hat{\Delta}$ that also satisfies

$$\begin{aligned}\hat{\Delta}(j\omega_1) &= \begin{bmatrix} v_{11} \\ \vdots \\ v_{n1} \end{bmatrix} (-\sigma_1)^{-1} \begin{bmatrix} u_{11}^* & \ldots & u_{n1}^* \end{bmatrix} \\ &= V\,\mathrm{diag}[(-\sigma_1)^{-1}, 0, \ldots, 0]\,U^*,\end{aligned} \tag{B.33}$$

and

$$\hat{\Delta}(j(\omega_1 + k\omega_s)) = 0 \quad \text{for } k = \pm 1, \pm 2, \ldots, \text{and } k \neq -2\omega_1/\omega_s. \tag{B.34}$$

The assumptions on W, and Δ, imply that a formula similar to (6.7) may be used to calculate $(FW\hat{\Delta}GH)_d$. Using (B.33) and (B.34) yields

$$\begin{aligned}(FW\hat{\Delta}GH)_d(e^{j\omega_1\tau}) = &-\frac{1}{\tau}F(j\omega_1)W(j\omega_1)V \\ &\times \mathrm{diag}[-\frac{1}{\sigma_1}, 0, \ldots, 0]\,U^*G(j\omega_1)H(j\omega_1),\end{aligned} \tag{B.35}$$

and therefore[1]

$$\begin{aligned}
&\det[I + S_d(e^{j\omega_1\tau})K_d(e^{j\omega_1\tau})(FW\hat{\Delta}GH)_d(e^{j\omega_1\tau})] \\
&\qquad = \det[I + S_d K_d \frac{1}{\tau} FWV \operatorname{diag}[(-\sigma_1)^{-1}, 0, \ldots, 0]\, U^* GH] \\
&\qquad = \det[I + V \operatorname{diag}[(-\sigma_1)^{-1}, 0, \ldots, 0]\, U^* \frac{1}{\tau} GHS_d K_d FW] \\
&\qquad = \det[I + V \operatorname{diag}[(-\sigma_1)^{-1}, 0, \ldots, 0]\, U^* Q(j\omega_1)] \\
&\qquad = [I + V \operatorname{diag}[-1, 0, \ldots, 0]\, V^*] \\
&\qquad = \det[V] \det[\operatorname{diag}[0, 1, 1, \ldots, 1]] \det[V^*] \\
&\qquad = 0.
\end{aligned}$$

Hence, (B.32) fails and so the perturbed system is unstable.

It remains to show that $\hat{\Delta}$ satisfying the required properties exists. We do this following a construction used in Chen & Desoer (1982). Consider

$$\hat{\Delta}(s) \triangleq \begin{bmatrix} \alpha_1(s) \\ \vdots \\ \alpha_n(s) \end{bmatrix} \left(-\frac{1}{\sigma_1}\right) f_q(s)^{k'} z(s) \left[\beta_1(s), \ldots, \beta_n(s)\right],$$

where k' is a natural number, and

$$\begin{aligned}
f_q(s) &\triangleq \frac{\omega_1 s}{q(s^2 + \omega_1^2) + \omega_1 s}, \quad q > 0, \\
\alpha_i(s) &\triangleq \frac{s}{\omega_1} \operatorname{Im} v_{i1} + \operatorname{Re} v_{i1}, \\
\beta_i(s) &\triangleq -\frac{s}{\omega_1} \operatorname{Im} u_{i1} + \operatorname{Re} u_{i1}, \\
z(s) &\triangleq \frac{H_{ZOH}(s - j\omega_1) H_{ZOH}(s + j\omega_1)}{\tau |H_{ZOH}(j2\omega_1)|} \eta(s), \\
\eta(s) &\triangleq \left(-\frac{s}{\omega_1} \sin(\sphericalangle H_{ZOH}(j2\omega_1)) + \cos(\sphericalangle H_{ZOH}(j2\omega_1))\right),
\end{aligned}$$

and where $H_{ZOH}(s)$ is the frequency response function of the ZOH, and $\sphericalangle$ denotes the phase of a complex number. It is then straightforward to verify that

(i) $\hat{\Delta}(j\omega_1)$ satisfies (B.33) and (B.34), and

(ii) by choosing both k' and q large enough, $\hat{\Delta}$ is exponentially stable and, for all $\omega \neq \pm\omega_1$, $\lim_{\omega\to\infty} \overline{\sigma}(\hat{\Delta}(j\omega)) \to 0$, i.e., $\|\hat{\Delta}\|_\infty < 1$ is satisfied.

□

[1] We suppress dependence on the transform variable when convenient and where the meaning is clear from the context.

B.2.3 *Proof of Lemma 6.2.5*

The proof follows the same lines of that of Lemma 6.2.4 after noting that we can alternatively write the perturbed discrete sensitivity function as

$$\begin{aligned}\tilde{S}_d &= \left[1 + K_d(F\tilde{G}H)_d\right]^{-1} \\ &= \left[1 + K_d(FGH)_d - K_d\left(\frac{F\Delta WGH}{1+W\Delta}_d\right)\right]^{-1} \\ &= \left[1 - S_d K_d\left(\frac{F\Delta WGH}{1+W\Delta}_d\right)\right]^{-1} S_d. \end{aligned} \tag{B.36}$$

That the nonsingularity of the term between brackets in (B.36) implies (6.25) may be shown by an argument by contradiction, similar to that used in the proof of Lemma 6.2.4, and is omitted to avoid repetition. □

Appendix C
The Laplace Transform of the Prediction Error

Throughout Chapter 10 we have used a modified Laplace transform that shifts the lower limit of integration from $t = 0$ to $t = -\tau$, i.e., the *shifted* Laplace transform of a function $h(t)$ was defined to be

$$H(s) \triangleq \int_{-\tau}^{\infty} e^{-st} h(t) dt.$$

Obviously, this modification leaves unchanged the transforms of signals starting at $t = 0$, but it over-evaluates the transform of the shifted state $x(t+\tau)$. Despite this overmeasure, the results given in Chapter 10 provide a good indication of the performance limits in prediction.

In this appendix, we will obtain the conventional Laplace transform (denoted by $\mathcal{L}$) of the estimation error and point out the difficulties of using this version in sensitivity analysis.

We recall the prediction error from (10.5) (for full-state prediction, i.e., $z = x$ in (10.1)),

$$\tilde{x}(t + \tau|t) = x(t + \tau) - e^{A\tau}\hat{x}(t).$$

Taking the Laplace transform, we have

$$\tilde{X}(s) = e^{s\tau}\left[X(s) - \int_0^{\tau} x(t)e^{-st} dt\right] - e^{A\tau}\hat{X}(s).$$

The second term above was not considered in our analysis in Chapter 10. We will see that this second term cannot be written as affine in the transforms of the input signals.

The contribution to the prediction error due to the process input ν is given by[1]

$$\tilde{X}_\nu(s) = e^{s\tau}\left[H_{x\nu}(s)V(s) - \int_0^\tau x_\nu(t)e^{-st}dt\right] - e^{A\tau}H_{\hat{x}\nu}(s)V(s),$$

where $x_\nu(t) = \int_0^t e^{A(t-\theta)}B\nu(\theta)d\theta$, $0 \le t \le \tau$, and $V = \mathcal{L}\nu$. The above equation can be expressed as

$$\begin{aligned}\tilde{X}_\nu(s) &= e^{s\tau}\left[H_{x\nu}(s)V(s) - \int_0^\infty x_\nu(t)h_\tau(t)e^{-st}dt\right] - e^{A\tau}H_{\hat{x}\nu}(s)V(s)\\ &= e^{s\tau}\left[H_{x\nu}(s)V(s) - \mathcal{L}[x_\nu(t)h_\tau(t)](s)\right] - e^{A\tau}H_{\hat{x}\nu}(s)V(s)\end{aligned} \tag{C.1}$$

where h_τ is the pulse function given by

$$h_\tau(t) = \begin{cases} 1 & \text{if } 0 \le t \le \tau,\\ 0 & \text{otherwise.}\end{cases}$$

Solving the transform of the real multiplication in (C.1), we further obtain

$$\tilde{X}_\nu(s) = e^{s\tau}\left[H_{x\nu}(s)V(s) - [H_{x\nu}(s)V(s)] \otimes H_\tau(s)\right] - e^{A\tau}H_{\hat{x}\nu}(s)V(s), \tag{C.2}$$

where $H_\tau = \mathcal{L}h_\tau$, and where '$\otimes$' denotes complex convolution (Gardner & Barnes 1949), i.e.,

$$\begin{aligned}[H_{x\nu}(s)V(s)] \otimes H_\tau(s) &\triangleq \int_{c-j\omega}^{c+j\omega} H_{x\nu}(\theta)V(\theta)H_\tau(s-\theta)d\theta\\ &= \int_{c-j\omega}^{c+j\omega} H_{x\nu}(\theta)V(\theta)\frac{1-e^{-(s-\theta)\tau}}{s-\theta}d\theta.\end{aligned} \tag{C.3}$$

The real constant c in the integration limits of (C.3) is chosen such that $\sigma_\nu < c < \infty$, where σ_ν is the abscissa of absolute convergence of $H_{x\nu}(s)V(s)$.[2]

From (C.2), we can define a "transfer function" (although not in the usual multiplicative form) from the process input to the prediction error as

$$H_{\tilde{x}\nu}(s) \triangleq e^{s\tau}\left[H_{x\nu}(s) - K_\tau[\cdot](s)\right] - e^{A\tau}H_{\hat{x}\nu}(s), \tag{C.4}$$

[1] In the sequel, the symbol H_{ba} will stand for the mapping from signal a to signal b.

[2] Since $H_\tau(s)$ is an entire function, the integral in (C.3) converges absolutely in the half plane $\mathrm{Re}(s) > \sigma_\nu$.

where the linear operator

$$K_\tau[v](s) \triangleq [H_{xv}(s)V(s)] \otimes H_\tau(s)$$

is clearly not a multiplication operator. With the expression for $H_{\tilde{x}v}$ given in (C.4), the corresponding prediction sensitivities would be defined as

$$P \triangleq e^{-s\tau} H_{\tilde{x}v} \left(H_{xv} - K_\tau[\cdot] \right)^{-1},$$

$$M \triangleq e^{-s\tau} H_{\hat{x}v} \left(H_{xv} - K_\tau[\cdot] \right)^{-1}.$$

It is clear from the above definitions that the derivation of interpolation and integral constraints for this version of the prediction sensitivities is nontrivial. Moreover, it cannot be performed with the tools developed in this book.

Appendix D

Least Squares Smoother Sensitivities for Large τ

In this appendix, we derive an expression for the smoothing sensitivities of Chapter 11 that holds for large values of the smoothing lag.

Consider, in system (11.1) of Chapter 11, that $D_1 = D_2 = 0$, i.e.,

$$\begin{aligned} \dot{x} &= Ax + Bv, \\ z &= C_1 x, \\ y &= C_2 x + w, \end{aligned} \tag{D.1}$$

where v and w are uncorrelated white noises with incremental covariances equal to Q and R, respectively.

Let the Kalman filter for the above system be given by

$$\begin{aligned} \dot{\hat{x}} &= \hat{A}\hat{x} + K_y y, \\ \hat{z} &= C_1 \hat{x}, \end{aligned} \tag{D.2}$$

where

$$\begin{aligned} \hat{A} &= A - K_y C_2, \\ K_y &= \Phi C_2' R^{-1}, \end{aligned} \tag{D.3}$$

and $\Phi \geq 0$ is a stabilizing solution of the Riccati equation (11.6), reproduced here for convenience

$$A\Phi + \Phi A' + BQB' - \Phi C_2' R^{-1} C_2 \Phi = 0.$$

Let F be the Laplace transform from y to $\hat{x}$ for the Kalman filter (D.2), i.e.,

$$F(s) = (sI - \hat{A})^{-1} K_y.$$

Using (D.3) and the matrix inversion lemma (e.g., Kailath 1980, p. 656) we can further write F as

$$F(s) = (sI - A)^{-1}K_y[I - C_2(sI - \hat{A})^{-1}K_y]. \tag{D.4}$$

Consider next the least squares smoother (11.7) of Example 11.1.1. We then have the following result.

Lemma D.0.3. Consider the least squares smoother of Example 11.1.1 for the system given in (D.1). Assume that the solution, Φ, of the Riccati equation (11.6) is positive definite. Then, for $\tau = \bar{\tau} \gg \tau_{max}(\hat{A})$, where $\tau_{max}(\hat{A})$ is the dominant time constant[1] of the Kalman filter, the smoothing sensitivities (11.17) can be approximated by

$$\begin{aligned} M_{[\bar{\tau}]}(s) &= H_{zv}(s)QH'_{\iota v}(-s)R^{-1}H_{\iota v}(s)H_{zv}^{-1}(s), \\ P_{[\bar{\tau}]}(s) &= I - H_{zv}(s)QH'_{\iota v}(-s)R^{-1}H_{\iota v}(s)H_{zv}^{-1}(s), \end{aligned} \tag{D.5}$$

where $H_{\iota v}$ is the transfer function from the process input, v, to the innovations process $\iota = y - C_2\hat{x}$ corresponding to the Kalman filter (D.2), and given by

$$H_{\iota v} = C_2(sI - \hat{A})^{-1}B.$$

Proof. For the least squares smoother of Example 11.1.1, the expression of the smoother function (11.10) is

$$\begin{aligned} H_s &= C_1\Phi(sI + \hat{A}')^{-1}\left[e^{\tau\hat{A}'} - e^{-s\tau}I\right]C_2'R^{-1} \\ &= C_1\Phi(sI + \hat{A}')^{-1}e^{\tau\hat{A}'}C_2'R^{-1} - e^{-s\tau}\Phi(sI + \hat{A}')^{-1}C_2'R^{-1}. \end{aligned}$$

Note that the expression $e^{\tau\hat{A}'}C_2'R^{-1}$ can be thought of as the state of the filter at time $t = \tau$, in response to the initial conditions given by the columns of $C_2'R^{-1}$. Then, for $\tau = \bar{\tau} \gg \tau_{max}(\hat{A})$, where $\tau_{max}(\hat{A})$ is the dominant time constant of the Kalman filter, $e^{\bar{\tau}\hat{A}'}C_2'R^{-1}$ will be negligible compared to the value of $C_2'R^{-1}$. Hence, for $\tau = \bar{\tau}$, we may neglect the first term on the RHS above and assume that H_s is of the form

$$H_{s[\bar{\tau}]} = -e^{-s\bar{\tau}}C_1\Phi(sI + \hat{A}')^{-1}C_2'R^{-1}. \tag{D.6}$$

For convenience, we will use the Riccati equation (11.6) to express $\hat{A}'$ in (D.3) as

$$\begin{aligned} \hat{A}' &= A' - C_2'R^{-1}C_2\Phi \\ &= -\Phi^{-1}(A + BQB'\Phi^{-1})\Phi \\ &\triangleq -\Phi^{-1}\bar{A}\Phi, \end{aligned}$$

[1]Recall that the dominant time constant of a stable system can be taken as the inverse of the real part of the eigenvalue with smallest magnitude for the real part.

with the obvious definition for $\bar{A}$. Hence

$$(sI + \hat{A}')^{-1} = \Phi^{-1}(sI - \bar{A})^{-1}\Phi. \tag{D.7}$$

Using (D.7) in (D.6), we have

$$H_{s[\tau]} = -e^{-s\tau}C_1(sI - \bar{A})^{-1}K_y, \tag{D.8}$$

where $K_y = \Phi C_2' R^{-1}$ is the Kalman filter gain given in (D.3).

Next, recall from (11.21) the expression for the smoothing complementary sensitivity, M

$$M = e^{s\tau}F_s H_{yv} H_{zv}^{-1},$$

where $H_{zv} = G_z$ and $H_{yv} = G_y$. Replacing F_s using (11.14), we have

$$M = [C_1F + e^{s\tau}H_s(I - C_2F)]H_{yv}H_{zv}^{-1}, \tag{D.9}$$

where F is given in (D.4). Using (D.4) and (D.8) in (D.9), we have, for $\tau \gg \tau_{max}(\hat{A})$,

$$M_{[\tau]} = C_1\left[(sI - A)^{-1} - (sI - \bar{A})^{-1}\right]K_y[I - C_2(sI - \hat{A})^{-1}K_y]H_{yv}H_{zv}^{-1}. \tag{D.10}$$

Next note that, using (D.3) and $H_{yv} = C_2(sI - A)^{-1}B$, we can write

$$\begin{aligned}[I - C_2(sI - \hat{A})^{-1}K_y]H_{yv} &= C_2(sI - \hat{A})^{-1}B \\ &= H_{\iota v},\end{aligned}$$

where $H_{\iota v}$ is the transfer function from the process input, v, to the innovations process, ι. Using this expression in (D.10), we have

$$M_{[\tau]} = C_1\left[(sI - A)^{-1} - (sI - \bar{A})^{-1}\right]K_y H_{\iota v} H_{zv}^{-1}.$$

Using $\bar{A} = A + BQB'\Phi^{-1}$ and replacing K_y from (D.3) in the above equation, and rearranging, yields

$$M_{[\tau]} = -C_1(sI - A)^{-1}BQB'\Phi^{-1}(sI - \bar{A})^{-1}\Phi C_2' R^{-1} H_{\iota v} H_{zv}^{-1}.$$

Using $H_{zv} = C_1(sI - A)^{-1}B$, and (D.7) again, we further get

$$\begin{aligned}M_{[\tau]} &= -H_{zv}QB'(sI + \hat{A}')^{-1}C_2'R^{-1}H_{\iota v}H_{zv}^{-1} \\ &= H_{zv}Q[C_2(-sI - \hat{A})^{-1}B]'R^{-1}H_{\iota v}H_{zv}^{-1} \\ &= H_{zv}(s)QH_{\iota v}'(-s)R^{-1}H_{\iota v}(s)H_{zv}^{-1}(s),\end{aligned} \tag{D.11}$$

where we have explicitly indicated the arguments s or $-s$ to avoid confusion. This establishes (D.5) for M. By complementarity, the result for P is also as given in (D.5). The result then follows. □

References

Anderson, B. (1969), 'Properties of optimal linear smoothing', *IEEE Trans. on Automatic Control* **14**, 114–115.

Anderson, B. & Chirarattananon, S. (1971), 'Smoothing as an improvement on filtering: A universal bound', *Electronics Letters* **7**(18), 524–525.

Anderson, B. & Doan, H. (1977), 'Design procedures for stable suboptimal fixed-lag smoothers', *IEEE Trans. on Automatic Control* **22**(6), 949–952.

Anderson, B. & Moore, J. (1979), *Optimal Filtering*, Prentice Hall, Englewood Cliffs, NJ.

Araki, M. (1993), Recent developments in digital control theory, *in* 'Proceedings of the 12th IFAC World Congress', Vol. IX, Sydney, Australia, pp. 251–260.

Araki, M., Ito, Y. & Hagiwara, T. (1993), Frequency response of sampled-data systems, *in* 'Proceedings of the 12th IFAC World Congress', pp. VII–289. Also *Automatica*, 32(4), pp. 483-497, 1996.

Astolfi, A. & Guzzella, L. (1993), Robust control of nonlinear systems. An H_∞ approach, *in* 'Proceedings of the 12th IFAC World Congress', Vol. IX, Sydney, Australia, pp. 281–284.

Åström, K. (1995), 'Fundamental limitations of control system performance', Preprint.

Åström, K. & Wittenmark, B. (1990), *Computer-Controlled Systems: Theory and Design*, 2nd edn, Prentice Hall, Englewood Cliffs NY.

Bernstein, D. & Haddad, W. (1989), 'Steady-state Kalman filtering with an H_∞ error bound', *Systems and Control Letters* **12**, 9–16.

Bhattacharyya, S. (1976), 'The structure of robust observers', *IEEE Trans. on Automatic Control* **21**, 581–588.

Björck, Å. & Golub, G. (1973), 'Numerical methods for computing angles between linear subspaces', *Math. Computation* **27**(123), 579–594.

Bode, H. (1945), *Network Analysis and Feedback Amplifier Design*, D. van Nostrand, New York.

Bower, J. & Schultheiss, P. (1958), *Introduction to the Design of Servomechanisms*, John Wiley & Sons.

Boyd, S. & Desoer, C. (1985), 'Subharmonic functions and performance bounds on linear time-invariant feedback systems', *IMA Journal of Math. Control and Information* pp. 153–170.

Braslavsky, J. (1995), Frequency domain analysis of sampled-data control systems, PhD thesis, Department of Electrical and Computer Engineering, The University of Newcastle, NSW 2308 Australia.

Braslavsky, J., Meinsma, G., Middleton, R. & Freudenberg, J. (1995), On a key sampling formula relating the Laplace and Z-transforms, Technical Report EE9524, Dept. of Elec. & Comp. Eng., The Univ. of Newcastle, Australia.

Braslavsky, J., Middleton, R. & Freudenberg, J. (1995), Frequency response of generalized sampled-data hold functions, *in* 'Proceedings of the 34th CDC', New Orleans, LO.

Braslavsky, J., Seron, M., Goodwin, G. & Grainger, R. (1996), Tradeoffs in multivariable filter design with applications to fault detection, *in* 'Proceedings of the 35th CDC', Kobe, Japan.

Caines, P. (1988), *Linear Stochastic Systems*, Wiley & Sons, New York.

Carlson, A. (1975), *Communication Systems. An Introduction to Signals and Noise in Electrical Communication*, 2nd. edn, McGraw-Hill Book Co.

Chen, J. (1995), 'Sensitivity integral relations and design trade-offs in linear multivariable feedback systems', *IEEE Trans. on Automatic Control* **40**(10), 1700–1716.

Chen, J. & Nett, C. (1995), 'Sensitivity integrals for multivariable discrete-time systems', *Automatica* **31**(8), 113–1124.

Chen, M. & Desoer, C. (1982), 'Necessary and sufficient condition for robust stability of linear distributed feedback systems', *Int. J. of Control* **35**(2), 255–267.

Chen, T. & Francis, B. (1991), 'Input-output stability of sampled-data systems', *IEEE Trans. on Automatic Control* **36**(1), 50–58.

Chen, T. & Francis, B. (1995), *Optimal Sampled-Data Control Systems*, Springer-Verlag.

Churchill, R. & Brown, J. (1984), *Complex Variables and Applications*, fourth edn, McGraw-Hill Book Co.

Conway, J. (1973), *Functions of One Complex Variable*, Springer-Verlag, New York.

Conway, J. (1990), *A Course in Functional Analysis*, 2nd edn, Springer-Verlag, New York.

Coron, J.-M., Praly, L. & Teel, A. (1995), Feedback stabilization of nonlinear systems: Sufficient conditions and Lyapunov and input-output techniques, *in* A. Isidori, ed., 'Trends in Control. A European Perspective', Springer-Verlag, London, 3rd. European Control Conference, Rome, Italy, pp. 293–348.

Cramér, H. (1946), *Mathematical Methods of Statistics*, Princeton University Press.

De Branges, L. (1968), *Hilbert Spaces of Entire Functions*, Prentice-Hall, Inc., Englewood Cliffs, N.J.

de Souza, C., Shaked, U. & Fu, M. (1995), 'Robust H_∞ filtering for continuous time varying uncertain systems with deterministic input signals', *IEEE Trans. on Signal Processing* **43**(3), 709–719.

Desoer, C. & Gündes, A. (1986), 'Decoupling linear multi-input multi-output plants by dynamic output feedback: An algebraic theory', *IEEE Trans. on Automatic Control* **31**(8), 744–750.

Desoer, C. & Vidyasagar, M. (1975), *Feedback Systems: Input-Output Properties*, Academic Press.

Dion, J. & Commault, C. (1988), 'The minimal delay decoupling problem: feedback implementation with stability', *SIAM J. Control and Optimization* **26**, 66–82.

Doyle, J., Francis, B. & Tannenbaum, A. (1992), *Feedback Control Theory*, Macmillan Publishing Company, New York.

Doyle, J., Georgiou, T. & Smith, M. (1993), 'The parallel projection operators of a nonlinear feedback system', *Systems and Control Letters* **20**, 79–85.

Doyle, J. & Stein, G. (1981), 'Multivariable feedback design: concepts for a classical/modern synthesis', *IEEE Trans. on Automatic Control* **26**.

Dullerud, G. & Glover, K. (1993), 'Robust stabilization of sampled-data systems to structured LTI perturbations', *IEEE Trans. on Automatic Control* **38**(10), 1497–1508.

Er, M. & Anderson, B. (1994), 'Discrete-time loop transfer recovery via generalized sampled-data hold functions based compensators', *Int. J. of Robust and Nonlinear Control* **4**, 741–756.

Feuer, A. (1993), Periodic control of linear time-invariant systems – a critical point of view, *in* 'Proceedings of the 12th IFAC World Congress', pp. VIII–145.

Feuer, A. & Goodwin, G. (1994), 'Generalized sample hold functions: Frequency domain analysis of robustness, sensitivity, and intersample difficulties', *IEEE Trans. on Automatic Control* **39**(5), 1042.

Feuer, A. & Goodwin, G. (1996), *Sampling in Digital Signal Processing and Control*, Birkhäuser.

Francis, B. (1991), Lectures on H_∞ control and sampled-data systems, *in* C. Foias, E. Mosca & L. Pandolfi, eds, 'H_∞-Control Theory: Lectures Given at the 2nd. Session of the CIME', Lecture notes in mathematics, Springer-Verlag, pp. 37–105.

Francis, B. & Georgiou, T. (1988), 'Stability theory for linear time-invariant plants with periodic digital controllers', *IEEE Trans. on Automatic Control* **33**(9), 820–832.

Francis, B. & Zames, G. (1984), 'On H_∞-optimal sensitivity theory for SISO feedback systems', *IEEE Trans. on Automatic Control* **29**(1), pp. 9–16.

Franklin, G., Powell, J. & Emami-Naeini, A. (1994), *Feedback Control of Dynamic Systems*, 3rd edn, Addison-Wesley, Reading, MA.

Franklin, G., Powell, J. & Workman, M. (1990), *Digital Control of Dynamic Systems*, 2nd edn, Addison-Wesley, Reading, MA.

Freudenberg, J. & Looze, D. (1985), 'Right half plane poles and zeros and design tradeoffs in feedback systems', *IEEE Trans. on Automatic Control* **30**(6), 555–565.

Freudenberg, J. & Looze, D. (1987), 'A sensitivity tradeoff for plants with time delay', *IEEE Trans. on Automatic Control* **32**(2), 99–104.

Freudenberg, J. & Looze, D. (1988), *Frequency Domain Properties of Scalar and Multivariable Feedback Systems*, Vol. 104 of *Lecture Notes in Control and Information Sciences*, Springer-Verlag.

Freudenberg, J., Middleton, R. & Braslavsky, J. (1994), Robustness of zero-shifting via generalized sampled-data hold functions, *in* 'Proceedings of the 33th CDC', Florida, pp. 231–235.

Freudenberg, J., Middleton, R. & Braslavsky, J. (1995), 'Inherent design limitations for linear sampled-data feedback systems', *International Journal of Control* **61**(6), 1387–1421.

Gardner, M. & Barnes, J. (1949), *Transients in Linear Systems*, John Wiley & Sons, Inc., New York.

Georgiou, T. & Smith, M. (1995*a*), Distance measures for uncertain nonlinear systems, *in* 'Proc. of the 3rd European Control Conference', Rome, Italy, pp. 1016–1020.

Georgiou, T. & Smith, M. (1995*b*), Metric uncertainty and nonlinear feedback stabilization, *in* B. Francis & A. Tannenbaum, eds, 'Feedback Control, Nonlinear Systems and Complexity', Vol. 202 of *Lecture Notes in Control and Information Sciences*, Springer-Verlag, pp. 88–98.

Georgiou, T. & Smith, M. (1995*c*), Remarks on robustness of nonlinear systems, *in* 'Proc. 34th Conf. Decision Control', New Orleans, Louisiana, pp. 1662–1663.

Gille, J.-C., Pelegrin, M. & Decaulne, P. (1959), *Feedback Control Systems*, McGraw-Hill Book Co.

Golub, G. & Van Loan, C. (1983), *Matrix Computations*, Johns Hopkins Univ. Press., Baltimore.

Gómez, G. & Goodwin, G. (1995), Vectorial sensitivity constraints for linear multivariable systems, *in* 'Proceedings of the 34th CDC', New Orleans, LO. An extended version published in *Automatica*, 32(4), pp. 499-518, 1996.

Goodwin, G. & Feuer, A. (1992), 'Linear periodic control: a frequency domain viewpoint', *Systems and Control Letters* **19**, 379–390.

Goodwin, G. & Gómez, G. (1995), Frequency domain design trade-offs for linear periodic feedback systems, Technical Report EE9544, Department of Electrical and Computer Engineering, The University of Newcastle, Australia.

Goodwin, G., Mayne, D. & Shim, J. (1995), 'Trade-offs in linear filter design', *Automatica* **31**(10), 1367–1376.

Goodwin, G. & Salgado, M. (1994), 'Frequency domain sensitivity functions for continuous time systems under sampled data control', *Automatica* **30**(8), 1263.

Goodwin, G. & Seron, M. (1995), Complementarity constraints for nonlinear systems, *in* 'Proc. of the IFAC Symposium on Nonlinear Control System Design', Tahoe City, California.

Hagiwara, T., Ito, Y. & Araki, M. (1995), 'Computation of the frequency response gains and H_∞-norm of a sampled-data system', *Systems and Control Letters* **25**, 281–288.

Hara, S. & Sung, H.-K. (1989), 'Constraints on sensitivity characteristics in linear multivariable discrete-time control systems', *Linear Algebra and its Applications* **122-124**, 889–919.

Hill, D. & Moylan, P. (1980), 'Dissipative dynamical systems: basic input-output and state properties', *J. of the Franklin Institute* **309**(Nr. 5), pp. 327–357.

Hoffman, K. (1962), *Banach Spaces of Analytic Functions*, Dover, New York.

Horowitz, I. (1963), *Synthesis of Feedback Systems*, Academic Press, New York.

Hung, N. & Anderson, B. (1979), 'Triangularization technique for the design of multivariable control systems', *IEEE Trans. on Automatic Control* **24**(3), 455–460.

Kabamba, P. (1987), 'Control of linear systems using generalized sampled-data hold functions', *IEEE Trans. on Automatic Control* **32**(9), 772–783.

Kailath, T. (1968), 'An innovations approach to least-squares estimation. Part I: Linear filtering in additive white noise', *IEEE Trans. on Automatic Control* **13**(6), pp. 647–655.

Kailath, T. (1974), 'A view of three decades of linear filtering theory', *IEEE Trans. on Information Theory* **20**(6), 145–181.

Kailath, T. (1980), *Linear Systems*, Prentice-Hall, Inc., Englewood Cliffs, NJ.

Kailath, T. (1981), *Lectures on Wiener and Kalman Filtering*, Springer-Verlag.

Kailath, T. & Frost, P. (1968), 'An innovations approach to least-squares estimation. Part II: Linear smoothing in additive white noise', *IEEE Trans. on Automatic Control* **13**(6), pp. 655–660.

Kalman, R. & Bucy, R. (1961), 'New results in linear filtering and prediction theory', *Trans. ASME, J. Basic Engrg.* **83-D**, 95–107.

Kalman, R., Ho, B. & Narendra, K. (1963), Controllability of linear dynamical systems, *in* 'Contributions to Differential Equations', Interscience. Vol.1.

Kaplan, W. (1973), *Advanced Calculus*, 2nd edn, Addison-Wesley, Reading, MA.

Khargonekar, P., Poolla, K. & Tannenbaum, A. (1985), 'Robust control of linear time-invariant plants using periodic compensation', *IEEE Trans. on Automatic Control* **30**(11), 1088–1096.

Kinnaert, M. & Peng, Y. (1995), 'Residual generator for sensor and actuator fault detection and isolation: a frequency domain approach', *Int. J. of Control* **61**(6), 1423–1435.

Kwakernaak, H. (1985), 'Minimax frequency domain performance and robustness optimization of linear feedback systems', *IEEE Trans. on Automatic Control* **30**, 994–1004.

Kwakernaak, H. (1995), Symmetries in control system design, *in* A. Isidori, ed., 'Trends in Control: A European Perspective', Springer-Verlag, pp. 17–51.

Kwakernaak, H. & Sivan, R. (1972), *Linear Optimal Control Systems*, Wiley-Interscience, New York.

Lee, S., Meerkov, S. & Runolfsson, T. (1987), 'Vibrational feedback control: zero placement capabilities', *IEEE Trans. on Automatic Control* **32**(7), 604–611.

Levinson, N. & Redheffer, R. (1970), *Complex Variables*, Holden-Day, Inc.

MacFarlane, A. & Karcanias, N. (1976), 'Poles and zeros of linear multivariable systems: a survey of the algebraic, geometric and complex-varible theory', *Int. J. of Control* **24**, 33–74.

Maciejowski, J. (1991), *Multivariable Feedback Design*, Addison-Wesley Publishing Company.

Markushevich, A. (1965), *Theory of Functions of a Complex Variable*, Vol. 1, Prentice-Hall, Inc., Englewood Cliffs, N.J.

Martin Jr., R. (1976), *Nonlinear Operators and Differential Equations in Banach Spaces*, John Wiley & Sons, Inc., New York.

Massoumnia, M., Verghese, G. & Willsky, A. (1989), 'Failure detection and identification', *IEEE Trans. on Automatic Control* **34**(3), 316–321.

Meyer, D. (1990), 'A parametrization of stabilizing controllers for multirate sampled-data systems', *IEEE Trans. on Automatic Control* **35**(11), 233–236.

Middleton, R. (1991), 'Trade-offs in linear control systems design', *Automatica* **27**(2), 281–292.

Middleton, R. & Freudenberg, J. (1995), 'Non-pathological sampling for generalised sampled-data hold functions', *Automatica* **31**(2).

Middleton, R. & Goodwin, G. (1990), *Digital Control and Estimation. A Unified Approach.*, Prentice-Hall, Inc.

Middleton, R. & Xie, J. (1995), Non-pathological sampling for high order generalised sampled-data hold functions, *in* 'Proceedings of the ACC', Seattle, WA.

Morari, M. & Zafiriou, E. (1989), *Robust Process Control*, Prentice-Halls, Englewood Cliffs.

Nagpal, K. & Kharghonekar, P. (1991), 'Filtering and smoothing in an H_∞ setting', *IEEE Trans. on Automatic Control* **36**, 152–166.

Naslin, P. (1965), *The Dynamics of Linear and Non-Linear Systems*, London: Blackie.

O'Reilly, J. (1983), *Observers for Linear Systems*, Academic Press.

O'Young, S. & Francis, B. (1985), 'Sensitivity tradeoffs for multivariable plants', *IEEE Trans. on Automatic Control* **30**(7), 625–632.

Postlethwaite, I. & MacFarlane, A. (1979), *A Complex Variable Approach to the analysis of linear Multivariable Feedback Systems*, Vol. 12 of *Lecture Notes in Control and Information Sciences*, Springer-Verlag.

Ramstad, T. (1984), Analysis/synthesis filterbanks with critical sampling, *in* 'Proc. IEEE Conf. DSP.'.

Ravi, R., Kharghonekar, P., Minto, K. & Nett, C. (1990), 'Controller parametrization for time-varying multirate plants', *IEEE Trans. on Automatic Control* **35**(11), 1259–1262.

Reynolds, O. (1883), 'An experimental investigation of the circumstances which determine whether the motion of water shall be direct or sinuous, and of the laws of resistance in parallel channels', *Trans. Royal Society, London* **174**.

Riesz, F. & Sz.-Nagy, B. (1990), *Functional Analysis*, Dover, New York.

Rosenbrock, H. (1969), 'Design of multivariable control systems using the inverse Nyquist array', *IEE Proceedings Part D* **116**(11), 1929–1936.

Rosenvasser, Y. (1995), 'Mathematical description and analysis of multivariable sampled-data systems in continuous-time: Part I', *Automation and Remote Control* **56**(4), 526–540. Part II, *Automation and Remote Control*, vol.56(5), pp. 684-697, 1995.

Rudin, W. (1987), *Real and Complex Analysis*, 3rd. edn, McGraw-Hill Book Co.

Sandberg, I. (1965), 'An observation concerning the application of the contraction mapping fixed-point theorem and a result concerning the norm-boundedness of solutions of nonlinear functional equations', *Bell Sys. Tech. J.* **44**, 1809–1812.

Seron, M. (1995), Complementary operators in filtering and control, PhD thesis, Department of Electrical and Computer Engineering, The University of Newcastle, NSW 2308 Australia.

Seron, M. & Goodwin, G. (1995), Design limitations in linear filtering, *in* 'Proceedings of the 34th CDC', New Orleans, LO.

Seron, M. & Goodwin, G. (1996), 'Sensitivity limitations in nonlinear feedback control', *Systems and Control Letters* **27**, 249–254.

Shamma, J. (1991), 'Performance limitations in sensitivity reduction for nonlinear plants', *Systems and Control Letters* **17**, 43–47.

Shamma, J. & Dahleh, M. (1991), 'Time-varying versus time-invariant compensation for rejection of persistent bounded disturbances and robust stabilization', *IEEE Trans. on Automatic Control* **36**(7), 838–847.

Shannon, C. (1948), 'A mathematical theory of communications', *BSTJ* **27**, 379–623.

Shenoy, R., Burnside, D. & Parks, T. (1994), 'Linear periodic systems and multirate filter design', *IEEE Trans. on Signal Processing* **42**(9), 2242–2256.

Sivashankar, N. & Khargonekar, P. (1993), 'Robust stability and performance analysis for sampled-data systems', *IEEE Trans. on Automatic Control* **39**(1), 58–69.

Smith, M. & Barnwell, T. (1987), 'A new filter bank theory for time-frequency representation', *IEEE Trans. on Acoustics, Speech, and Signal Processing* **35**(3), 314–327.

Sontag, E. (1990), *Mathematical Control Theory. Deterministic Finite Dimensional Systems*, Springer-Verlag.

Sule, V. & Athani, V. (1991), 'Directional sensitivity tradeoffs in multivariable feedback systems', *Automatica* **27**(5), 869–872.

Sung, H.-K. & Hara, S. (1988), 'Properties of sensitivity and complementary sensitivity functions in single-input single-output digital control systems', *Int. J. of Control* **48**(6), 2429–2439.

Truxal, J. (1955), *Control System Synthesis*, McGraw-Hill, New York.

Vetterli, M. (1987), 'A theory of multirate filter banks', *IEEE Trans. on Acoustics, Speech, and Signal Processing* **35**(3), 356–372.

Vetterli, M. (1989), 'Invertibility of linear periodically time-varying filters', *IEEE Trans. on Circuits and Systems* **36**(1), 148–150.

Vidyasagar, M. (1985), *Control System Synthesis, A Factorization Approach*, MIT Press, Cambridge.

Wall, Jr., J., Doyle, J. & Harvey, C. (1980), Trade-offs in the design of multivariable feedback systems, *in* 'Proc. of the 18th Allerton Conference', pp. 715–725.

Weller, S. (1996), Sensitivity limitations for multivariable linear filtering, *in* 'Proceedings of the 35th CDC', Kobe, Japan.

Weller, S. & Goodwin, G. (1993), Controller design for partial decoupling for linear multivariable systems, *in* 'Proceedings of the 32nd CDC', pp. 833–834.

Widder, D. (1961), *Advanced Calculus*, 2nd edn, Prentice-Hall, Inc., Englewood Cliffs, N.J.

Willems, J. (1971*a*), *The Analysis of Feedback Systems*, The MIT Press, Cambridge, Massachusetts.

Willems, J. (1971*b*), 'Least-squares stationary optimal control and the algebraic Riccati equation', *IEEE Trans. on Automatic Control* **16**, 621–634.

Xie, J. & Middleton, R. (1996), Inherent design limitations for linear siso output multirate sampled-data control, Technical Report EE9628, Dept of E&CE, The University of Newcastle, Australia.

Yaesh, I. & Shaked, U. (1989), Game theoretic approach to optimal linear estimation in the minimum H_∞ norm sense, *in* 'Proc. 28th Conf. Decision Control', Tampa, Florida.

Yamamoto, Y. & Araki, M. (1994), 'Frequency responses for sampled-data systems - Their equivalence and relationships', *Linear Algebra and its Applications* **206**, 1319–1339.

Yamamoto, Y. & Khargonekar, P. (1996), 'Frequency response of sampled-data systems', *IEEE Trans. on Automatic Control* **41**(2), 166–176.

Zadeh, L. (1950), 'Frequency analysis of variable networks', *Proceedings of the I.R.E.* **38**, 291–299.

Zames, G. (1966*a*), 'On the input-output stability of time-varying nonlinear feedback systems. Part I: Conditions derived using concepts of loop gain, conicity and positivity.', *IEEE Trans. on Automatic Control* **11**, 228–238.

Zames, G. (1966*b*), 'On the input-output stability of time-varying nonlinear feedback systems. Part II: Conditions involving circles in the frequency plane and sector nonlinearities', *IEEE Trans. on Automatic Control* **11**, 465–476.

Zames, G. & Bensoussan, D. (1983), 'Multivariable feedback, sensitivity, and decentralized control', *IEEE Trans. on Automatic Control* **28**(11), 1030–1035.

Zarantonello, E. (1967), 'The closure of the numerical range contains the spectrum', *Pacific Journal of Mathematics* **22**(3), 575–595.

Zhang, J. & Zhang, C. (1994), Robustness analysis of control systems using generalized sample hold functions, *in* 'Proceedings of the 33th CDC', Lake Buena Vista, FL.

Zhang, Z. & Freudenberg, J. (1993), 'Discrete-time loop transfer recovery for systems with nonminimum phase zeros and time delays', *Automatica* **29**(2), 351–363.

Index